Plant Aging
Basic and Applied Approaches

NATO ASI Series

Advanced Science Institutes Series

A series presenting the results of activities sponsored by the NATO Science Committee, which aims at the dissemination of advanced scientific and technological knowledge, with a view to strengthening links between scientific communities.

The series is published by an international board of publishers in conjunction with the NATO Scientific Affairs Division

A	**Life Sciences**	Plenum Publishing Corporation
B	**Physics**	New York and London
C	**Mathematical**	Kluwer Academic Publishers
	and Physical Sciences	Dordrecht, Boston, and London
D	**Behavioral and Social Sciences**	
E	**Applied Sciences**	
F	**Computer and Systems Sciences**	Springer-Verlag
G	**Ecological Sciences**	Berlin, Heidelberg, New York, London,
H	**Cell Biology**	Paris, and Tokyo

Series A: Life Sciences

Plant Aging
Basic and Applied Approaches

Edited by

Roberto Rodríguez and R. Sánchez Tamés

University of Oviedo
Oviedo, Spain

and

D.J. Durzan

University of California, Davis
Davis, California

Plenum Press
New York and London
Published in cooperation with NATO Scientific Affairs Division

Proceedings of a NATO Advanced Study Institute
on Molecular Basis of Plant Aging,
held July 2–15, 1989,
in Ribadesella, Spain

Library of Congress Cataloging in Publication Data

Plant aging: basic and applied approaches / edited by Roberto Rodriguez and
  R. Sánchez Tamés, and D. J. Durzan.
       p.   cm.—(NATO ASI series. Series A, Life sciences; vol. 186)
    "Published in cooperation with NATO Scientific Affairs Division."
    "Proceedings of a NATO Advanced Study Institute on Molecular Basis of
Plant Aging, held July 2–15, 1989 in Ribadesella, Spain"—Copr. p.
    Includes bibliographical references.
    ISBN 0-306-43518-7
    1. Plants—Aging—Congresses. 2. Plant propagation—Congresses. I.
Rodriguez, Roberto. II. Sánchez, Tamés, R. III. Durzan, D. J. IV. North Atlantic
Treaty Organization. Scientific Affairs Division. V. NATO Advanced Study In-
stitute on Molecular Basis of Plant Aging (1989: Ribadesella, Spain) VI. Series:
NATO ASI series. Series A, Life sciences; v. 186.
QK762.5.P58   1990   90-6996
582′.0372—dc20   CIP

© 1990 Plenum Press, New York
A Division of Plenum Publishing Corporation
233 Spring Street, New York, N.Y. 10013

Printed in the United States of America

ORGANIZING AND SCIENTIFIC COMMITTEE

RODRIGUEZ, R. Lab. Fisiología Vegetal, Dpto. B.O.S.
Facultad de Biología, Universidad de
Oviedo. Arias de Velasco, s/n.
33005 Oviedo. Spain

DURZAN, D. Department of Environmental Horticulture. University of California.
Davis, CA 95616 USA

PAIS, M.S. Departamento de Biologia Vegetal
Facultade de Ciencias de Lisboa,
Boco C2, Campo Grande 1700. Lisboa
Portugal

WITHERS, L.A. International Board for Plant Genetic Resources. c/o Food and Agricul-
ture Organization of the United Na-
tion. Via delle Terme di Caracalla
00100 Rome. Italy

TRAN THANH-VAN, K. Institut de Physiologie Végétale.
CNRS, 91190 Gif sur Yvette. France.

SANCHEZ TAMES, R. Lab. Fisiología Vegetal, Dpto. B.O.S
Facultad de Biología, Universidad de
Oviedo. Arias de Velasco, s/n.
33005 Oviedo. Spain

SABATER, B. Dpto. de Biología Vegetal, Universi-
dad de Alcalá de Henares. Apdo. 20,
Alcalá de Henares. 28871-Madrid.
Spain

NORTON, C.R. Faculty of Environmental Studies
Meriot-Watt University
Edinburgh, E43 9 DF. Scotland

PASQUALETO, P.L. Azienda Agricola Meristema SRL.
Laboratorio di micropropagazione,
Via Martiri della liberta, n. 13,
56030 Cascine di Buti (Pisa), Italy

PREFACE

For many, the terms aging, maturation and senescence are synonymous and used interchangeably, but they should not be. Whereas senescence represents an endogenously controlled degenerative programme leading to plant or organ death, genetic aging encompasses a wide array of passive degenerative genetic processes driven primarily by exogenous factors (Leopold, 1975). Aging is therefore considered a consequence of genetic lesions that accumulate over time, but by themselves do not necessarily cause death. These lesions are probably made more severe by the increase in size and complexity in trees and their attendant physiology. Thus while the withering of flower petals following pollination can be considered senescence, the loss of viability of stored seeds more clearly represents aging (Norden, 1988). The very recent book "Senescence and Aging in Plants" does not discuss trees, the most dominant group of plants on the earth. Yet both angiospermic and gymnospermic trees also undergo the above phenomena but less is known about them. Do woody plants senesce or do they just age? What is phase change? Is this synonymous with maturation?

While it is now becoming recognized that there is no programmed senescence in trees, senescence of their parts, even in gymnosperms (e.g., needles of temperate conifers last an average of 3.5 years), is common; but aging is a readily acknowledged phenomenon. In theory, at least, in the absence of any programmed senescence trees should live forever, but in practice they do not. Is death simply a response to external environmental factors?. As trees grow older, their growth rates decrease, and although they apparently never stop growing, very old trees increase in size rather insignificantly. Under natural conditions, a plot of the cumulative height and diameter growth by dominant trees in a stand exhibits a sigmoidal growth curve, with the maximum annual increment occurring relatively early in the age of the tree (Assmann, 1970, quoted in Greenwood, 1988). The time at which the maximum increment occurs appears to be species-specific; and for the same species occurs later on relative poor soils, but at the same relative height and diameter. While there is little doubt that the maximum size a tree can attain is primarily a function of its genetics, is incremental growth more a function of size rather than chronological age? Is the phase-change process a consequence of the amount of growth that has occurred or is it the result of the physiological consequences of increased size (Greenwood, 1988)?

There are other equally intriguing questions. Do the above changes arise from the accumulation of harmful metabolites with time, or from the physical, biophysical and physiological consequences of trees having to sustain such a large mass, the requeriments to move large volumes of water and nutrients, and photosynthates over great distances? Although the debate as to whether meristems themselves age continues, more and more researchers have come to accept this idea, as first suggested by Shaffalitsky de Muckadell (1959); so much so that today we accept the fact that the oldest part of a tree chronologically is the youngest part developmentally, and attempt to utilize this fact in clonal propagation of woody plants.

The organizers of the 1989 NATO Advanced Study Institute (ASI) on the "Molecular Basis of Plant Aging" recognized these and other dilemmas we face in understanding the behaviour of woody plants. They explicitly state that "Aging is a phenomenon of considerable theoretical importance in relation to morphogenic control, differentiation and determination in plant development. It also has practical importance due to its implication in several economical areas such as flowering, fruit set, "in vitro" tree manipulation, genetic improvement, etc".

Although many theories have been advanced to explain aging, the mechanisms involved in this process still are among the least understood factors in biology. Thus, to increase our knowledge in this scientific field will be very useful for a better understanding of plant biology.

Due to the loss of morphogenic competence, aging is one of the most critical barriers to capture special genetic traits of selected trees. In order to fully exploit the possibilities that tissue culture technologies provide for forest tree improvement, it is imperative to establish first the conditions to maintain plant tissues in their most reactive phase.

Thus I was delighted to accept the invitation of Prof. Roberto Rodríguez of the University of Oviedo, to participate in this ASI, which was held in Ribadesella, Asturias, Spain from June 25 to July 8, 1989. The objective of such ASI is to disseminate advanced knowledge not yet in University curricula and foster international scientific contacts through high level teaching courses. The organizers of this ASI accepted this mandate, and rose to the challenge of providing a detailed description of the process which occurs during aging. Thus, some 70 participants from 15 countries, consisting of graduate students, post doctoral fellows, recently and long established researchers from Universities, Government Research Institutions and Private Industry assembled in ribadesella to address their charge.

The programme was divided into five sections, namely, 1. Juvenility, maturation and rejuvenation, 2. Vegetative propagation, advantages and limitations, 3. Ultrastructural, genetic and biochemical characteristics of aging and senescence, 4. Modulators of aging and maturation, and 5. Genetic manipulations. Information was presented in the form of major lectures (review and primary data), posters and short presentations. The work of each section ended with a round-table discussion. The results of these deliberations are captured in this volume.

It is not easy to summarize in a few sentences the gist of a meeting with so wide a theme and with such a large number of presentations, both oral and by posters. Nevertheless, in a very superficial way, one can divide the meeting in two very broad areas. These were: 1) problems and practices of woody plant tissue culture, and 2) biochemical and molecular aspects of aging. Any such division is bound to be arbitrary and several papers in this volume fall into neither area, and perhaps this division structuralizes the ASI, which has led to the production of this very important volume.

A number of papers dealt with the culture and manipulation of woody plants, particularly as these were affected by the maturation process. Some technology, such as use of the vibratome for explant production, thin layer methodology and micrografting techniques, was also presented. Factors affecting culture establishment and multiplication, problems of vitrification, polyphenol production, generation of somaclonal variation, the use of minimal culture media, including the possible harmful effects of cytokinins, and the gaseous environment on the cultures all received attention. Because mature woody species are less responsive in culture, not only was the nature of the phase-change discussed, but empirical approaches to reactivate or reinvigorate woody explants for ease of manipulation in culture were also described. Some physiological and biochemical aspects of regeneration were also covered.

The second major focus was on the genetic biochemical and molecular basis of phase change, and aging in woody species. The use of information from senescent herbaceous systems served as a model for understanding aging and its related phenomena, as many of the manifestations of aging are similar to senescence. Nitrogen metabolism, oxidative stress, free radical generation, and the role of phytohormones were among the topics covered. One view that seemed to emerge is that the rate of results from aging, a balance between oxidative stress damage and regenerative repair capacity. Lastly, the potential genetic manipulation of forest trees by traditional and molecular approaches to generate the forest tree of tomorrow highlighted the last session.

The value of NATO ASIs, such as this one, cannot be overestimated. It brought together citizens of various countries, scientists from different disciplines, people of both sexes and of different ages and experiences for a period of two weeks to think on the common topic. Different perspectives were brought to bear on the subject and the format allowed for the free and unfettered exchange of ideas. Although many questions remain uncovered, we are all a little clearer in our understanding of aging and maturation in woody plants, of the role that stress plays in tissue cultures, etc. This volume allows us to share our experiences with a much wider audience. A trust that all of you who read and study their articles will gain as much knowledge and pleasure in doing so, as we who participated in this ASI had during our two weeks in Ribadesella.

The organizers and Scientific Committee under the Directorships of Roberto Rodríguez and Don Durzan deserve our sincere gratitude for job well done.

Trevor A. Thorpe
Professor of Botany, Department of
Biological Sciences; and Associate
Dean (Research) Faculty of Science,
Univ. of Calgary, Alberta, Canada.

REFERENCES

GREENWOOD, M. 1988. The effect of phase change on annual growth increment in Eastern Larch. INRA-IUFRO International Symposium on Forest Tree Physiology, Nancy, France, Sept. 25-30.

LEOPOLD, A.C. 1975. Aging, senescence and turnover in plants. BioScience 25: 654-662.

NORDEN, L.D. 1988. The phenomena of senescence and aging. In: "Senescence and Aging in Plants" (Norden LD and Leopold, AC, eds.). pp. 1-50. Academic Press, New York.

NORDEN, L.D. and LEOPOLD, A.C. 1988. Senescence and Aging in Plants. Academic Press, New York. 526 pp.

SCHAFFALITSKY DE MUCKADELL, M. 1959. Investigations on aging of apical meristems in woody plants its importance in silviculture. Torstl. Torsgsv. Danm. 25: 310-455.

CONTENTS

SECTION V:
GENETIC MANIPULATION

SELECTED POSTERS

SECTION I

AGING, MATURATION AND REJUVENATION

AGING OF MERISTEMS AND MORPHOGENETIC POTENTIALITIES

Victorio S. Trippi

Laboratorio de Fisiología Vegetal, Univ. Nac. de Córdoba

Casilla de Correo 395, 5000-Córdoba, Argentina

AGING OF MERISTEMS

The term "aging" refers to chronological changes in plants. In many plants requiring a maturation period, morphogenetic potentialities of meristems may change during aging until flowering and later on, until death. These changes induce "heteroblastic development" (Goebel, 1889). As plants display different morphogenetic qualities along the shoot, an ascendent gradient of aging has been suggested. Molisch (1938) called this phenomenon "topophysis" (from the Greek topos = location and physis = nature) to point out the different morphogenetic potentialities of meristems located at different places. The same phenomenon was referred later on as "physiological aging of meristems" (Schaffalitzky, 1959). The aging of meristems can be defined as the qualitative changes induced by internal and/or environmental factors on morphogenetic potentialities resulting in flowering and the loss of proliferation capacity which leads to senescence.

GROWTH HABIT

Heteroblastic development is the morphological expression of progressive changes during the aging of meristems, as shoot and leaf forms depend on meristem function. These changes are observed from the base to the apex of the shoot, which is finally transformed into flower.

Plagiotropic shoot growth is a juvenile character of plants like Hedera helix capable of transforming their growth habit from plagiotropic to orthotropic in the adult stage which normally bears flowers. A similar heteroblastic development has been observed in Ficus repens (Robbins, 1957). In woody plants orthotropic branches are typical of young plants and a change in the growth habit of branches along the stem has been pointed out. The stability of the growth habit depends on the species. In Aegopodium podagraria autonomous plagiotropic growth of stalk pieces does not last long; in Mentha rotundifolia and M. acuatica stoloniferous growth depends

Plant Aging: Basic and Applied Approaches
Edited by R. Rodríguez *et al.*
Plenum Press, New York, 1990

on the lenght of the shoot. In Epilobium hirsutum plagiotropic growth may last for two years. In Phyllanthus lathyroides shoots may remain plagio- tropic growth for many years, as was also observed in Araucaria excelsa.

In woody plants the stability of the growth habit seems to be kept by endogenous factors, however in herbaceous plants it appear to be the result of exogenous regulation. Light intensity is a powerful factor in the regulation of the growth habit in Cynodon dactylon. While darkness promotes orthotropic growth, high light intensity promotes plagiotropic growth (Montaldi, 1969). In Stachys silvatica new horizontal shoots become orthotropic under SD and remain plagiotropic under LD. An opposite behav- ior is observed in Proserpinaca palustris: while SD determine plagiotropic growth or juvenile form, LD induce orthotropic growth and flowering.

Gibberellic acid (GA) is capable of modifying the growth habit, but its effects depend on the species. While in Cynodon, Trifolium and Proserpinaca GA promotes orthotropic growth, in Hedera helix it induces plagiotropic growth. Plagiotropic growth induced by 0.3 M sucrose in Cynodon can be reversed by GA (Montaldi, 1973).

In Stachys plagiotropic growth is induced by flowering and depends on the leaves. In this case leaves can only be replaced by ABA, but not by IAA, GA nor Kn (Pfirsch, 1978).

Morphogenetic potentialities can be propagated by cuttings in plants showing an endogenous control of forms, as observed for leaf form in tomato (Trippi, 1964), for growth habit in woody plants like Araucaria (Wochting, 1904) and for other characters in Robinia pseudoacacia and Fagus silvatica (Schaffalitzky, 1959). However, in other plants, which are sensitive to environmental conditions, morphogenesis can be exogenously controlled such as in Cynodon and Proserpinaca palustris.

LEAF FORM

Heterophylly is a common aspect of the aging of meristems and it has been described in a number of species (Allsopp, 1966). Typical plants showing changes in leaf form are Hedera helix which displays lobated leaves in juvenile forms and lanceolate leaves in the adult form and Passiflora coerulea which changes from entire leaves in the juvenile stage to heptalobate in the adult form. In Lycopersicon esculentum the first entire leaves are followed by tri-, penta- and hepta-lobate leaves, the latter accompanying flowering. The stability of meristem changes can be observed by ablation of the stem at different nodes (Trippi, 1964).

Generally, basal sprouts of tomato and Passiflora display same heterophylly like the main shoot, but after flowering and in senile plants the change is faster than in vegetative plants. This fact suggests that morphogenetic substances accumulate in the basal zone of plants. Defolia- tion in young plants prevented heterophylly development in Ipomea and tomato and induced reversion from adult to juvenile forms in Ipomea and woody plants. Adventitious buds in woody and herbaceous plants induced reversion to juvenile form in apple and pear trees (Wellensieck, 1952). In decapitated plants of tomato buds formed on the callus at the ablation zone and resulting leaves showed typical juvenile forms (Trippi, 1964) (Fig. 1). Similar rejuvenation was observed by Montaldi et al. (1963b) in Passiflora when adventitious buds were formed on the ablation zone of rooted leaves. In Passiflora buds formed on roots, also produced juvenile leaf forms. Pruning has also been described to induce reversion in Hedera helix (Doorembos, 1954).

Morphogenetic potentialities of meristems are modulated by light intensity, photoperiod and temperature. Goebel (1889) reported that high light intensity is necessary for inducing adult forms in Campanula rotundifolia. Passiflora coerulea shows juvenile trilobate leaves under low light intensity. Furthermore, when plants showing adult forms (5-lobate leaves) are moved to low light intensity, a reversion to 3-lobate leaves is observed (Montaldi et al., 1963a).

In Ipomea coerulea the progression from entire to 3-lobate leaves was different in plants under photoperiod of 16 h and 8 h respectively (Ashby, 1950). In Ulex europaeus, SD of 8 h induce 3-foliolate juvenile type of leaves and LD of 16-20 h the simple entire adult form of leaves (Millener, 1961). A similar behavior has been observed in Gaillardia pulchella (Trippi, 1965). A temperature effect has been described in Ranunculus hirtus, Ipomea coerulea and Hedera canariensis. All these plants showed juvenile forms at high temperature (Trippi, 1982).

Fig. 1. Aging of meristems in tomato showed as changes in the leaf form and morphogenetic potentialities along the stem.

Hormonal control of morphogenetic potentialities has been shown particularly for GA. In some plants, GA induces juvenile leaf forms, i.e. Hedera helix, Ipomea coerulea and Acacia melanoxylon, but in others such as in Eucalyptus, Gaillardia pulchella, Proserpinaca palustris and Xanthium GA induces adult forms. According to Njoku (1958), IAA, NAA, 2,4-D and TIBA delayed the development of adult leaf form in Ipomea.

Experimental evidence that carbohydrate supply is capable of inducing adult form in the fern Marsilea has been obtained by Allsopp (1965). Allsopp (1964) pointed out that environmental factors capable of regulating heteroblastic development like light intensity, defoliation, etc. correlate with the carbohydrate level. However, a detailed investigation on this subject is still lacking.

Changes in other characters such phyllotaxis, leaf size and anatomical features have also been mentioned to be the result of the aging of meristems (Trippi, 1982).

ROOTING

The aging of meristems may lead either to their transformations into flowers or to senescence. In polycarpic plants shoots can be used as cuttings. These cuttings may exhibit reduced rooting capacity, in agreement with a loss of cell proliferation capacity. This phenomenon has been also attributed to an aging of meristems, that is, essentially to senescence.

Knight (1926) pointed out that the position of the cutting in the mother plant determines its rooting capacity. In pine, olive and blueberry the apical zone bearing adult characters (flowers) shows lower rooting capacity than the base (Trippi, 1982).

Rooting capacity has been reported to be regulated by light intensity, photoperiod, hormones and sugars. Cotyledonar cuttings from Sinapis alba under 8000 Lux produce a higher number of roots than under 16000 Lux (Lovell and Moore, 1969). Light inhibited rooting in Pisum sativum (Leroux, 1973) and a pre-treatment with high irradiance reduced rooting as compared to low irradiance (Hansen and Eriksen, 1974).

In Anagallis arvensis SD stimulate rooting and LD prevent rhizogenesis and induce flowering and senescence (Trippi and Brulfert, 1973). Similar effects of SD were observed in Bryophyllum tubiflorum (Nanda et al., 1967). Rooting of leaves was inhibited by sucrose in Anagallis arvensis, but IAA treatment restored rooting capacity (Larrieu et Trippi, 1979). In Sinapis alba (Lovell et al., 1972) and in Pisum sativum (Leroux, 1973) the inhibitory effects of sugar depend on light.

THE IMPLICANCE OF SUGARS IN MERISTEMS AGING

As Allsopp (1964) suggested, regulation of juvenile and adult forms by different factors are in line with carbohydrate contents. Conditions determining low carbohydrate content associate with juvenile forms and increased carbohydrate content with adult forms.

The implicance of sugars in flowering has been experimentally shown by Nitsch (1968) in Plumbago indica. In this plant, explants flower only when provided with high sugar concentrations (sucrose, glucose, fructose, maltose) which exclude vegetative organ formation. Studies on Anagallis arvensis have shown that sucrose modulates rooting, shoot and leaf growth, flowering and fruiting, including the change from two to three leaves per verticil which is associated with flowering. While low sucrose concentration (1%) induces root and shoot formation, concentrations of 3% or above induce flowering and fruit formation and inhibit shoot growth in lenght, leaf surface growth and rooting. These effects, as in other cases, were modulated by an interaction with light intensity. Studies on Anagallis are particularly relevant because sucrose concentration modulates morphogenesis, namely a high sugar concentration stimulates flowering while simultaneously inhibiting vegetative growth and rooting (Silvente and Trippi, 1986). All these morphogenetic events can be seen in plants of Anagallis under LD conditions leading to the aging of meristems and to senescence in the whole plant. Tuber formation (Mes and Menge, 1954) and the induction of resting stages in Spirodella (Henssen, 1954) are linked to a halting of vegetative growth and can be controlled by the sugar supply.

Control of aging of meristems and related morphogenetic potentialities by carbohydrates can be easily accepted in plants displaying an endogenous or genetic determinism for maturation and determinate growth (quantitative photoperiodics ?), but the idea becomes hard to accept when dealing with photoperiodic qualitative plant responses. As in Anagallis sucrose level may determine all morphogenetic responses. Under LD and an illumination of 6 W m^{-2} no growth could observed without the addition of sugar. However, photoperiodic effects could not be replaced by sucrose. It is likely that sugar controls internal hormone balance, i.e. a low sucrose content promotes rooting and a high sucrose content promotes flowering and inhibits rooting. We consider photoperiod to be like a key (a biophysic photoreaction) linked to the genetic determination conditioning photoperiodic response, with no function other than the openning of a "door" to sugars.

THE GENERAL SCHEME

As plants display an open organization, senescence affecting the whole plant is rather difficult to accept, unless provoked by external factors. We can imagine that plants grow in suboptimal environmental conditions. High light intensity and CO_2 concentration in an oxygenated atmosphere may induce a progressive alteration in the proportion of chemical constituents, namely an increase in carbohydrate content as compared to nitrogen compounds. The idea is supported by the well known increase of C/N ratio during aging, necessarily linked to an O_2-effect on SH-enzymes, among them nitrate reductase (Kenis and Trippi, 1986). While carbohydrate content is low, indeterminate growth takes place (morphogenesis of roots and shoots), but with progressive increase in sugar content a modulating effect on morphogenetic potentialities begins and it is reflected in heteroblastic development (growth habit, heterophylly). The highest sugar levels induce flowering, growth cessation, dormancy and senescence (Fig. 2).

Thus, flowering as an index of meristem aging seems to be a reaction to environmental stresses, tending to restore embryonic/vegetative qualities as an expression of the homeostasis of living plants. As senescence is already present in indeterminate plants, conditions determining flowering and reproductive structures should increase senescence phenomena in the plant.

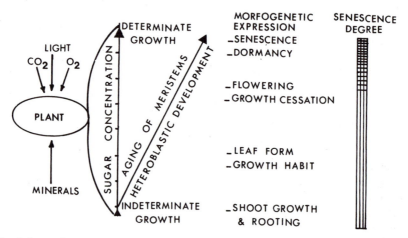

Fig. 2. Schematic representation of aging of meristems in a general proposed scheme of senescence.

Concerning genetic determination plants of indeterminate habit may have genes capable of preventing environmental stresses, which would be lacking/suppressed in those with determinate growth and senescence. Stress condition may induce flowering in Lemna (Hillmann, 1961) and other plants.

REFERENCES

Allsopp, A., 1964, The metabolic status and morphogenesis, Phytomorphology 14: 1.
—— 1965, Heteroblastic development in cormophytes, in: "Encyclop. Plant Physiol.", Springer-Verlag.
—— 1966, Heteroblastic development in vascular plants, Adv. in Morphogenesis 5: 127.

Ashby, E., 1950, Studies in the morphogenesis of leaves. VI. Some effects of lenght of day upon leaf shape of Ipomea coerulea, New Phytol. 49: 375.

Doorembos, J., 1954, Rejuvenation of Hedera helix in graft combinations, Proc. Kon. Ned. Akad. Wetensch. Ser. C 57: 99.

Goebel, K., 1889, Ueber die Jugendzudstände der Pflanzen, Flora 72: 1.

Hansen, J., and Eriksen, E. N., 1974, Root formation of pea cuttings in relation to irradiance of the stock plants, Physiol. Plant. 32: 170.

Henssen, A., 1954, Die Dauerorgane von Spirodella polyrrhiza (L.) Schalid., Flora 141: 523.

Hillmann, W. S., 1961, The lemnaceae, or duckweeds, The Bot. Rev. 27: 221.

Kenis, J. D., and Trippi, V. S., 1986, Regulation of nitrate reductase in detached oat leaves by light and oxygen. Physiol. Plant. 68: 387.

Knight, R. C., 1926, The propagation of fruit tree stocks by stem cuttings, Journ. Pomology 5: 248.

Larrieu, C., et Trippi, V. S., 1979, Etude du viellissement d'Anagallis arvensis L.: Action régulatrice de la lumiere, du saccharose et de l'acide indolyl-acetique sur la capacité rhizogénique des feuilles et des rameaux axillaires, Biol. Plantarum 21: 336.

Leroux, R., 1973, Contribution a l'etude de la rhizogenese de fragments de tiges de Pois (Pisum sativum L.) cultivés in vitro, Rev. Cytol. et Biol. Veg. 36: 1.

Lovell, P. H., and Moore, K. G., 1969, The effect of light and cotyledon age on growth and root formation in excised cotyledons of Sinapis alba L., Planta 85: 351.

——— , Illsey, A., and Moore, K. G., 1972, The effects of light intensity and sucrose on root formation, photosynthetic ability and senescence in detached cotyledons of Sinapis alba L. and Raphanus sativus L., Ann. Bot. 36: 123.

Mes, M. G., and Menge, I., 1954, Potato shoot and tuber cultures in vitro, Physiol. Plant. 7: 637.

Millener, L. H., 1961, Day-lenght as related to vegetative development in Ulex europaeus L. I. The experimental approach, New Phytol. 60: 339.

Molisch, H., 1938, "The longevity of plants", H. Fulling, ed., New York.

Montaldi, E. R., 1969, Gibberellin-sugar interaction regulating the growth habit of Bermudagrass (Cynodon dactylon (L.) Pers.), Experientia 25: 91.

——— 1973, Epinasty in Cynodon plectostachyum R. Pilger induced by sucrose and its reversion by gibberellic acid and nitrogen compounds, Experientia 29: 1031.

——— , Caso, O. H., y Lewin, I., 1963a, Algunos factores que afectan la morfología de las hojas de una planta de desarrollo heteroblástico, Rev. Inv. Agric. 17: 321.

——— 1963b, Rejuvenecimiento de plantas por diferenciación de yemas en callos, Rev. Inv. Agric. 17: 441.

Nanda, K. K., Purohit, A. N., and Bala, A., 1967, Effect of photoperiod, auxins and gibberellic acid on roting of stem cuttings of Bryophyllum tubiflorum, Physiol. Plant. 20: 1096.

Nitsch, C., 1968, Induction in vitro de la floraison chez une plante de jour courts. Plumbago indica L., Ann. Sci. Natur. Bot. Serie 12 9: 1.

Njoku, E., 1958, Effect of gibberellic acid on leaf form, Nature 182: 1097

Pfirsch, E., 1978, Induction de la croissance plagiotrope chez le Stachys silvatica L. Role de l'acide abscissique dans le mécanisme autorépétitif, Bull. Soc. Bot. France 125: 231.

Robbins, W. J., 1957, Gibberellic acid and reversal of adult Hedera to a juvenile state, Amer. J. Bot. 44: 743.

Schaffalitzky de Muckadell, M., 1959, Investigations on aging of apical meristems in woody plants and its importance in silviculture, Det Forstlige Forsogsv. Danmark 25: 309.

Silvente, S., and Trippi, V. S., 1986, Sucrose-modulated morphogenesis in Anagallis arvensis L., Plant Cell Physiol. 27: 349.

Trippi, V. S., 1963, Studies on ontogeny and senility in plants. VI. Reversion in Acacia melanoxylon and morphogenetic changes in Gaillardia pulchella, Øyton 20: 172.

—— 1964, Studies on ontogeny and senility in plants. VII. Aging of the apical meristems, physiological age and rejuvenation in tomato plants Øyton 21: 167.

—— 1965, Studies on ontogeny and senility in plants. XI. Leaf shape and longevity in relation to photoperiodism in Gaillardia pulchella, Øyton 22: 113.

—— 1982, "Ontogenia y Senilidad en Plantas", Univ. Nac. Córdoba, Córdoba

—— and Brulfert, J., 1983, Organization of the morphophysiologic unit in Anagallis arvensis L. and its relation with the perpetuation mechanism and senescence, Amer. J. Bot. 60: 641.

Wellensiek, S. J., 1952, Rejuvenation of woody plants by formation of sphaeroblasts, Koninkl. Nederl, Akademie van Wetenschappen, Proc. Serie G 55: 567.

Wochting, H., 1904, Uber die regeneration der Araucaria excelsa, Jb. Wiss. Bot. 40: 144.

MATURATION AND SENESCENCE: TYPES OF AGING

Victorio S. Trippi

Laboratorio de Fisiología Vegetal, Univ. Nac. de Córdoba

Casilla de Correo 395, 5000-Córdoba, Argentina

INDETERMINATE AND DETERMINATE GROWTH

Plants are organisms of indeterminate growth because of their external meristems. However under natural conditions, this is not always true,
due to environmental influences. Certainly there are plants showing indeterminate growth like those capable of vegetative reproduction through
stolons, rhizomes, etc· i.e. Cynodon, Chlorophytum, Saxifraga, etc. These
plants may also flower but vegetative reproduction always takes place.
Other type of plants show determinate growth. Most of these plants flower
and fructify but rarely propagate vegetatively i.e. Triticum, Xanthium,
etc.

TYPE OF GROWTH AND ONTOGENETIC MORPHOGENESIS

Ontogenetic morphogenesis can be defined as the development of forms
ending with the formation of reproductive structures. In plants with
indeterminate growth like Musa, Pothos, Phylodendron, etc. ontogenetic
morphogenesis is very simple and agamic reproductive structures are
limited to rooted branches. These are the simplest forms of propagation.
Other species show more specialized agamic reproductive structures such
as stolons, tubers, bulbs, gemiferous roots, etc. but also flowers which
can appear simultaneously. These plants showing agamic and sexual reproductive structures may be considered as displaying a more complicated
ontogenetic morphogenesis, i.e. Solanum, Cynodon, Chlorophytum, etc.

In most cases determinate growth is observed in plants showing only
sexual reproduction like in monocarpics (wheat, maize, etc.). But in some
cases it also occurs in plants with vegetative reproduction like Lemna
(Wangermann, 1965). The counterpart also is observed i.e. Anagallis arvensis under experimental condition may show indeterminate growth while
having flowers and fruits. This is the case when plants under SD are
subjected to a LD periodically i.e. one LD every 15 days (Trippi and
Brulfert, 1973) and LD of low light intensity.

Plant Aging: Basic and Applied Approaches
Edited by R. Rodríguez *et al.*
Plenum Press, New York, 1990

MATURATION

Maturation refers to the qualitative changes that allow a plant or organ to express a potentiality. In flowering plants maturation can be defined as the changes that must take place for the plant to be able to form flowers. In both indeterminate and determinate plants this period of maturation varies from days (ephemerals) to years (perennials), depending on both genetic determination and environment.

Maturation should not only be connected with flower formation but also with the formation of other reproductive structures such as tubers, bulbs, etc. The lenght of the period of maturation is genetically determined but it can be modulated by the environment. While some plants like Pharbitis nil can be induced to flower at a cotyledonary stage, the case of perennials which have a long juvenile phase during which they can not react to photoperiod, suggests that endogenous factor/s is/are sometimes involved. Certainly, maturation may not be necessary for flowering in plants very sensitive to environment but in plants with heteroblastic growth, internal changes (or maturation) in meristems (aging) seem to be necessary.

THE TYPE OF GROWTH AND SENESCENCE

Senescence affects organs or organisms always associated with determinate growth. In true indeterminate-growth-plants senescence affects only organs but not the whole plant. Only in determinate-growth-plants senescence affects the whole plant. In this manuscript the term "plant senescence" is defined as the senescence of the whole plant.

The degree of senescence in indeterminate-growth-plants seems related to the lenght of the period during which agamic reproductive structures remain attached to the mother plant, senescence being observed always in the oldest part of the body. The new growth may remain attached to older parts for a long period and senescence and growth may occur simultaneously and continuously. In a multicaulinar system like Sequoia, Chlorophytum, Cynodon, etc, senescence affects only the older parts of the body before flowering; after flowering senescence spreads out from the apex to the base of the whole flowering branch, but never affects the whole plant. In these cases chronological senescence and that derived from the occurrence of flowers and fruits can be added up. In determinate plants, reproductive structures remain frecuently attached to the mother plant for a short time and then become independent. In these cases senescence and growth are not simultaneous phenomena and plant senescence results of both the chronological gradient (base to apex) and that originated in flower and seed formation (apex to base). Most woody plants may be classified as determinate-growth-plants, although they may look like indeterminate ones, because , once they have achieved the biggest size, growth stops and senescence affects progressively the whole plant, from the apex to the base (Fig. 1).

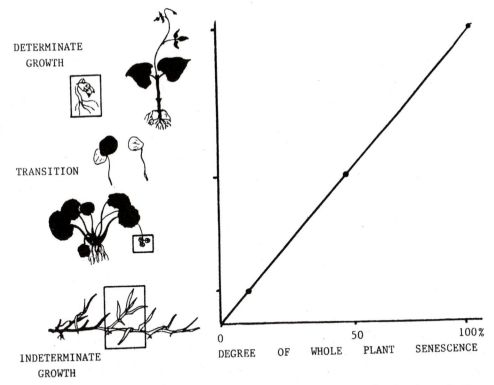

DETERMINATE
GROWTH

TRANSITION

INDETERMINATE
GROWTH

0   50   100%
DEGREE   OF   WHOLE   PLANT   SENESCENCE

Fig. 1. Senescence affects organs in indeterminate-growth-plants and the
whole plant in determinate growth ones.

GROUPS OF PLANTS

While organ senescence has probably an endogenous origin plant senescence seems to depend on environmental conditions. Regarding the influence
of environment on plant senescence, two or perhaps three groups of plants
can be considered.

A first group, almost insensitive to environment, in which indeterminate growth is genetically determined. These plants may form flowers and
fruits but maintain their capacity for indeterminate growth. In these
plants this genetic determination prevents senescence of the whole plant.
Cynodon, Saxifraga, Chlorophytum, Sequoia, etc. can be included in this
group.

A second group shows strong sensitivity to environment and weak
genetic determination to both determinate or indeterminate growth. These
plants may form different kinds of reproductive structures and senescence
affects either organ or the whole plant according to environmental
conditions. Whole plant senescence and determinate growth are induced by
the environment, especially by light intensity, photoperiod and/or
temperature.

Anagallis is a good example of plant senescence and type of growth as controlled by the environment. Anagallis arvensis, grown under 9 h-SD (14000 Lux, $\pm$ 140 W m^{-2}) is an indeterminate-growth-plant and a "open organization". Under these conditions senescence affects only organs (leaves) from the base to the shoot apex, and growth is homoblastic, with opposite leaves. Reproduction is only vegetative through natural rooting of plagiotropic branches (Trippi and Brulfert, 1973). The plant has expanded bush habit, tending towards indefinite projection in absence of external limiting factors.

Under LD conditions growth is determinate. Senescence affects progressively the regenerative capacity and the activity of apical meristems until the whole organism dies. Lateral branching is limited and branches loose rooting capacity.

Complementary observations showed that plants receiving just one inductive LD of 24 h complete its ontogenetic morphogenesis including flowering (Brulfert, 1965) but plagiotropic shoots still maintain their regenerative capacity. When Anagallis is transferred from LD to SD conditions plants revert to indeterminate vegetative growth (Trippi and Brulfert, 1973). Results suggest that LD determine flowering and senescence of the whole plant. Thus, LD can be considered a stress condition.

A third group of plants showing strong genetic determination to determinate growth, flowering and senescence can be suspected. This could be the case of plants growing as heterotrophic such as potatoes, peas and beans, which apparently can flower in complete darkness (Leopold, 1949). However senescence under such conditions has not been studied. These observations and others by Nitsch and Nitsch (1967) in the SD Plumbago indica and in the LD Anagallis arvensis (Silvente and Trippi, 1986) suggested that plant organization and ontogenetic morphogenesis are strongly dependent on sugars, which seem capable of modulating morphogenesis and ontogenesis. Complementary experimental evidence on this subject and on plant behavior under light intensities lower than those used in photoperiod studies seems necessary for accepting that entire plant senescence is in some cases genetically controlled.

ORGAN AND PLANT SENESCENCE

An internal determination of senescence can be recognized in the chronological gradient affecting organs, leaves and shoots, from the base to the apex in both indeterminate and determinate plants. Compensatory growth (Ashby, 1948), regulation of abscission (Rosseter and Jacobs, 1953) and chemical senescence parameters like chlorophyll, protein, DNA, etc. contents also regulated by elimination of parts (shoots, leaves) of the plants (Walkley, 1940; Mothes, 1960) or increase of longevity in rooted leaves (Chibnall, 1954), can be considered experimental evidence on internal regulation of senescence. Then senescence in organs like leaves can be considered as genetically determined since form and organization of plant depend on genes. However, it must be said that any alteration of correlative effects by elimination of flowers, fruits, etc. or isolation and leaf rooting, increases longevity but does not prevent senescence suggesting that another factor determining senescence exist.

A second gradient from the apex to the base of shoots, associated with the formation of flowers and seeds has also been recognized. This gradient affects organs (shoots) in plants showing indeterminate growth i.e. Cynodon, etc. Many experimental evidence confirms that elimination of flowers and fruits can delay senescence of either part or the whole plant (Molisch, 1938; Kelly and Davies, 1988). Whether or not this second senescence gradient affects only organs or, in determinate plants (monocarpics), it also affects the whole plant is a matter of discussion.

We consider senescence as a degradative process originated in the internal organization of unity, but flowering and sexual reproductive structures production seems rarely to be originated by an internal or genetic determination but induced by particularly light intensity, photoperiod, temperature and other environmental factors and substitution from one by others. Therefore, elucidation of whole plant senescence originated in reproductive structures effects, seems necessary to be studied in plants forming flower and fruit by internal determination. The use of photoperiodic or light sensitive plants makes easily possible of dealing with endogenous and environmental factors simultaneously.

In many plants such as perennials, it seems clear senescence is not strongly determined by the presence of flowers and other reproductive structures. Flowers and fruits generally result from internal readjustment derived from vegetative growth (when plants need a maturation period) and/or from environmental influences (light, temperature) which stimulate determinate growth or growth cessation. Therefore, flowering could result from stress conditions originated either in the internal plant organization (vegetative growth) and/or in environmental influences and can be considered an index of plant senescence (Fig. 2).

Summing up, whole plant senescence could result from the sum of three kinds of stress: one originated in vegetative growth, other originated in reproductive growth to which the environmental effect can be added.

MECHANISM INVOLVED

In both chronological gradient of senescence (from base to apex), the one originated in vegetative growth and that induced by reproductive structures, the relation source-sink fits well with Molisch's idea (1938) of growing points inducing senescence in older parts, as well as with the observed growth cessation and the experimental evidence available today.

The source-sink relation is controlled by hormones-induced translocation of organic (Mothes, 1960) and inorganic ions (Lewin and Bukovac, 1965). Hormone balance may induce senescence or rejuvenation irrespective of the age of the organ (Leopold and Kawase, 1964) and may act by taking

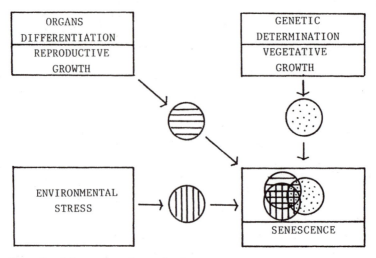

Fig. 2. Schematic view of the origin of senescence in
plants. Vegetative growth itself has the
property of bringing about senescence, according to a chronological pattern. Organ differentiation (reproductive growth) and environmental stress add their senescence determining
effects.

the place of reproductive structures as the sink site of the plant
(Seth and Wareing, 1967; Woolhouse, 1983). Hormonal regulation of membrane
permeability and metabolism can be responsible of determining the behavior
of a tissue or organ as a sink or a source.

The source-sink relation can also be influenced by chemicals like
sugars, capable of increasing membrane permeability in light conditions
(Luna and Trippi, 1986), of decreasing photosynthesis (Lovell et al.,
1972; Nafziger and Koller, 1976) and of inducing an oxidative stress
leading to cell degradation.

Other questions remain to be elucidated. How is genetic determination
connected to senescence? Which are the endogenous signals?.

Comparing "source" and "sink" parts of the plant it becomes clear
that the source (at least in most cases) is the oldest part having cells
larger than those of the sink or youngest part which has meristematic
cells. It can be said that senescence and rejuvenation seem to depend on
cell size. When yeasts divide in the presence of O_2 it produces a large
and older mother cell which will senesce after a few sprouts and a small
and younger daughter cell which will continue growing. Is there a genetic
control of unequal cell division? Is there an environmental determining
factor?

REFERENCES

Ashby, E., 1948, Studies on the morphogenesis of leaves. II. The area, cell size and cell number of leaves of Ipomea in relation to their position on the shoot, New Phytol. 47: 117.

Brulfert, J., 1965, Physiologie de la mise a fleurs d'Anagallis arvensis L. et du developpement vegetatif en retour des meristemes floraux, Bull. Soc. Franc. Physiol. Vég. 11: 247.

Chibnall, A. C., 1954, Protein metabolism in rooted runner-bean leaves, New Phytol. 53: 31.

Kelly, M. O., and Davies, P. J., 1988, The control of whole plant senescence, CRC Critical Reviews in Plant Sciences 7: 139.

Leopold, A. C., 1949, Flower initiation in total darkness, Plant Physiol. 24: 359.

—— and Kawase, M., 1964, Benzyladenine effects on leaf growth and senescence, Amer. J. Bot. 51: 294.

Lewin, I. J., y Bukovac, M. J., 1965, Efecto de la N Bencil adenina sobre la absorción y el transporte de P32 aplicado por vía foliar a plántulas de poroto (Phaseolus vulgaris L.), Rev. Inv. Agrop. Serie 2 2: 63.

Lovell, P. H., Illsley, A., and Moore, K. G., 1972, The effect of light intensity and sucrose on root formation, photosynthetic ability and senescence in detached cotyledons of Sinapis alba and Raphanus sativus, Ann. Bot. 36: 123.

Luna, C. M., and Trippi, V. S., 1986, Membrane permeability-regulation by exogenous sugars during senescence of oat leaf in light and darkness, Plant Cell Physiol. 27: 1051.

Molisch, H., 1938, "The longevity of plants", H. Fulling, ed., New York.

Mothes, K., 1960, Uber das Altern der Blatter und die Moglichkeit ihrer Wiederjungung, Naturwissenschaften 47: 337.

Nafziger, E. D., and Koller, H. R., 1976, Influence of leaf starch concentration on CO_2 assimilation in Soybean, Plant Physiol. 57: 560.

Nitsch, C., and Nitsch, O. P., 1967, The induction of flowering in vitro in stem segments of Plumbago indica L. II. The production of reproductive buds, Planta 72: 371.

Rosseter, F. N., and Jacobs, W. P., 1953, Studies on abscission: The stimulating role of nearby leaves, Amer. J. Bot. 40: 276.

Seth, A. K., and Wareing, P. F., 1967, Hormone-directed transport of metabolites and its possible role in plant senescence, J. Exp. Bot. 18: 65.

Silvente, S., and Trippi, V. S., 1986, Sucrose-modulated morphogenesis in Anagallis arvensis L., Plant Cell Physiol. 27: 349.

Trippi, V. S., and Brulfert, J., 1973, Organization of the morphophysiologic unit in Anagallis arvensis and its relation with the perpetuation mechanism and senescence, Amer. J. Bot. 60: 641.

Walkley, J., 1940, Protein synthesis in mature and senescent leaves of barley, New Phytol. 39: 362.

Wangermann, E., 1965, Longevity and ageing in plants and plant organs, in "Enc. Plant Physiol.", Rhuland, ed., Springer-Verlag.

Woolhouse, H. W., 1983, Hormonal control of senescence allied to reproduction in plants, in "Strategies of Plant Reproduction", W. J. Meudt, ed., U.S. Dept of Agriculture, Washington.

ADULT vs. JUVENILE EXPLANTS: DIRECTED TOTIPOTENCY

D. J. Durzan

Department of Environmental Horticulture
University of California
Davis, CA 95616 USA Fax (916) 752-1819

Introduction

The exploitation of stages of the life cycle of a woody perennial through cell and tissue culture is examined in terms of explant selection and discrete developmental outcomes. For mass propagation, explants from adult trees require rejuvenation, dedifferentiation, induction and redifferentiation. These processes traverse phases of the life cycle and involve cells having cell-line replicator activity. The components of the rejuvenation process, replicator monitoring and the concept of directed totipotency are illustrated with *Prunus* sp. and with some conifers. The result leads to some general notions for diagnostic reasoning in ontogeny under artificial conditions and different genetic backgrounds.

Rejuvenation and Phase Changes

By rejuvenation we normally mean to "make young or youthful again." "Reinvigorate" is also used to mean "restore to an original or new state of development." Explants represent physiological gradients in the tree (*cf.* Bonga, 1982). Gradients represent gene expressions that have positional, developmental and microclimatic histories (Durzan, 1984a). These histories should be removed as cells pass through juvenile to embryonic phases of development. Judgements on how this is best done are based on genotypic characteristics and model-references. In practice, the successful and predictable control of the rejuvenation process represents an enormous theoretical and practical problem. The problem has to be decomposed into all traversed phases of the life cycle, i.e., we have to sequentially and logically recover sets of physiological states in regularly recurring, reversible cycles of change. Each characteristic phase should have well-defined developmental set points associated with an appropriate histogenic algorithm. Rejuvenation or reinvigoration should also be "phased," i.e., adjusted so as to be in a synchronized condition or scheduled as required.

The expression of totipotency in rejuvenated cells should recapitulate without constraint the normal phases of the life cycle. Brink (1962) has provided evidence for the genetic basis for phase changes. Brink's concepts have been extended to the molecular level to include biochemical genetics, metabolic control theory and molecular phenogenetics (Durzan, 1990ab).

In pomology, juvenility impedes fruit cultivar development. Prospects for reducing the length of the juvenile period have been outlined by Hansche (1983). The genes influencing the length of the juvenile phase are primarily additive (Visser, 1976). Mass selection should be effective in genetically reducing the juvenile phase.

Plant Aging: Basic and Applied Approaches
Edited by R. Rodríguez *et al.*
Plenum Press, New York, 1990

Evidence for how embryony can facilitate propagation of adult ideotypes is not yet available. However, attempts at rendering histogenic algorithms and the appropriate process controls for the recovery of embryos have begun. Current emphasis is placed on developing true-to-type models that include polyembryony and the origin of meristems (primary and secondary) that contribute to juvenile and adult forms. Sphaeroblasts in cell suspensions develop uniquely with a cambial-like meristem (Durzan, 1982). Sphaeroblasts are organogenic, i.e. from the cambial-like cells, new primary meristems (root, shoot) are restored. We now have a way to generate at will either primary or secondary meristems through a callus phase (Figure 1).

The adult phase is characterized by its reproductive habit, developed primary and secondary meristems, morphological and anatomical features and by the reallocation of resources to foster the fruiting habit. Cells in most adult trees are sufficiently variable to encourage us that as explants, especially from meristems and new flushes of growth, will indeed undergo reversible and directed phase changes.

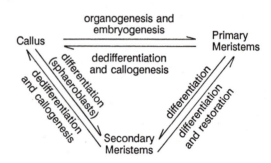

Figure 1. The sequence of alternatives in restoring meristematic expressions in cells explanted from different phases of the life cycle of a tree.

Cell Line Replicators and Mitotic Events in Reversible Phase Changes

The term "replicator" characterizes an essential property of genes that relates to regeneration (Durzan, 1988b; Dawkins, 1982). This term usually refers to the genetic replicators (Dawkins, 1982), and represents several types. Cells are best not considered as replicators, because they are vehicles in which replicators express themselves. Replicators are DNA molecules of which copies are made. Copies are made in daughter cells as cells grow, divide, differentiate and pass through the sequences in Figure 1. An <u>active</u> replicator has some influence over its probability of being copied in a daughter cell. A <u>passive</u> replicator has no influence over its probability of being copied. A <u>promeristem-cell-line</u> replicator, which may be active or passive, is potentially the ancestor of an indefinitely long line of descendent replicators recoverable from daughter cells, i.e. it is potentially immortal. A gene encoding the meristematic process in a cell from adult trees may be a replicator. A <u>dead-end</u> replicator, which also may be active or passive, is copied a finite number of times. It may give rise to a short chain of descendents and is not a potential ancestor of an indefinitely long line of descendents. Short chains of descendents tend to become highly differentiated and specialized. Most cells in adult trees may be considered as being determined by dead-end replicators. Through sorting, searching and testing, replicator genes whose product is "acetocarmine reactive" may be selected for. This replicator product is found in cells that have embryogenic and polyembryogenic potential. Since acetocarmine is not always specific, a more specific probe will eventually be required for detecting macro-molecular products of replicator activity.

In the restoration of embryony through phase changes we need to consider 1) non-nuclear inheritance and the role of organelles (contributed by the paternal parent) in regenerating embryos, 2) the rejuvenation of somatic cells from adult trees having replicators of various types, and 3) the globally regulated expression of developmental genes as daughter cells pass through consecutive cell cycles.

Gene expressions linked to replicator activity are based on: a) self-similarity preserving expressions (codes) as expressed in polyembryogenesis (multiple embryo output), b) difference-preserving gene expressions, as evident in the differentiation of a proembryonal cell to an embryonal tube in somatic embryogenesis (single embryo output), and c) callus in which daughter-cell growth patterns and gene expressions are comparatively unorganized and redundant.

Given the recombinant DNA and restriction fragment length polymorphism (RFLP) technologies, we can expect to define basic units of gene activity needed for rejuvenation, phase change and embryogenesis. Gene expression must link with the activity of promeristematic-cell-line replicators and meristematoids of various types. Each combination of replicator activity enables the expression of histogenic algorithms through gene promoters that direct future ontological events in a reversible way.

For gene expressions having discrete phase specificity, it is now possible to diagnose and probe information flow as in Figure 2.

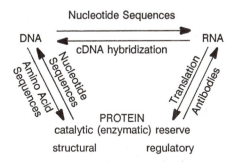

Rejuvenation *In Vitro*

So far, we have cited the need to sort, search, test and select for cells throughout the life cycle having an active promeristem replicator activity based

Figure 2. Molecular bases to diagnose information flows among macromolecules in cells expressing reversible phase changes.

on the production of acetocarmine-reactive products. Ontogenetic expressions pass normally through a feedback control sequence as follows:

Explants from the adult phase have a history based on the fate of cell lineages determined by at least three components: a) topological position in a system of correlations, b) a developmental history based on genomic rules, and c) an environmental fix on the physiological state based on the nuclear-cytoplasmic cycle of determination (Brink 1962).

All developmental outcomes relate to the activities of vegetative <u>and</u> reproductive meristems. Replicator activity leads to discrete adult ideotypes. Outcomes are deterministic and somewhat stochastic, especially where genotype × environment interactions are selective factors. In the juvenile phase, most of the growth and vigor of the tree goes into wood formation. This activity is dominated by replicators in the secondary cambial meristem and crown surface.

In the <u>embryonic phase</u>, replicator activity occurs in acetocarmine-reactive proembryonal cells (pE) that reconstitute multiple proembryos by polyembryogenesis. In hardwoods and fruit trees, similar acetocarmine-reactive cells are seen, but the histogenic algorithms are not well understood, nor are plants always reliably regenerable from these cell lines.

To rejuvenate and restore cells in explants from the adult phase and to recover the full potential for embryony, we have to 1) remove (i.e. subtract) the history of the adult and juvenile phases and, to some extent, the later stages of the embryonic phase

$$\begin{matrix} \text{Rejuvenation based} \\ \text{on selection and} \\ \text{recovery of promeristematic} \\ \text{cell line replicators} \end{matrix} = \alpha \begin{pmatrix} \text{Adult-Adult history} \\ \text{topophysis} \\ \text{cyclophysis} \\ \text{periphysis} \end{pmatrix} - \beta \begin{pmatrix} \text{Juvenile-} \\ \text{Juvenile} \\ \text{history} \end{pmatrix} - \gamma \begin{pmatrix} \text{Embryo-Late} \\ \text{Embryonal} \\ \text{history} \end{pmatrix}$$

21

and 2) reset development to recapitulate histogenic algorithms based on active pro-embryonal cell line replicators. Here α, ß and γ are the probabilities of achieving each reversible step in the rejuvenation and restoration process. Each step involves tactical, structural, functional, developmental, positional, nutritional and environmental input parameters. Moreover, since many biological parameters are not easily dealt with in Euclidean dimensions, we may have to describe the phenomenal and process controls in more nondimensional fractal terms (Mandelbrot, 1982), using the appropriate model references (Durzan, 1988a).

For polyembryogenesis in dedifferentiated cells, a cell mass balance equation for a batch run over time t has been formulated (Durzan and Durzan, 1990). This balance equation is based on pE cells embodying active DNA replicators with promoters and other basic genetic elements. The use of restriction fragment length polymorphism techniques will emerge as a useful tool to define and locate DNA fragments that can probe the rejuvenation of explants aimed at the reconstitution of somatic embryos.

Replicator Penetrance and Expressivity

Replicator activity can be traced by the informational sequences shown in Figure 2. It has historically been described at various levels of gene expression. The polymerase chain reaction (PCR) is a powerful technique for amplifying specific rare sequences from genomic DNA. At least a 10^5-fold amplification can be achieved in a few hours. The amplified fragments are easily visualized on ethidium bromide-stained agarose gels, so that genomic sequences can be rapidly detected without hybridization analysis. Amplification enables the study of the rejuvenation process, disease, mutations, DNA sequences, rare mRNAs for cloning and sequence analysis, and the segregation of introduced genes in transgenic plants. The location of foreign genes, replicators, promoters and other genetic elements could possibly be mapped by RFLP methods (e.g., Tanksley et al., 1989).

Directed Totipotency: Somatic Parthenocarpy

Reversible phase changes in woody perennials enable the isolation of specific expressions of the life cycle. For example, with a feedback approach, explants from mature trees may be rejuvenated for clonal propagation (Gupta and Durzan, 1987) and possibly embryogenesis (Durzan, 1988b), provided the propagation is massive and the gene expression is true-to-type. In a feedforward approach, the precocious expression of genes in the mature tree may be facilitated in explants that are not fully rejuvenated or already programmed for aspects of mature development. An example of precocious feedforward development is seen in the recovery of micropropagated shoots of Douglas fir by rejuvenation from 60-year-old trees. Some shoots continue to develop female cones *in vitro* (Gupta and Durzan, 1987).

With fruit trees, the totipotency of cells of suspension cultures of petiole callus from a mature (28-year-old) cherry rootstock (*Prunus cerasus* cv. Vladimir) can be directed to alternative fates (Durzan, 1988c). Cell suspensions (initially screened to 60-300 μ cell dia.) can be programmed with plant growth regulators (PGRs) for the production of roots or for low frequency somatic embryogenesis (rejuvenation) (Hansen and Durzan, 1986). By contrast, cells from the same suspension, when shifted onto a different medium, act like dead-end replicators and restore cell masses that resemble the ripening mesocarp. The whole suspension turns red like the cherry fruit, produces strong cherry fragrances and deteriorates after 2 to 3 weeks (maturation, ripening and senescence). These "directed-phase shifts" illustrate the importance of PGRs and cultural conditions in the expression of totipotency along divergent paths.

In fruit having a climacteric respiratory rise, an irreversible phase shift occurs (*cf.* Steward, 1968). The irreversibility marks the final expression of dead-end replicators. The forcing of a precocious expression of fruit characteristics in cell suspensions is useful in screening genotypes before propagules from these cell populations produce fruit under field conditions.

Temporal Mapping of Regeneration

The temporal mapping of "explant regenerative potential" is possible as follows:

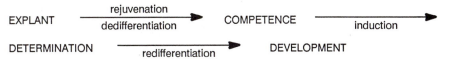

Temporal mapping is based on the optimal time that explants spend on dedifferentiation, induction and redifferentiation media. Media should be optimized for each discrete initial state and developmental outcome. In this scheme, dedifferentiation is the reversion of specialized cells to a more competent or primitive condition as a preliminary step to further major changes. Induction is the sum of processes by which the fate of competent cells is determined. The competence and determination of a cell or tissue refers to a whole range of physiological states during development. These states are assayed by the response of cells to external stimuli, e.g., PGRs and media changes.

Competent cells tend to be dedifferentiated and respond to an inductive agent, i.e. they become "induced." Competent cells change from state n to state n+1 upon exposure to the inductive signal. If a second exposure to the same inductive signal maintains the n+1 state without change, the cells cannot be judged to have been competent by this criterion alone. If the second exposure to the inductive signal causes the cells to change to a n+2 state, then the cells are judged competent for the n+1 to n+2 induction step. The end result is a causal series of inductions that determines a discrete developmental outcome.

Differentiation is the sum of processes whereby apparently generalized or primitive cells, tissues and structures attain their adult form and function (Glock and Gregorius, 1984). Differentiation leads to division of labor whereby tissues in their development become unlike one another. The differentiation of dedifferentiated cells is usually referred to as redifferentiation.

The sequential simplification of culture media used by FC Steward (1968) indicates that shifts from heterotrophic to autotrophic conditions are also needed to regenerate plants from cells, i.e. temporal mapping of developmental stages should include effects of media enrichment and starvation under uniform and controlled conditions. Events leading to final outcomes may be displayed by physiological state-network maps that derive from the sequences shown in Figures 2 and 3 (Durzan, 1990ab).

Plant Growth Regulators in Phase Specificity

Metabolic behavior and the resultant thresholds in experimentally determined PGR-directed phase specificity can be depicted as "signal surfaces" by computer graphics (Durzan, 1987b, 1990ab). These metabolic spreadsheets have been used for diagnostic reasoning in the developing zygotic embryo of the pistachio, for the developing epi- and mesocarp of cling peach, and for characterization of the bud-failure syndrome in the almond. As each phase is traversed, the emerging specific phenotypes, as directed by PGR activities, have several measurable properties. Metabolite fluxes are used to characterize the emerging phenotypes. The departure of fluxes from equilibrium or mean values reveals maxima and minima that "time-stamp" the response surface. Each threshold helps to characterize the physiological state (n, n+1, ...) and carries with it an intensity (rate and magnitude) as a function of the state of the system and of the metabolic networks. The sequential appearance of thresholds represents gene expression in terms of penetrance and expressivity (Durzan, 1990b). Some thresholds may represent discrete signals that feedback or feedforward to help control the nuclear-cytoplasmic cycle of cell determination (Brink 1962). Now that the behavior of metabolites can be portrayed over development for most phase changes, how are the PGRs involved in controlling these processes?

Diagnostic Reasoning in Ontogeny

Diagnostic reasoning for the study of phase shifts is based on genetic algorithms, critical signals, and dynamic thresholds in the life cycle (*cf.* Figures 1, 2 and 3).

Variations in metabolic phenotypes at various levels (Fig. 1) and under the influence of PGRs and specific cultural parameters can now be diagnosed in terms of pattern, threshold values, costs for biosynthesis, stability, bifurcation, global relatedness, engrams, noise, percent misadjustment, and process control. Biosynthetic and catabolic capacities of explants should be correlated with a wide range of PGR-directed phase specificities. For microorganisms, a rapid redox chemical method has been devised to measure global phenotypes using microwell plates and computerization (Bochner, 1989). Carbon source utilization is revealed by a color change.

Diagnostic evaluations involve quantitative genetic definition, gene expression and evolutionary strategies (Wright, 1968) with the following PGR-interactive parameters:

As for the PGR-directed strategies inherent in phenotypic dynamics, we can postulate that for any cell line the net effect or response portrayed by the signal-surface pattern is unbeatable under natural selection. In phasic development, the signal surfaces should represent true-to-type gene expression, even though patterns are an artificial geometrical representation of events. As for

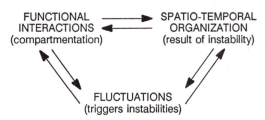

Figure 3. Interactive dynamic behaviors among large data sets help to understand and characterize phenotypic expression. Fluctuations are represented as state-network maps (Durzan, 1987) and their diagnostic variations (Durzan, 1990ab).

cultural practices, we should look for the change in responsiveness of the signal surface upon the removal or introduction of constraints. This will enable us to evaluate performance and irritability of the phenotype through the behavioral geometry of responsive parameters. Where trees are cloned by *in vitro* methods, the developmental stability, origin of somaclonal aberrations, and points of phenotypic bifurcation need to be evaluated in terms of PGR-directed process control for each specific phase.

Over the long run, more quantitative and diagnostic approaches could reduce the time for progeny field-testing and assist in establishing juvenile-mature correlations. Metabolic phenotypes could be useful, especially to determine how PGRs program plant development and how totipotent cells could be selectively and genetically modified.

We appear to be at least a decade or so away from applications of models with PGRs in understanding directed totipotency and phase changes. Phenotype expression derived from metabolic control theory should lead to the development of common metabolic command languages for exploitation. More comprehensive signal surfaces for gene expressions, based on current dogma and improved algorithms, will be useful to further our understanding of growth, differentiation and morphogenesis. "Process control systems" regulating mass clonal propagation should enhance future genetic approaches seeking to introduce new and useful genetic variation.

References

Bochner, B.R., 1989, Sleuthing out bacterial identities, *J. Genet.*, 65:45-54.

Bonga, J.M., 1982, Vegetative propagation in relation to juvenility, maturity and rejuvenation, in: "Tissue Culture in Forestry," J.M. Bonga and D.J. Durzan, eds., Martinus Nijhoff, Dordrecht.

Bonga, J.M., von Aderkas, P., and James, D., 1988, Potential application of haploid cultures of tree species, in: "Genetic Manipulation of Woody Plants," J.W. Hanover and D.E. Keathley, eds., Plenum Press, N.Y., pp. 57-77.

Brink, R.A., 1962, Phase change in higher plants and somatic cell heredity, *Q. Rev. Biol.*, 37:1-22.

Dawkins, R., 1982, "The Extended Phenotype," WH Freeman Co., San Francisco, CA.

Durzan, D.J., 1982, Somatic embryogenesis and sphaeroblasts in conifer cell suspensions. Proc. 5th Intl. Cong. Plant Cell Tissue Culture, July 11-16, 1982, Tokyo, Japan, 113-114.

Durzan, D.J., 1984a, Special problems: Adult vs. juvenile explants, Chap. 17, in: "Handbook of Plant Cell Culture, Crop Species," W.R. Sharp, D.A. Evans, P.V. Ammirato, and Y. Yamada, eds., MacMillan Publishing Co., N.Y. 2:471-503.

Durzan, D.J., 1984b, Potential for genetic manipulation of forest trees: Totipotency, somaclonal aberration, and trueness to type, in: "Proceedings International Symposium on Recent Advances in Forest Biotechnology," July 10-13, 1984, Michigan Biotechnology Institute, Traverse City, Michigan, pp. 104-125.

Durzan, D.J., 1987a, Plant growth regulators in cell and tissue culture of woody perennials, *Plant Growth Regulation*, 6:95-112.

Durzan, D.J., 1987b, Physiological states and metabolic phenotypes in embryonic development, in: "Cell and Tissue Culture in Forestry," J.M. Bonga and D.J. Durzan, eds., Martinus Nijhoff/Dr. W. Junk, Dordrecht. Vol. 2, pp. 405-439.

Durzan, D.J., 1988a, Somatic polyembryogenesis and plantlet regeneration in selected tree crops, *Biotech. Gen. Eng. Revs.* 6:339-376.

Durzan, D.J., 1988b, Process control in somatic polyembryogenesis, in: "Molecular Genetics of Forest Trees," J.-E. Hällgren, ed., Frans Kempe Symp., Umea, Sweden, pp. 147-186.

Durzan, D.J., 1988c, Applications of cell and tissue culture in tree improvement, in: "Applications of Plant Cell and Tissue Culture," Ciba Foundation Symp. 137, John Wiley & Sons, N.Y. pp. 36-49.

Durzan, D.J., 1990a, Performance criteria in response surfaces for metabolic phenotypes of clonally propagated woody perennials, in: "Application of Plant Biotechnology in Forestry," V. Dhawan, ed., Plenum Press, N.Y. (in press).

Durzan, D.J., 1990b, Molecular phenogenetics as an aid to fruit breeding, *Acta Hort.* (in press).

Durzan, D.J. and Durzan, P.E., 1989, Future technologies: Model-reference control systems for the scale-up of embryogenesis and polyembryogenesis in cell suspension cultures, in: "Micropropagation," P. Debergh and R. Zimmerman, eds., Martinus Nijhoff, Dordrecht. (in press).

Glock, H. and Gregorius, H.-R., 1984, Differentiation -- A consequence of ideotype-environment interaction, *BioSystems*, 17:23-34.

Gupta, P.K. and Durzan, D.J., 1987, *In vitro* establishment and multiplication of Juvenile and mature Douglas-fir and sugar pine, Symposium: "*In Vitro* Problems Related to Mass Propagation of Horticultural Plants," September 16-20, 1985, Gembloux, Belgium, *Acta Hort.*, 212:483-487.

Hansche, P.E., 1983, Response to selection, in: "Methods in Fruit Breeding," J.N. Moore and J. Janick, eds., Purdue Univ. Press, West Lafayette, Indiana pp. 154-171.

Hansen, K.C. and Durzan, D.J., 1986, Somatic embryogenesis and morphogenesis in *Prunus cerasus*, Proc. XXII Intl. Hort. Cong., August 10-18, 1986, University of California, Davis, CA. Abstr. No. 1130.

Mandelbrot, B.B., 1982, "The Fractal Geometry of Nature," W.H. Freeman, N.Y.

Steward, F.C., 1968, "Growth and Organization in Plants," Addison-Wesley Publ. Co., Reading, Mass.

Tanksley, S.D., Young, N.D., Paterson, A.H., and Bonierbale, M.W., 1989, RFCP mapping in plant breeding: New tools for an old science, *Bio/Tech.*, 7:257-264.

Visser, J., 1976, A comparison of apple and pear seedlings with reference to the juvenile period, II. Mode of inheritance, *Euphytica*, 25:334-342.

Wright, S., 1968, "Evolution and the Genetics of Populations," Vol. 1, Univ. Chicago Press.

RECOVERY OF TRANSIENT JUVENILE CAPACITIES DURING

MICROPROPAGATION OF FILBERT*

Díaz-Sala, C. Rey, M.; Rodríguez, R.

Lab. Fisiología Vegetal, Dpto. B.O.S. Facultad de
Biología. Universidad de Oviedo
Arias de Velasco, s/n; 33005 Oviedo; Spain

In the last few years several reports on maturation and
rejuvenation have appeared (Greenwood, 1987; Bonga and
Durzan, 1987; Monteuuis, 1988) but discussions on this
controversial area are common and still represent a
phenomenon which is only partially understood due in part to
the complexity of the physiological processes involved and of
course due to the difficulties found in timing this period in
the plant cycle. However most authors agree that maturation
and aging seem to be responsible for the morphogenic
potential decline found in most   woody species. As is
common, micropropagation pathways found in filbert (Corylus
avellana L.) are strongly linked to the chronological age and
of course to the ontogenical and physiological changes
related; situations that may be at least partialy exogenously
manipulated and probably reversed.

Due to the enormous contradictions that actually exist,
in this paper it was decided to avoid the terms rejuvenation
or reinvigoration, the Title proposed may be a little
dogmatic but it seems to be more realistic, since the
exogenous manipulation carried out in the tested tissues
facilitates **the manifestation of morphogenic patterns in
mature tissues which are easily expressed in juvenile ones.**
These effects are not permanent and mainly depend on the
culture media used.

The possibilities offered by in vitro culture techniques
have been used for characterizing the phase-changes that
occur during culture, and for determining the properties of
Mature, Juvenile and Embryonic clones. The First differences
found between the tissues dealt with in vitro culture:

-----
* Granted by Junta Nacional de Investigación Científica y
  Técnica Poj. nº

I in vitro culture responses of embryonic, juvenile and mature tissues.

In the experiments carried out during the last three years working with filbert (Rodríguez et al., 1989) of different varieties and tissues and explants of various chronological ages, the current status of tissue culture patways can be summarized as follows.

- Cotyledonary nodes taken from seedling were able to yield a good rate of shoot proliferation and callus induction in a defined media. By simply changing the cytokining/auxin ratio embryoid induction with further plantlet regeneration was also achieved (Fig. 1a, b, c, d).

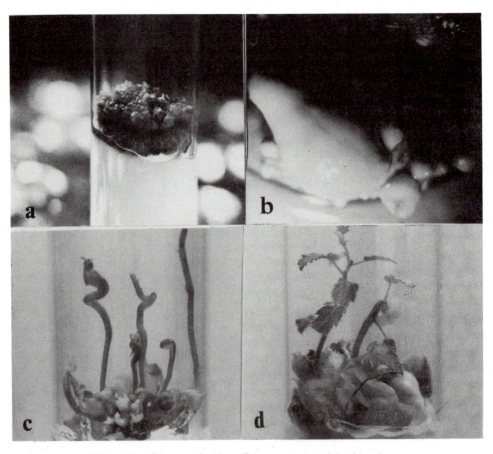

fig. 1. Some of the Responses obtained from cotyledonary nodes taken from 30 day old seedlings: a) callus induc--tion, b) embryoid formation and c), d) different stages of further plantlet regeneration.

-Shoot proliferation and rooting rates counted in explants taken from different locations along the main shoot of the seedlings seem to be correlated to the intraclonal zonation or topoclonal concepts. The most responsive tissues were always those close to the roots.

– When juvenile or adult tissues were used, superficial and endogenous contamination were always the roadblocks to efficient process for micropropagation. The use of primary explants removed after forced-outgrowth in field grown branches allow in vitro culture initiation.

– With mature tissues in the morphogenic pathways was a reduction always present, making it possible to induce shoot proliferation and low rates of callus induction, but not embryoid formation. As tissues became older a more complex media with regard to growth regulators was necessary. The impact of a high concentration of cytokinin during the first subcultures greatly favoured the establishment of cycloclonal lines.

– Dealing with mature tissues the effect of a double-phase culture system greatly enhanced the elongation of the arrested shoots, also reducing the length of the culture period. (fig. 2 and 3).

fig. 2. Shoot-bud induction on the basal end of mature – explants. Note the poor elon-gation, that occurs on the – single-phase culture system.

fig. 3. Shoot proliferation and elongation achieved by the double-phase culture sys-tem. Morphogenic responses – obtained from derived ex- – plants are similar to those in fig. 1.

– Subsequent sulcultures may favour proliferation and

rooting rates on mature tissues. Furthermore it was also demostrated that tissues coming from mature explants which have already been in several subcultures may behave in vitro quite similarly to those derived from seedlings. These tissues underwent increases not only in shoot-bud proliferation, but also in callus and embryoid induction.

Therefore it seems clear that: 1) there is a decline of morphogenic capacities with age, but, 2) the reappeareance of potential juvenile capacities by means of tissue culture is possible.

Since techniques which allow the shortening of the interval between phase-changes have immense practical applications; biochemical research could provide a very promising opportunity to progress in the study of maturation and relevant rejuvenation possibilities. Thus the methods available from in vitro culture have been used to obtain responses from seedlings, juvenile and mature tissues, which were analyzed with these aims in mind.

II Molecular analyses of the phase-changes in plant tissues.

Efforts have been made by several authors, for example the architectural approach done by Barthelemy et al in 1989, dealing with aging in trees; or the studies related to the differences between dormancy and active growth in woody species carried out by several authors including Champagnat in 1989.

However the molecular approach to this subject still remains quite proplematic (Arnoud et al, 1988), in this sense despite the number of studies, general information in this area is fairly limited and sketchy.

J. Allemand et al. described in 1985 and 1988, the polyphenolic and enzymatic characterization of aging and rejuvenation in hybrid walnut trees, indicating that aging seems to be linked with an accelerated and unsynchronized functioning of GDH on the one hand and the pentose phosphate pathway and PAL on the other. Moreover adult shoots were characterized by a late accumulation of myricetine-derived compounds. Rejuvenation modified this pattern by inducing a synchronization between GDH, pentose phosphate pathway, and PAL during maximum growth acceleration, coinciding with a specific accumulation of polyphenols, probably p- coumaric derived compounds. The results described indicated that the analysis of phenolic compounds could show a possible biochemical expression of the rejuvenation phenomena.

The first study done in filbert to determine biochemical causes related to different morphogenic potential, was carried out using apical, basal and cotyledonary nodes removed from in vitro germinated seedlings. All the biochemical data quantified revealed the existence of a specific-physiological state for those tissues with higher plasticity, namely cotyledonary nodes (fig. 4 a, b, c, d).

These tissues also showed the highest activity level in the biochemical parameters measured: - respiration level,

soluble and reducer sugar content as well as nucleic acids,
peroxidase activity and proteins. Results which may be
interpreted as the correlation found between intraclonal

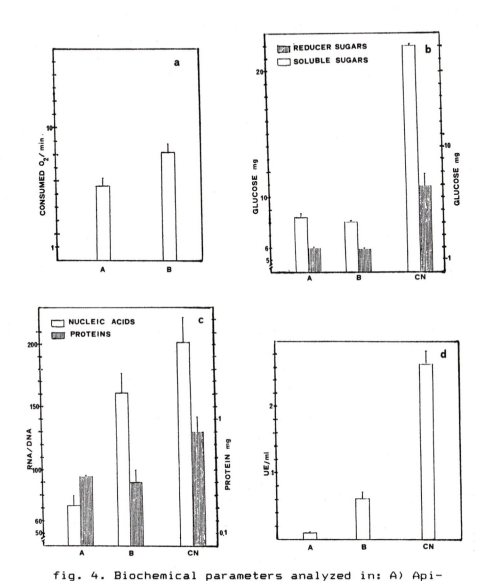

fig. 4. Biochemical parameters analyzed in: A) Apical, B) Basal and CN) Cotyledonary nodes of 30 days
old of seedlings.
Data expressed by fresh weight are: a) res-piration
level, b) soluble and reducer sugars c) nucleic acid
and protein content and d) peroxidase activity.

zonation and future behaviour in vitro.

All the proteins analyzed on a 5-20% polyacrylamide
linear gradient slad gels revealed the existence of different
patterns band, making it possible to divide the general

zimograms into two main parts, those lower than 66 kd and those of a higher molecular weight.

By densitometric analysis of coomassie blue stained proteins bands, the next patterns were obtained (fig. 5, a, b, c, d, e). By using several molecular weight marker polipeptides, it was possible to make a general analysis in the following terms:

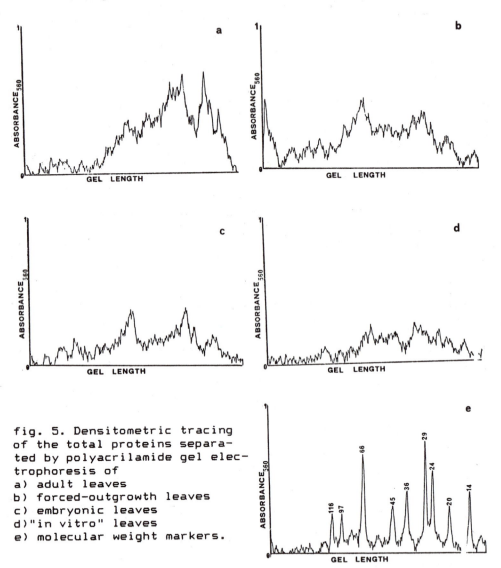

fig. 5. Densitometric tracing of the total proteins separated by polyacrilamide gel electrophoresis of
a) adult leaves
b) forced-outgrowth leaves
c) embryonic leaves
d)"in vitro" leaves
e) molecular weight markers.

- There is a great similarity between densitometric profiles found in in vitro forced-outgrowth and embryonic tissues, but those of mature samples showed significant differences.

- There is a range 40-60 kd, which is common among in vitro, forced-outgrowth and embryonic tissues.

- The range between 24-36 kd. are similar in mature, in

32

vitro and forced-outgrowth tissues.

- The ranges around 14.2 and 20.1 kb. seems to be specific to mature tissues.

Although the differences observed could be attributed to varying growth intensities, the preliminary results described may indicate the existence of specific protein markers.

Genomic DNA from leaves of different chronological and physiological ages has been purified to relate age with specific genomic DNA methylation pattern.

DNA purification from filbert leaves: embryonic, juvenile, mature trees, "in vitro" plantlets from mature trees and forced outgrowth under controlled conditions were tested tryed by using protocols (Murray and Thompson, 1980; Dellaporta et al. 1983; Rogers and Bendich, 1985), to purify DNA without using cesium chloride gradients, unfortunatly limitations were also present.

Methods based on Dellaporta et al. (1983) procedures do not allow us to obtain successful results. The analysis of purified DNA by spectrophotemetry and agarose gel electrophoresis showed a high content of contaminants which interfere with the quantification of nucleic acids by spectophotometric methods and do not allow the DNA to run into the agarose gel (fig. 6).

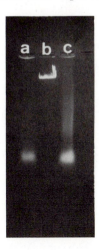

fig. 6. Analysis of purified DNA using Dellaporta et al. (1983) procedure in agarose gel.
a) DNA from mature leaves
b) DNA from λ cI 8571
c) DNA from junvenile samples

Techniques based on the CTAB nucleic acid extraction methods of Rogers and Bendich (1985) and Murray and Thompson (1980) allow obtain small amounts of DNA when 1:15 (w/v) extraction buffer was added (fig. 7). Proteinase K (100 ug/ml) treatment was necessary for at least there hours.

The DNA obtained presented 50 pairs of kb (using DNA from lambda, cI 8571), as marker) (fig. 8 a, b), it was DNA-ase free (fig. 8d) and there was no inhibition for enzyme restriction BamHI (fig. 8c) in any of the material tested.

However, the yield was low; from 7-9 ng in juvenile and mature leaves to 2-2.5 ug in "in vitro" and forced outgrowth leaves per gram of tissue.

Limited cellular breakdown as a result of using dry ice to grind the tissue into a fine powder may be a factor

fig. 7. Genomic DNA purified from mature samples using Murray and Thompson (1980) method.
a) DNA from mature leaves
b) DNA from λcI 8571

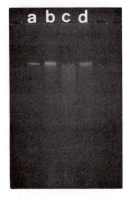

fig. 8. Analysis of DNA quality
a) DNA from λcI 8571
b) DNA from mature samples
c) DNA from mature samples incubated with BamHI for 1 h. at 37º
d) DNA from mature samples incubated withaut BamHI.

affecting the yield; more drastic methods for breaking the cell wall being necesary.

Although the qualitative pattern of polyamines was very similar among all the materials tested, PAS seems to be a good marker. There was a discrepancy in the Put and DAP titer. Put was not detected in adult and in vitro leaves, and conversely, DAP was not detected in the forced and embryonic shoots. Perhaps, this is a problem of identification, because polyamine titer was performed for the first time in filbert (Rey et al., in this volume). Additional assays are in progress, to arrive at a definitive identification of filbert polyamines.

In the free polyamine fraction (S fraction) Spd and Spm titer were quite different between adult and in vitro, embryonic or forced shoots, among which the differences, although present, were not clear (Table I). Therefore, polyamine titer may reflect a transient juvenile status achieved by in vitro culture, Also, the DAP/Spd+Spm ratio was higher in the in vitro than in the adult material. Other authors reported different polyamine content in relation to the age of the material (Palavan and Galston, 1982); our data give more evidence on this view, supporting the hypothesis of a rejuvenation process during filbert in vitro culture.

give more evidence on this view, supporting the hypothesis of a rejuvenation process during filbert in vitro culture.

TABLE I

PLANT MATERIAL			
	FORCED SHOOTS	EMBRYONIC	ADULT	IN VITRO
Put	23.28 ±0.75	67.62 ±0.12	N.D.	N.D.
Spd	22.35 ±0.99	36.69 ±2.13	236.36 ±42.42	76.20 ±0.00
Spm	18.96 ±0.54	22.41 ±0.09	180.02 ±28.28	59.03 ±0.67
DAP	N.D.*	N.D.	72.02 ±1.41	40.32 ±0.00
Put/PAs	0.56	1.68	----	----
DAP/PAs	----	----	0.17	0.30

Values are means ±S.E.   (*) N.D.: Not detected.
nmol/g. fresh weight

REFERENCES

ARNAUD, Y.; FOURET, Y.; LARRIEU, Ch.; TRANVAN, H.; FRANCLET, A.; MIGINIAC, E. (1988).
Reflexion sur les modalities d´appreciation du rajeunissement "in vitro" chez la Sequoia sempervirens.
INRA-IUFRO Int. Symp. of Forest Tree Physiology. Nancy. France.

BARTHELEMY,D.; EDELIN, C.; HALLE,F. (1988).
Approche architucturele du vieilliseement chez les vegetaux ligneus.
INRA-IUFRO Int. Symp. on Forest Tree Physiology. Nancy. France.

BONGA, J.M.and DURZAN, D.J. (1987).
Cell and Tissue Culture in Forestry. Vol. I. Martinus Nijhoff Publishers.

CHAMPAGNAT, P. (1988).
Repos et activite chez les bourgeons vegetatifs des vegetaux ligneux.
INRA-IUFRO Int. Symp. on Forest Tree Physiology. Nancy. France.

DELLAPORTA, S.L.; Wood, J.; Hicks, J.B. (1983).
A plant DNA minipreparation: Version II.
Plant Mol. Biol. Rep. 1:19-21..

GREENWOOD, M.S. (1987).
Rejuvenation in Forest Trees.
Plant Growth Regulation 6:1-12.

JAY- ALLEMAND, Ch.; CORNU, D.; MACHEIX, J.J. (1985).
Les marqueurs biochemiques de la rejuvenilisation chez le noyer. Relation avec les possibilités de multiplication vegetative.
5º Colloque sur les recherches frutieres. Bordeaux. France.

JAY-ALLEMAND, Ch.; CORNU, D.; MACHEIX, J.J. (1988)
Biochemical attributes associated with rejuvenation of walnut tree.
Plant Physiol. Biochem. 26:2:139-144.

MONTEUUIS, O. (1988).
Maturation concept and possible rejuvenation of arborescent species.
Limits and promises of shoot apical meristems to ensure sucessful cloning.
IUFRO Symposium. Thailand. (In Press).

MURRAY, H.G. and Thompson, W.F. (1980).
Rapid isolation of high molecular weight DNA.
Nucleic Acids Res. 8:4321-4325.

PALAVAN, N. and GALSTON, A. W. (1982).
Polyamine biosynthesis and titer during various developmental stages of
Phaseolus vulgaris .
Physiol. Plant. 55:438.

RODRIGUEZ, R.; RODRIGUEZ, A.; GONZALEZ, A.; PEREZ, C. (1989).
Hazelnut. p.127-160. In: Bajaj, Y.P.S. (Ed). Biotechnology in Agriculture
and Forestry. Vol. 5: Trees II. Springer-Verlag.

ROGERS, S.O. and BENDICH, A.J. (1985).
Extraction of DNA from Milligram amounts of fresh, herbarium and mummified
plant tissues.
Plant Mol. Biol. 5:69-74.

JUVENILITY AND MATURITY OF WOODY SPECIES IN NEW ZEALAND

J.L. Oliphant

Cyclone Flora
14 Clifton Road
Takapuna, Auckland, New Zealand 9

The Origins of the New Zealand Flora

Millions of years ago, so it has been argued, New Zealand was once part of a great southern continent, Gondwanaland, in close contact with Australia and Antarctica and so linked through to South America.[1] The animal and plant species that evolved on this continent were inherited by New Zealand. During the Jurassic period (190 million years ago) in the age of the gymnosperms New Zealand acquired the ancestral podocarps. Later, in the Cretaceous period (120 million years ago) the newly developed angiosperms were received, the Magnolia, Protea and Fushia families, but predominantly the Fagaceae. At this time amphibians and reptiles roamed the land and the first birds were evolving.

Shortly afterwards, (100 million years afterwards) the greater New Zealand continent eroded, the Tasman Sea developed and New Zealand was cast adrift, three little islands in the ocean, on which the animal and plant life was destined to continue to evolve in splendid isolation.

There was a complete absence of land mammals, but the birdlife flourished even to the extent of evolving flightless birds e.g. kiwi, moa and takahe, all quite secure in their ground nesting habits.

The plant life although closely related to late Cretaceous and early Tertiary ancestors, owes most of its formation to the glaciations and climatic changes of the last two million years. At each glaciation period the forest shrank in size, and in some instances only small patches of forest survived in the more benign areas. In the interglacial period the remnants exploded to repopulate the moraines, the scrub, and the tussock land. In this resurgence the trees were aided by the birds as seed distributors particularly the podocarps.[2]

Today, New Zealand (a country the size of Britain) has settled down to rest between latitudes 35°-47°S which as you will realise is the antipodeal equivalent of Spain and Portugal. The forests survive as temperate rainforests. At lower altitudes there are evergreen mixed forests, the podocarps dominating over a subcanopy of tree ferns, with a miscellany of trees, shrubs and vines. In the warmer north the auracarian Agathis holds sway, while further south Dacrydium, Dacrycarpus, Prumnopitys or Podocarpus may take the lead role.

Plant Aging: Basic and Applied Approaches
Edited by R. Rodríguez *et al.*
Plenum Press, New York, 1990

At higher altitudes in the North Island and in much of the South Island the evergreen southern beech forest contains mainly the Nothofagus spp. This genus Nothofagus is found throughout the Pacific, Australia, New Caledonia, New Guinea and South America. It is thought to have originally come from stock which also produced the northern hemisphere Fagaceae, the beeches, oaks, and chestnuts.

New Zealand was first colonised 1,000 years ago by the Polynesian people, and during this time 1/3 of the forest disappeared. The European peoples arrived 150 years ago and a further 1/3 of the forest was removed, the timber milled and the land cleared for agricultural and horticultural farming.

Today there are 6 million hectares left in native forest.[3] Afforestation has occurred and over one million hectares of the fast growing Pinus radiata have been planted for pulp and paper production and the treated timber is used for building construction.

Eucalyptus sp have also been planted and to a much lesser extent some native species.

Heteroblastic Woody Plants in the New Zealand Flora

In the native forests of New Zealand, there are a large number of heteroblastic plants. There are 200 species of seed plants involving 30 different families and 50 genera. Of these heteroblastic plants 80% are woody plants. The differences between the juvenile and the adult stages involve both leaf construction and branching habit, often in combination.[4]

As examples of heteroblasty in the New Zealand flora some of the plants are listed below with their differences in form.

Pseudopanax crassifolius - the deflexed xerophytic juvenile leaves and unbranched trunk lasts 15-20 years followed by branching to form a crown of smaller mesophytic leaves, either simple or trifoliate.

Weinmannia racemosa - the seedling leaves are simple, followed by simple to 5-foliate juvenile leaves and then by simple adult leaves.

Phyllocladus glaucus - the seedling bears the only true leaves which are narrow and needle-like, after which time cladodes appear. In mature plants the leaves are reduced to scales or are absent. Under "in vitro" conditions the terminal buds produce needle-like leaves, the cladodes forming more often after the plantlets have rooted.

Knightia excelsa - the young trees have an upright form but later secondary trunks diverge to form a wider crown and lateral branches of fastigiate branching.

Agathis australis - the young trees have a narrow based conical shape usually with a single trunk, but the mature form has spreading limbs.

Dacrydium cupressinum - the young trees have a weeping form with pendulous branches, while the adult spreads out to form a crown.

The following plants all show divaricate branching in the juvenile form, and the leaves are always smaller than in the adult.
Dacrydium dacrydioides, Elaeocarpus hookerianus, Carpodetus serratus, Podocarpus spicatus, Plagianthus regius, Prumnopitys ferruginea, Sophora microphylla, Pennantia corymbosa.

Why should New Zealand have so many heteroblastic plants?

A reduction in the amount of water available may have lead to some of the xerophytic juvenile forms. The divaricating habit may have been an adaptation which enabled the plant to resist damage from wind abrasion, frost and dessication in a highly changeable climate, from the glacial past ages to the present day.[5]

Alternatively the divaricating form may have developed by natural selection in response to browsing by moas (large flightless birds, now extinct, that once roamed the land in prehistoric times). The much branched juvenile form with small leaves and tough stems may have deterred browsing, while the transition to adult foliage occurs between 3-4 metres, just out of reach of even the tallest moas.[6]

The importance of this phase-change syndrome in horticultural research, and the understanding of the transition from juvenile to mature forms, requires, as Wesley Hackett has pointed out, the need to find "a woody plant experimental system in which the transition to the mature phase is rapid, can be induced at will, and is easily identified by rapid, controlled induction of flowering or associated morphological characteristics."[7] Perhaps amongst the heteroblastic plants of New Zealand there will be found such a system.

Rejuvenation and associated problems in the micropropagation of woody plants

In the micropropagaion of woody plants where explants are taken from mature trees, rejuvenation often occurs under the "in vitro" conditions. This is a temporary phase. After deflasking, the plant returns to the adult state. It is becoming increasingly important commercially to hasten this return to the flowering condition. The retail ornamental trade demands that a plant be sold in flower so they can capitalise on impulse buying, and also so the uneducated public know what they are getting. It is essential that we find methods to jog the memory of the plant back from its "in vitro" rejuvenation to the mature state as quickly as possible. In New Zealand there is a population of only 3.5 million people and the research dollar is spent mainly to assist the horticultural exports of apples, pears and kiwifruit. There is little left for the ornamental trade.

However, some of the most popular ornamental plants are the Metrosideros sp and although the leaf change from glabrous, to tomentose in the adult state is not regarded by some to be truly heteroblastic, it nevertheless gives a good visual indication of the phase change.[8]

The Metrosideros sp are difficult to root by normal cutting propagation, and several superior forms have been micropropagated over the last 3-4 years. The explants were taken from cutting grown plants which in turn were taken from the selected tree. It was important to acclimatise the plants in a growing room at 25°C with low light conditions of 1-2000 lux for a 16 hr/day. The soft new growth 1-2cm in length was used as explant material.

Metrosideros collina 'Tahiti' (formerly known as M.villosa) is an evergreen shrub that has been introduced to New Zealand from Tahiti. Seedling plants have a juvenile phase of 1-2 years and flowering occurs in 4-5 years. Cuttings from adult plants flower in 1-2 years. Explants from cutting grown plants with adult foliage are rejuvenated "in vitro". After deflasking, the tomentose adult leaves begin to appear after six months, and flowering occurs in the second year.

Metrosideros _carminea_ is endemic to New Zealand. A forest climbing woody vine the juvenile form has obtuse leaf tips while the adult has acute leaf tips. A seedling maintains the juvenile climbing phase for 12-15 years before branching out into a shrubby habit and flowering. Cuttings taken from the shrubby adult form, flower in the second year. Explants taken from cutting grown plants with adult foliage are once again rejuvenated "in vitro". After deflasking, the pointed mature leaves begin to form after six months and flowering occurs in the second year. The plant still maintains some of the characteristics of its climbing habit, but pruning of the younger downward spreading shoots stimulates the growth of the more upright mature foliage.

Metrosideros _excelsa_ is endemic to New Zealand. These are large evergreen trees found in coastal situations. Seedlings have a juvenile form with glabrous leaves, lasting 3-4 years before developing tomentose leaves and flowering after 7-8 years. Cuttings taken from adult plants flower at the earliest in the second year. Explants taken from cutting grown mature plants are rejuvenated "in vitro". After deflasking, the foliage begins to show change not after six months as was first indicated but after 12-18 months, depending on the clonal parent. Flowering is expected after this the third year. The pruning of shoot tips in this instance encourages the growth of further juvenile foliage. While this may be detrimental in production time, the young shoots root readily and may be used as a basis for further increase in stock. At six months after deflasking one cutting may be taken, at 12 months at least another two, so that 500 plants may be derived from an initial 100 in one year, a 5-fold increase.

Commercially it is important to speed the change to maturity, not only to hasten flowering but also because the juvenile foliage is susceptible to chermid or psyllid attack which spoils the cosmetic appearance of the plant in the eyes of the consumer.

This year we have begun a two fold approach to this problem of hastening the mature foliage on rejuvenated plants.

1. Physical: by regular root pruning after deflasking.[9] This method was chosen for investigation because of its practicality. The average plant nursery in New Zealand would propagate 1-200,000 plants per annum and this would include 2-300 varieties of plants. Sometimes it would be simpler to root prune 10 trays of tubes amongst another 1000 trays in a glasshouse or shade house than it would be to spray.

2. Chemical: by the inclusion of abscisic acid in the rooting medium prior to deflasking; and by the use of growth regulators after deflasking. The effects of abscisic acid and growth retardants on the heteroblastic development of other plants is well documented.[10,11,12]

Ideally the inclusion of abscisic acid at the "in vitro" stage would be simple, and cost effective in an attempt to induce maturation. However, spraying with growth retardants is also being trialled.

The results of this work will not be known for some months.

In a similar area, ie. the phase change of juvenile to mature forms after micropropagation, investigation has begun on a selected form of _Eucalyptus_ _ficifolia,_ the flowering gum. Since it is not possible to root cuttings from mature foliage the following approaches were taken.

1. The lower branches were pruned to stimulate coppice growth.
2. The base of the trunk was wounded to provide coppice growth.

In both instances the new growth was used as explant material for micropropagation. In this new growth the leaves were opposite as in seedlings, or alternate as in the mature form.

Under "in vitro" conditions the leaves became rounded, and opposite, rejuvenation had occurred.

Multiplication of the shoots has been satisfactory although there appears to be a faster growth rate of the more juvenile explants from the base of the trunk. Root elongation has occurred within two weeks and has shown a similar growth rate from either explant source.

The plantlets have been deflasked and we now await with interest to compare the foliage growth and flowering time between the two different explant areas. Will this be a further vindification of the ontogenetic basis for phase change and once again will we be looking for methods to hasten the growth of mature foliage and flowering?

In concluding this address I would emphasise that there are a host of New Zealand heteroblastic plants all steeped in the history of Gondwana-land, waiting to be assayed for their value to plant physiology.

I and many like me, are limited almost entirely to handling species of horticultural value, but surely there are less mercenary workers who can spend some time studying the economically useless heteroblast.

With this thought in mind I welcome you all to New Zealand to find a heteroblastic plant which can expand the tale of juvenility, maturity and rejuvenation started so many years ago with embryoidal carrot tissue.

ACKNOWLEDGEMENTS

I wish to thank Dr. J.D. Ferguson, D.S.I.R., Lincoln, New Zealand and Dr John Herbert, F.R.I., Rotorua, New Zealand, for reviewing the manuscript.

REFERENCES

1. Graeme Stevens, "Lands in Collision", Science Information Publicity Centre, D.S.I.R., Wellington, New Zealand (1985).
2. G.R. Stevens, M. McGlone and B. McCulloch, "Prehistoric New Zealand", Heinemann Redd, Auckland, New Zealand (1988).
3. Carolyn King, "Immigrant Killers", Oxford University Press, Auckland, New Zealand (1984).
4. E.J. Godley, Paths to Maturity. New Zealand Journal of Botany, 23:687-706 (1985).
5. M.S. McClone, C.J. Webb. Selective forces influencing the evolution of divaricating plants, New Zealand Journal of Ecology, 4:20-28 (1981).
6. R.M. Greenwood, I.A.E. Atkinson. Evolution of divaricating plants in New Zealand in relation to moa browsing. Proceed. of the New Zealand Ecological Society 24:21-33 (1977).
7. W.P. Hackett, Phase change and intraclonal variability. Hort.Science 18(6):840-844 (1983).
8. J.L. Oliphant, Considerations of juvenility and maturity in the micropropagation of the Metrosideros species. Acta Hort 227:482-484 (1988).
9. W.W. Schwabe, Applied aspects of juvenility and some theoretical considerations. Act Hort 56:45-55 (1976).
10. M.A. Kane, L.S. Albert. Abscisic acid induces aerial leaf morphology and vasculature in submerged Hippuris vilgaris Aqua.Bot. 28:81-88 (1987)
11. H.Y. Ram Mohan and S. Rao. In vitro induction of aerial leaves and of precocious flowering in submerged shoots of Limnophila indica by

abscisic acid. <u>Planta</u> 155:521-523 (1982).

12. C.E. Rogler and W.P. Hackett. Phase change in <u>Hedera</u> <u>helix</u>: stablisation of the mature form with abscisic acid and growth retardants. <u>Physiol.Plant</u> 34:148-152 (1975).

DEVELOPMENT OF REJUVENATION METHODS FOR IN VITRO ESTABLISHMENT,

MULTIPLICATION AND ROOTING OF MATURE TREES

A. Ballester, M.C. Sánchez, M.C. San-José,
F.J. Vieitez and A.M. Vieitez

Plant Physiology, CSIC, Apartado 122
15080 Santiago de Compostela, Spain

INTRODUCTION

The utility of the micropropagation of forest trees is limited by the efficiency with which selected trees can be reproduced vegetatively. As is well known, there are several woodland species for which satisfactorily successfull establishment in vitro has still not been achieved, at least for adult trees. Juvenile trees are in general readily cloned by conventional techniques, but the ease of propagation of many trees tends to diminish as they approach a size sufficient to allow reliable evaluation of their crop potential (Bonga, 1987).

The transition from juvenile to adult often occurs gradually, and some parts of an otherwise adult tree may retain juvenile or transient characteristics for many years. Material of this kind includes epicormic shoots, root suckers and stump sprouts, all of which are generally easier to establish in vitro than material from the upper branches of the same tree. In our own work, for example (Vieitez et al., 1983; 1985), we have achieved in vitro regeneration of chestnut and oak using both tissues from juvenile trees and stump sprouts from adult trees, but it was very difficult to establish in vitro cultures of material from the upper branches of adult trees. However, adult trees do not always feature material with juvenile properties, and in such cases adult tissue must be rejuvenated in some way to allow its establishment in vitro.

Pre-culture rejuvenation methods reported in recent years include severe pruning, spraying with cytokinins or gibberellins (Magalewski and Hackett, 1979; Bouriquet et al.,1984), serial rooting or grafting on juvenile rootstocks (Franclet, 1983). Serial subculture of explants from mature trees (Gupta et al., 1983), and meristem culture (Margara, 1982; Monteouis, 1987) also results in a gradual increase in juvenility. In our laboratory, however, shoot multiplication cultures derived from 80-100 year-old specimens of the difficult-to-root species Camellia reticulata still fail to root after 2 years of serial subculture.

In this paper we describe our results on the partial rejuvenation of oak, chestnut or camellia by three other methods: grafting on juvenile rootstocks and subsequent spraying with 6-benzylaminopurine (BAP); partial etiolation of starting material; and successive horizontal reculture of decapitated shoots in vitro.

Plant Aging: Basic and Applied Approaches
Edited by R. Rodríguez *et al.*
Plenum Press, New York, 1990

In this work, a degree of rejuvenation was deemed to have taken place if in vitro establishment and shoot multiplication rates approached those obtained with material of juvenile origin, or (especially) if the in vitro rooting rates of recalcitrant mature clones rose.

GRAFTING ON JUVENILE ROOTSTOCKS

The rootstocks used were 2-week-old seedlings obtained by germinating seeds of a 30-year-old selected Castanea sativa x C. crenata hybrid (Clone HV). The same tree provided scions consisting of 4-5 cm of upper branch bearing 2-3 buds, which were collected in winter (January) and stored at 4ºC until grafted in April. The rootstocks were prepared for grafting by cutting out the epicotyl at the level of the cotyledon petioles, and the scions were then inserted into the split hypocotyl as shown in Fig. 1 (Vieitez and Vieitez, 1981). The proximity between the scion and the root system in this kind of grafting must be emphasized.

The grafts were kept for 5 weeks in a growth chamber to flush. During the last two weeks of this period, some of the grafts were sprayed with 50 mg/l BAP solution twice a week. Newly grown shoots were isolated and sterilized by successive immersion in 70% alcohol for 30 sec and 15% commercial bleach (40 g/l chlorine) for 10 min, after which they were rinsed thrice in sterile distilled water. Nodes and terminal buds were excised, and half of those derived from unsprayed grafts were placed in 25 mg/l BAP solution for 2 h before transfer to culture (pulse treatment).

Terminal branches collected and stored with the scions were placed in water for flushing in the same growth chamber as the grafts. Very few of these branches flushed: the buds produced by those that did were established in vitro using the same sterilization procedure as for the grafts.

For in vitro culture, the explants were placed individually in 20x150 mm test tubes containing 15 ml of the initial culture medium, which contained the macronutrients, micronutrients and vitamins of Gresshoff and Doy's (1972)

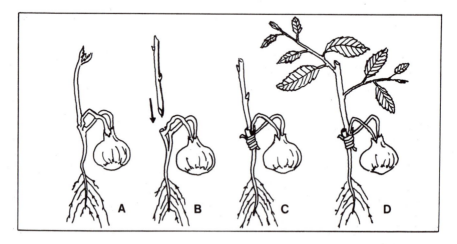

Fig. 1. Juvenile grafting. (A) chestnut seedling; (B) preparation of the rootstock: removal of epicotyl and longitudinal excision of the hypocotyl; (C) insertion of the scion into the split hypocotyl; (D) new shoots on grafted plants after 5 weeks in a growth chamber.

medium, the Fe-EDTA of Murashige and Skoog's (1962) medium, 0.5 mg/l BAP, 3% of sucrose and 0.6% of agar (pH had been brought to 5.5-5.6 before auto-claving). For shoot multiplication cultures the same culture medium was used, but with the BAP concentration reduced to 0.2 mg/l.

Table 1 lists the results obtained so far. Explants from ungrafted upper branches failed totally to take in vitro, whereas establishment in culture was achieved by 22% of the explants from grafts given no prior BAP treatment, 43% of those treated with BAP for 2 h immediately before culture (pulse), and 100% of those grafts had been sprayed with BAP during their last 2 weeks of flushing. The cultures from sprayed grafts also had the best shoot multipli-cation rate after 1 year's culture.

Table 1. Effect of juvenile grafting on in vitro establishment of plant material collected from an adult, 30-year-old (Clone HV) chestnut tree

Treatment	In vitro establishment (%)	Nº shoots/culture (2nd subculture)	Nº shoots/culture (after 1 year in culture)
Control HV	0	–	–
Grafting	22	2.0	1.8
Grafting + BAP (25 mg/l) Pulse	43	2.7	2.4
Grafting + BAP (50 mg/l) Spray	100	2.4	2.6

The above results show that tissue from the upper branches of adult chestnut trees can be established and proliferated in vitro after a single graft, and that treatment of the graft with BAP spray vastly improves the results. These findings support those of Reynoird (1983), who reported that spraying adult Pseudotsuga material with BAP had the same rejuvenating effect as several successive grafts. McGranahan et al. (1987) tried a similar method in walnut trees but mature cultivars had been difficult to establish in culture inspite of those treatments. For chestnut, however, the combina-tion of juvenile grafting and BAP spraying is a very efficient alternative to the laborious, time-consuming cascade grafting technique.

PARTIAL ETIOLATION

In our study, partial etiolation experiments were carried out with a mature tree of Castanea sativa x C. crenata (Clone HV) and a centenarian oak Quercus robur (Clone SL1). In late May, 10-15 cm of each of several crown branches with new year's growth was stripped of leaves and wrapped in aluminium foil. In this way the shoots acomplishes the initial growth in the light and then had a portion of its stem covered with an opaque material, the rest of the branche maintaining active leaves growing under ligth. Cuttings were taken in September from both the etiolated portions of banded shoots and from unetiolated shoots, and were stored at 4ºC until the following April, when they were placed in water in a growth chamber to flush. After 10-15 days the new shoots were collected and sterilized in the same way as the grafts of the previous section, and 5-mm-long shoot tips and nodes were excised and placed in the same culture medium as was used for the graft

Fig. 2. Initial cultures from etiolated material: (A) chestnut; (B) oak

explants. Established cultures were segmented and subcultured after 6 weeks.

Preliminary results shown that the rate of bud break and new shoot growth among chestnut cuttings was 77% for shoots from etiolated branches as against only 27% for those of unetiolated origin (Ballester et al., 1989, submitted for publication). Furthermore, the new shoots produced by etiolated material were 2-3 cm long and thus allowed both nodal segments and shoot tips to be used as initial explants, whereas only shoot tips could be obtained from the unetiolated material, whose shoots were all less than 1 cm in length. The difference between the responses of material of etiolated and unetiolated origins to in vitro culture was even greater, with a 79% establishment rate for the former as against 22% for the latter (Fig. 2A); and after 4 subcultures, the corresponding multiplication coefficients (the proportion of explants forming axillary shoots x the mean number of new 5-mm segments per explant) were 3.3 for material of etiolated origin and 1.1 for material of unetiolated origin.

With regard to the experiments with oak, we have failed to establish any in vitro cultures of material of unetiolated origin, but have achieved a 50% establishment rate for tissues of etiolated origin (Fig. 2B).

The above results show that partial etiolation can greatly facilitate the establishment in vitro and subsequent multiplication of mature chestnut and oak material, presumably by reducing the physiological aging. The in vitro response of the material of etiolated origin in this study was in fact similar to that of chestnut and oak stump sprouts, which conserve juvenile characteristics and whose successful use for establishment and proliferation in vitro has already been reported (Vieitez et al., 1983; Vieitez et al., 1985).

Although etiolation has been found to be a useful aid to in vivo and in vitro rooting, very little research has been done on its potential as a pretreatment for the establishment of in vitro cultures. Bennet and Davis (1986) and Hansen and Lazarte (1984) reported there to be no advantage in etiolating seedling stock plants of Quercus shumardii and Carya illioniensis respectively as a pre-treatment for in vitro culture; but since seedlings, being juvenile, ought to be quite easy to establish anyway, it is perhaps not surprising that etiolation produced no marked improvement.

The physiological effects of etiolation are poorly understood, but it is known that etiolation reduces lignification, alters anatomy and increases the concentration of IAA-oxidase inhibitors. Changes in phenols have also been reported: Vieitez et al. (1987) isolated two root-inhibiting phenolic compounds from unetiolated chestnut stump sprouts that were not present in etiolated sprouts.

The partial etiolation method described here may prove useful not only for pre-treatment of recalcitrant mature tissues, but also to facilitate basic research on the effects of etiolation on the physiology of woody plants.

SUCCESSIVE HORIZONTAL RECULTURE IN VITRO

Horizontal culture was used to improve the rootability of poorly rooting in vitro oak and camellia clones. Two Quercus robur clones of adult origin were employed (Clone 6 and Scer-fast), and a clone derived from an 80-year-old specimen of Camellia reticulata cv. Captain Rawes. The establishment and multiplication culture conditions of these clones have described elsewhere (San-José et al., 1988; San-José et al., 1989 submitted for publication). At the time horizontal culture experiments reported here were performed, shoot multiplication cultures had been maintained for over a year by vertical subculture of 1 cm segments every 4-6 weeks.

Shoots 20-25 mm long were excised from shoot multiplication cultures, 2 mm were removed from their tips, and the decapitated shoots were placed horizontally in glass jars containing multiplication culture medium. After 4 weeks for oak clones or 6 for camellia, all the new shoots produced on the horizontal explants were excised for rooting and the mother shoots were transferred to fresh medium (recultured) to obtain new crops of shoots. The mother shoots were recycled 3 or 4 times in all.

The rooting of the harvested shoots was induced by dipping their bases in 1 g/l indolebutyric acid solution for 2 min (for oak) or 15 min (for camellia). The treated shoots were then transferred to basal medium without growth regulators for 1 month. Camellia shoots spent the first two weeks of this month in darkness.

The shoots produced by the recultured explants in the second and third cycles differed from those of the first cycle in that they elongated faster to give longer internodes, and generally had a more seedling-like appearance; the leaves of the second- and third-crop oak shoots, for example, were less deeply lobed than those of the first crop. Whereas explants from vertical multiplication cultures had little or no rooting capacity, the rooting capacity of explants harvested from horizontal mother shoots increased with the latter's reculture to peak in the second or third crop at rooting rates of over 50% even in the worst case (Table 2). The decline in the rootability of subsequent crops parallels the decline in the multiplication coefficients and quality observed in harvested oak shoots after the second or third crop (San-José et al., 1988).
Horizontal culture has been used to improve the soot propagation coefficients of pear (Lane, 1979), rhododendron (Anderson, 1984) and apple (Yae et al., 1987), but as far as we know it has scarcely been employed expressly to improve the in vitro rooting of difficult-to-root species, though Economou and Read (1986) reported increased rooting rates in rhododendron cuttings harvested from repeatedly recultured horizontal explants.

According to Franclet (1980), reculturing is a similar process to

Table 2. Percentage rooting of shoots harvested from vertical subcultures and successive horizontal recultures of oak and camellia. Rooting was recorded after 4 weeks in rooting medium.

Plant Material	Vertical Subculture	Horizontal recultures			
		1st	2nd	3rd	4th
Oak (Clone 6)	50	83	96	–	–
Oak (Scer-fast)	4.2	8.3	42	58	33
Camellia	0	0	58	17	–

repeated drastic pruning, which is known to induce the development of shoots with more juvenile characteristics. At the same time, it is known (Brown and Leopold, 1973) that in woody species mechanical stress such as that involved in the horizontal position of the mother shoots induces the production of ethylene. The availability of ethylene, regulated endogenously, is essential to the release bud on decapitated plant in order to sustain its subsequent development into a lateral shoot possibly acting by reducing auxin transport (Yeang and Hillman, 1984). Increased ethylene production may therefore explain the high shoot multiplication rates achieved by horizontal cultures, though it provides no immediate explanation of increased rooting rates. Be that as it may, the combination procedures involved in horizontal reculturing clearly seem to retard or reverse the aging of the cultured tissues.

Perhaps related to this phemomenon is the fact in Hedera helix the mature growth ussually appears to be associated with vines which have been able to growth vertical from some time, and are infrequent on horizontally growing vines which may be the same age and size as those which have grown vertically (Greenwood, 1987).

REFERENCES

Anderson, W.C., 1984, A revised tissue culture medium for shoot multiplication of rhododendron, J. Amer. Soc. Hort. Sci., 109: 343.

Bennet, L.K., and Davies Jr., F.T., 1986, In vitro propagation of Quercus shumardii seedlings, HortScience, 21: 1045.

Bonga, J. M., 1987, Clonal propagation of mature trees: problems and possible solutions, in: "Cell and Tissue Culture in Forestry, vol. 1", J.M. Bonga and D.J. Durzan, eds., Martinus Nijhoff Publishers, Dordrecht.

Bouriquet, R., Tsogas, M., and Blaselle, A., 1984, Essais de rajeunissement de l'épicéa par les cytokinines, Ann. AFOCEL, 173.

Brown, K.M., and Leopold, A.C., 1973, Ethylene and the regulation of growth in pine, Can. J. For. Res., 3: 143.

Franclet, A., 1980, Rajeunissement et propagation végétative des ligneux, Ann. AFOCEL: 12.

Franclet, A., 1983, Rajeunissement, culture in vitro et pratique sylvicole, Bull. Soc. Bot. Fr., 130 Actual. Bot., 2: 87.

Economou, A.S., and Read, P.E., 1986, Microcutting production from sequential reculturing of hardy deciduous azalea shoot tips, HortScience, 21: 137.

Greenwood, M.S., 1987, Rejuvenation of forest trees, Plant Growth Regulation, 6: 1.

Gresshoff, P.M., and Doy, C.H., 1972, Development and differentiation of haploid Lycopersicon esculentum, Planta, 107: 161.

Gupta, P.K., Metha, U., and Mascarenhas, A.F., 1983, A tissue culture method for rapid multiplication of mature trees of Eucalyptus torelliana and E. camaldulensis, Plant Cell Rep., 2: 296.

Hansen, K.C., and Lazarte, J.E., 1984, In vitro propagation of pecan seedlings, HortScience, 19: 237.

Lane, W.D., 1979, Regeneration of pear plants from shoot meristem tips, Plant Sci. Lett., 16: 337.

Margara, J., 1982, "Bases de la multiplication végétative", INRA, Versailles.

Magalewski, R.L., and Hackett, W.P., 1979, Cutting propagation of Eucalyptus ficifolia using cytokinin-induced basal trunk sprouts, Proc. Int. Plant Prop. Soc., 29: 118.

McGranahan, G.H., Driver, J.A., and Tulecke, W., 1987, Tissue culture of Juglans, in: "Cell and Tissue Culture in Forestry, vol. 3", J.M. Bonga and D.J. Durzan, eds., Martinus Nijhoff Publishers, Dordrecht.

Monteouis, O., 1987, In vitro meristem culture of juvenile and mature Sequoiadendron giganteum, Tree Physiology, 3: 265.

Murashige, T., and Skoog, F., 1962, A revised medium for rapid growth and bioassays with tobacco tissue cultures, Physiol. Plant., 15: 473.

Reynoird, J.P., 1983, "Effet de prétraitment par greffages successifs ou par pulvérisation de cytokinine sur la réactivation de Douglas agé (Pseudotsuga menziesii) en vue de la propagation végétative. Incidence de quelques facteurs du milieu de culture sur la croissance in vitro d'individus juvéniles", DEA Université de Paris VI, Paris.

San-José, M.C., Ballester, A., and Vieitez, A.M., 1988, Factors affecting in vitro propagation of Quercus robur L., Tree Physiology, 4: 281.

Vieitez, A. M., Ballester, A., Vieitez, M.L., and Vieitez, E., 1983, In vitro plantlet regeneration of mature chestnut, J. Hortic. Sci., 58: 457.

Vieitez, A.M., San-José, M.C., and Vieitez, E., 1985, In vitro plantlet regeneration from juvenile and mature Quercus robur L., J. Hortic. Sci., 60: 99.

Vieitez, J., Kingston, D.G.I., Ballester, A., and Vieitez, E., 1987, Identification of two compounds correlated with lack of rooting capacity of chestnut cuttings, Tree Physiology, 3: 247.

Vieitez, M.L., and Vieitez, A.M., 1981, Injerto en hipocótilo de plántulas de castaño, Anal. Edafol. Agrobiol., 40: 647.

Yae, B.W., Zimmermann, R.H., Fordham, I., and Ko, K. Ch., 1987, Influence of photoperiod, apical meristem and explant orientation on axillary shoot proliferation of apple cultivars in vitro, J. Amer. Soc. Hort. Sci., 112: 588.

Yeang, H.Y., and Hillman, J.R., 1984, Ethylene and apical dominance, Physiol. Plant., 60: 275.

AGING IN TREE SPECIES: PRESENT KNOWLEDGE

Andreas Meier-Dinkel and Jochen Kleinschmit

Niedersächsische Forstliche Versuchsanstalt
Abt. Forstpflanzenzüchtung
D-3513 Staufenberg-Escherode
Federal Republic of Germany

INTRODUCTION

A comprehensive publication which is still relevant concerning the topic of this paper are the proceedings of the meeting on "Juvenility in woody perennials" held 14 years ago in College Park/Maryland and Berlin (ZIMMERMAN, 1976). Since that meeting, notable progress has been achieved in some applied fields of research like in vitro rejuvenation and flower induction techniques which are important for the vegetative and generative propagation of trees. Recent reviews on juvenility, aging, and rejuvenation of woody plants are those of BONGA (1982), FRANCLET (1983), HACKETT (1985, 1987), and GREENWOOD (1987).

Aging is a complex phenomenon with far reaching consequences for vegetative and generative reproduction, especially in the case of woody plants.

According to FORTANIER & JONKERS (1976), aging in plants has three aspects: a chronological, an ontogenetical, and a physiological one.

Chronological aging only means "getting older" and refers to the period of time since the germination of a seedling.

Physiological or somatic aging refers to the loss of vitality during the life of a tree ending with death and is mainly caused by an increasing complexity of the tree.

Ontogenetical aging or maturation means the genetically programmed process of phase change resulting in different phases of development: the embryogenetic phase, the juvenile or seedling phase, the transition phase and the mature phase.

Ontogenetical aging in woody plants is still poorly understood. The characteristics and the behaviour of juvenile and mature trees can be described which is very important for breeding and for propagation of all kinds of woody plants like forest trees, trees producing raw material, fruit trees and ornamentals. However, the molecular basis of plant aging, the topic of this meeting, is largely unknown.

Plant Aging: Basic and Applied Approaches
Edited by R. Rodríguez *et al.*
Plenum Press, New York, 1990

The process of aging is accompanied by
- increase of size
- increase of complexity
- physiological differentiation
- functional differentiation
- differential gene activities in the different morphological parts.

Chronological, physiological, and ontogenetical aging interact upon each other. These interactions occur on different levels of influence (Tab. 1). The latter for their part show a close dependence among themselves.

CHARACTERISTICS OF THE GROWTH PHASES AND THEIR CONSEQUENCES ON THE PROPAGATION OF TREES

Embryogenetic Phase

This phase lasts from the formation of the zygote until the formation of a mature, dormant embryo within the seed. Embryogenic tissue has a high regeneration capacity and is therefore often used as explant source for in vitro cultures. Usually, zygotic embryos are used for the induction of adventitious bud formation, especially in conifers e.g. *Picea abies* (VON ARNOLD, 1982), *Pinus radiata* (AITKEN & REILLY, 1978), *Pinus pinaster* (RANCILLAC, 1979), *Pseudotsuga menziesii* (BOULAY, 1979) etc.

Up to now, somatic embryogenesis could mostly be induced on explants of embryogenic tissue or of the very early seedling stage in hardwoods and in conifers. HAKMAN & FOWKE (1987) obtained somatic embryos and plants from immature embryos of *Picea glauca* and *Picea mariana*. In *Liriodendron tulipifera*, somatic embryogenesis from isolated immature zygote embryos was achieved by MERKLE & SOMMER (1986). Primary leaves of *Aesculus hippocastanum* seedlings gave rise to a high frequency of somatic embryos (DAMERI et.al., 1986)

Juvenile Phase

The juvenile phase begins with the germination of the seedling and is followed by the transition phase. According to DOORENBOS (1965), the juvenile phase of woody plants is characterized apart from its morphological properties by a greater readiness to form adventitious roots and an inability to form flowers.

The average duration of the juvenile growth phase of woody plants is varying from species to species. Forest trees with a comparatively short

Tab. 1. Interactions between chronological, physiological, and ontogenetical aging occur on different levels of influence.

Ecological level (competition, light, soil, nutrition)
Species level (e.g. difference between conifers and hardwoods)
Individual level (genetic program of the individual tree)
Structural level (root system, stem, crown)
Organ level (leaves, buds, branches, flowers)
Tissue level
Cellular level
Molecular level

juvenile phase are birches (5 - 6 years), whereas oak and beech stay juvenile for some decades (CLARK, 1983). However, these dates are only referring to the period until flowering occurs first. In single individuals, the length of the juvenile phase is influenced by factors such as genotype, environment and nutritional status (DOORENBOS, 1965).

The genetic control of the length of the juvenile phase was impressively demonstrated by STERN (1961) who selected early-flowering lines of birch in three generations using repeated controlled crossings. He succeeded in selecting lines which were able to form flower catkins during the first growth period. However, these birches retained juvenile morphological characteristics. This shows that different characters of a plant may change at different times during the ontogenesis and must not necessarily change simultaneously.

In fruit trees like apple, pear and peach it is also known that the length of the juvenile phase is inheritable (VISSER, 1965; HANSCHE, 1986).

The duration of the juvenile phase is also affected by environmental factors. Low light intensity reduces the rate of development. Beech seedlings which are growing in the shade of old trees stay longer juvenile than others standing free and in full sunlight.

In the case of oak seedlings, it is possible to induce early flowering by accelerating growth under continuous light and GA₃ treatment. 18 months old plants were 3 m high and flowered already (SCHWABE, personal communication).

The juvenile stage of woody plants may be characterized by morphological or physiological features as the shape or physiology of leaves, by the occurance of certain types of shoots or phyllotaxis or by the inability to flower under natural conditions. Some examples of species with a clear juvenile phase are *Hedera helix* with creeping or climbing, not flowering shoots with three to five lobes in the leaves, *Vitis vinifera* with a spiral phyllotaxy (2/5) and the abscence of tendrils (MULLINS et al., 1979), fruit trees with a different shape and size of leaves and a greater rooting capacity of cuttings (PASSECKER, 1943), *Eucalyptus* and *Acacia* with typical juvenile leaves (DIELS, 1906) and *Fagus sylvatica* with the ability to retain most of the withered leaves during the winter (SCHAFFALITZKY DE MUCKADELL, 1954).

The distance between the roots and the shoot apex seems to be a controlling factor in the expression of adult or juvenile characters (SCHWABE & AL-DOORI, 1973; FRYDMAN & WAREING, 1973; PATON et al., 1981). With increasing size of a tree, the shoot apex becomes more remote from the roots and the level of endogenous gibberellin in the apex declines (WAREING and FRYDMAN, 1976). Juvenile apices contain higher levels of gibberellin than adult ones. Moreover, rejuvenation could be induced by GA-treatment in adult *Hedera canariensis* (ROBBINS, 1957) and in *Acacia melanoxylon* (BORCHERT, 1965). BORCHERT supposed that the rejuvenating effect of GA is only secondary, resulting from the increased growth induced by the acid.

A contrary effect is common in conifers. Here treatments with gibberellins (GA 4/7) are used to induce flower formation.

Experiments with grafts of juvenile seedlings of apple (FRITSCHE, 1948) showed that a short distance between root and shoot apex and a resulting high GA concentration cannot be the only factor which maintains the juvenile state. Juvenile seedling scions grafted to flowering branches of adult trees did not flower earlier than seedlings. These results show that the juvenile scions remained juvenile over a distinct period of time, which may genetically be fixed.

Transition Phase

This phase is characterized by the <u>gradual change</u> of morphological, anatomical, physiological, biochemical and growth characteristics from the juvenile to the mature state. These different characters of a plant can chronologically change quite differently.

Attainment and maintenance of the ability or potential to flower is the only consistent criterion available to assess the begin of the mature growth phase. Although flowering occurs only after 25 - 40 years in some forest tree species like *Quercus*, *Fagus*, *Abies*, or *Picea*, this does not mean that the trees are completely juvenile before the first flower onset. Other characteristics, especially those which are affecting the potential for vegetative propagation, change much earlier. The ability of cuttings to form adventitious roots decreases quite early in the development of seedlings. Whereas oak cuttings from 2- and 3-year-old ortets gave high rooting percentages over 80 % in all tested clones, cuttings from 6- to 150-year-old ortets rooted gradually poorer. The variation between clones increased dramatically with the age of the mother tree and was 0 to 95 % in the cuttings from 8-year-old trees (Fig. 1), (SPETHMANN, 1986).

Many morphological traits also change gradually , e.g. branching type and foliar morphology as reported for *Pinus taeda* (GREENWOOD, 1984).

Mature Phase

Maturity is reached when the reproductive potential is attained and maintained. The mature growth phase is relatively stable. Under natural conditions, a reversion to the juvenile condition generally requires sexual reproduction, i.e. meiosis. In some tree species, like apple, apomictic seed formation gives rise to juvenile seedlings which have the same genotype as the mother tree and are therefore good subjects for studies on maturation and rejuvenation.

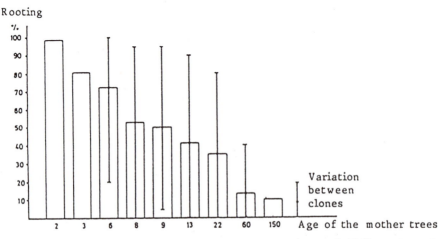

Fig. 1. Mean rooting percentage of Oak cuttings of different aged mothertrees

Tissue culture techniques allow the induction of somatic embryos which should ontogenetically be completely juvenile. Somatic embryogenesis has been induced in 60 species and numerous cultivars of woody plants (TULECKE, 1987) and this number continues to grow. Among them are several woody plants of the temperate zone, like *Ilex aquifolium* (HU & SUSSEX, 1971), *Corylus avellana* (RADOJEVIC et al., 1975), *Paulownia tomentosa* (RADOJEVIC, 1979), *Liquidambar styraciflua* (SOMMER & BROWN, 1980), *Malus* spec. (JAMES et al., 1984), *Castanea sativa* (GONZALEZ et al., 1985), *Juglans regia* (TULECKE & MCGRANAHAN, 1985), *Aesculus hippocastanum* (DAMERI et al., 1986), *Liriodendron tulipifera* (MERKLE & SOMMER, 1986) and a *Prunus* rootstock (HANSEN & DURZAN, 1986, cited by BOULAY, 1987).

However, in calli derived from mature tissue, somatic embryogenesis has never been observed (GREENWOOD, 1987). In order to achieve a true ontogenetic rejuvenation of adult trees which is a prerequisite for a large scale vegetative propagation, it is very desirable to induce somatic embryogenesis in explants from mature trees.

According to MEYER (1983), the epigenetic change from maturity to juvenility may take place in some flower tissues before the formation of an embryo.

There are several examples which indicate that the flower may be a suitable explant source for formation of juvenile tissue needed for clonal propagation of woody plants. Somatic embryo formation was induced in isolated unfertilized ovules of a non-apomictic plant - the grapevine (*Vitis vinifera*) - by MULLINS & SRINIVASAN (1976). Adventive embryos were initiated on micropylar halves of the nucellus in *Malus domestica* cv. 'Golden Delicious' by EICHHOLTZ et al. (1979). Recently JOERGENSEN (1989) of our lab succeeded in the induction of somatic embryogenesis in filament callus of *Aesculus hippocastanum*.

The results obtained so far with somatic embryogenesis in woody plants indicate that only juvenile tissue - obtained from juvenile or mature organisms - is capable of somatic embryogenesis. This is of course a question of definition: Is a filament - a part of the flower which is the most striking sign for maturity - juvenile?

There is no evidence that the reversion to the juvenile growth phase, the rejuvenation, can occur during the process of somatic embryogenesis itself in mature tissues. It seems that the phase change has to take place before, e.g. in flower tissues.

Flowers are also a suitable explant source regarding adventitious shoot formation. BONGA (1984) has shown that somatic tissues of female strobili of *Larix decidua* collected at, or near, the time of meiosis can form adventitious shoots in vitro on a reproducible basis. The fact that during meiosis somatic tissues near the site of meiosis gain organogenetic competence, i.e. act like juvenile tissue, suggest that they are activated by the same factors that initiate rejuvenation in the meiocytes. If this is correct, there are two possibilities. Either the activating factors are produced in the meiotic cells and permeate the surrounding tissues, or vice versa, they are primarily produced in the somatic tissues and are translocated to the sexual cells (BONGA, 1984).

By means of vegetative reproduction, a complete ontogenetic rejuven-ation of all characters does not seem to be possible in all species. There are examples like *Araucaria excelsa*, where once induced differentiation is maintained forever even within vegetative propagules. *Abies* species show a strong induction of morphological differentiation in vegetative propagules, too. In some hardwood species, this induction seems to be much less strict and a more or less complete rejuvenation is possible.

There are several examples how plants with juvenile characters can be produced. An often cited method is the formation of sphaeroblasts and adventitious buds (WELLENSIEK, 1952). Sphaeroblasts are round woody bodies which occur spontaneously in the bark of many trees or can be induced by cutting back and disbudding. They may arise from buds, either dormant or adventitious, or by an independent adventitious formation. In an experiment with apple, DERMEN (1948) demonstrated that induced sphaeroblasts were of true adventitious origin, since they appeared between nodes.

The sphaeroblasts induced by WELLENSIEK (1952) on apple and pear produced shoots from adventitious buds, in several cases up to 8 or 9 on one sphaeroblast. These could be rooted, contrary to one-year-old branches of old trees which did not root at all. However, the shoots originating from sphaeroblasts and their rooted cuttings did not show morphologically juvenile characteristics, whereas only after stooling shoots with juvenile characteristics arose from the cuttings. On the one hand, this could mean that the shoots on sphaeroblasts are not completely rejuvenated and need the influence of the roots for a complete reversion to juvenility. The influence of the root could be the production of gibberellic acid which is reported to be necessary for the maintenance of the juvenile phase (WAREING & FRYDMAN, 1976). On the other hand, it is possible that the shoots are juvenile, but the expression of juvenility is suppressed by correlative influences of the adult mother tree on which the sphaeroblasts were formed.

According to BONGA (1982), adventitious shoots are juvenile. This would mean that the process of adventitious bud and shoot formation itself may have a rejuvenating effect.

When interpreting the results of rejuvenation techniques, it has to be considered that the phase change from juvenile to mature is a gradual process and therefore the reversion will also be gradual which means that some techniques may only cause a partial rejuvenation. It is also important to distinguish between true ontogenetic rejuvenation and physiological rejuvenation (=invigoration). The interpretation is complicated because these two processes are not seperated, but influence one another.

Because of these difficulties, there is a need to find clear biochemical or genetical markers valid for all (woody) plant species.

From the literature and own experiments it is evident that in vitro propagation can result in rejuvenated plants. 10 years ago, MULLINS et al. (1979) already reported the formation of juvenile plants after several multiplication cycles via axillary bud formation on BAP containing medium in a very old cultivar of *Vitis vinifera*.

In blueberry cultivars, rejuvenation was also achieved by serial subculture of shoots derived from mature plants (LYRENE, 1981). In this case, adventitious shoots with juvenile characters were produced after some subcultures.

We made experiments with mature birch hybrids (*Betula platyphylla var. japonica* x *B. pendula*) which were more than 30 years old (MEIER-DINKEL, 1989). Explants consisting of the meristem, a few leaf primordia, and some stem tissue were prepared from dormant winter buds. From these explants, adventitious shoots arose whereas the initially formed axillary shoot died back. Then the adventitious shoots were multiplied by axillary shoot formation on BAP containing media for several subcultures. Microcuttings of 6 tested clones rooted well and were transplanted to a greenhouse and later to our nursery. The plants resembled seedlings showing a fast juvenile growth. The ability of cuttings to form adventitious roots, a very important trait for the clonal propagation of forest trees, was tested with cuttings from two-year-old in vitro propagated plants as well as from the adult mother trees. The cuttings from the mother trees did not root at all, whereas those from in vitro propagated plants formed roots at an average percentage of 70 to 90%.

A similar result was obtained by STRUVE & LINEBERGER (1988) with *Betula papyrifera*.

For practical application, the rejuvenation of mature selected forest trees has to fulfil several conditions:

- the fast and orthotropic juvenile growth has to be restored completely,
- the ability to form adventitious roots has to be restored in order to make an economical large scale vegetative propagation possible,
- no precocious flowering should occur compared to the mother tree, because this would reduce the vegetative growth earlier.
- the rejuvenated plants have to be true to type, i.e. if in vitro techniques such as induction of somatic embryos or adventitious shoots are involved, one should pay attention to somaclonal variation.

The rejuvenating effect of in vitro propagation techniques achieved in several deciduous woody plants contrast with the results obtained with *Pseudotsuga menziesii* and *Pinus taeda*. MAPES et al. (1981) propagated Douglas fir from isolated embryos by induction of axillary shoots in the axils of the cotyledons and of adventitious shoots at the base of the epicotyl. Plantlets were rooted and established in soil. While growing further on, the plants assumed a plagiotropic growth habit which is a typical characteristic of cuttings derived from ortets of Douglas-fir older than 3 years (Vieitez et al., 1977). MCKEAND (1985) compared seedlings and tissue cultured plantlets derived from cotelydons of *Pinus taeda*. After two growing seasons, the plantlets had numerous growth and morphological characteristics of mature trees. Additionally, early production of female strobili has been observed on many plantlet clones (FRAMPTON et al., 1985).

The observed more mature characteristics of tissue culture plantlets originating from juvenile embryonic tissue may be explained by environmental stress during the in vitro culture or the acclimatization process.

Whereas juvenility is required for vegetative propagation of forest trees, an early induction of flowers is an important aim in tree breeding programs. Cross breeding with trees is hindered by late and irregular flowering of most species.

The induction of flowers in plants is a problem of gene regulation. Factors affecting the flower onset in trees are temperature, photoperiod, light intensity, nutrition, grafting, mechanical treatments (girdling, strangulation, pruning of roots and crowns), treatments with growth regulators (phytohormones, inhibitors) and of course the degree of maturation.

Among the treatments with phytohormones, gibberellins have successfully been used in order to induce flowers in a number of conifer species, e.g. *Thuja plicata* and *Cupressus* (PHARIS & MORF, 1967), *Metasequoia glyptostroboides* (LONGMAN, 1970), *Cryptomeria japonica* (HASHIZUME, 1962), *Pinus taeda*, *Pinus contorta*, *Pseudotsuga menziesii* (PHARIS et al., 1976) and *Picea abies* (BLEYMÜLLER, 1976). We induced flowers on 2-year-old *Sequoia giganteum* seedlings by spraying with GA 4/7. In some of these species it was possible to induce flowers in adult plants as well as in juvenile material. Although flowers appeared in juvenile plants, other characteristics of juvenility were maintained. This means that gibberellins do not shorten or terminate the juvenile phase, but induce the shoot meristems to change from vegetative towards generative differentiation, for example by gene activation.

PHYSIOLOGICAL AND ONTOGENETICAL AGING OF A TREE

During the ontogenetic development, the size of a tree, the degree of specialization of the single organs and the complex interdependence of the single parts increase. This process is necessarily associated with physiological changes of the total system and of it parts. Ontogenetical aging is genetically programmed and can be explained by differential (selective) gene activities, that means different gene complexes are active in different phases of development. This can also indirectly be concluded by the fact that even small parts of a differentiated conifer, e.g. a Douglas fir explant that has been put under completely new physiological conditions in tissue culture, retains the differentiation status. There exists an interaction with the physiological conditions, too. This can be demonstrated in Douglas fir cuttings which slowly recover from branch habit to orthotropic growth. This process can take more than 10 years in some cases.

According to SUSSEX (1976), a coordinated gene activation of a number of genes which are required for the expression of an ontogenetical phase might be possible if a product of a gene that is activated early serves as the activator for a second gene, and if its product activates a third gene in a cascade manner.

Ontogenetic aging seems to be located in the terminal meristem of a plant and is carried over onto its lateral meristems when these are formed (ROBINSON & WAREING, 1969). With this theory it can be explained that the basal parts of a tree are juvenile whereas the top, which is chronologically the youngest part, is ontogenetically most mature. The first flowers of a tree, especially in conifers, are formed on the uppermost branches.

This is very important for the breeding work and propagation of forest trees. If trees are grafted for generative purposes e.g. seed orchards or controlled crosses, only mature scions from the upper crown should be used. If the aim is vegetative propagation by cuttings or in vitro culture, only the most juvenile material available like root suckers, stump sprouts, basal epicormic shoots or lower branches, in this range, should be used.

PHYSIOLOGICAL AND BIOCHEMICAL BASIS OF AGING

Investigations on physiological and biochemical basis of aging of woody plants are rare. Up to now, no general marker (s) has been found to define either the juvenile or the adult growth phase of trees and it may be doubted if such a general marker exists for all tree species.

However, some attempts were made to identify biochemical characteristics associated with the juvenile or adult growth phase. MILLIKAN & JANKIEWICZ (1966) analysed mineral elements and proteins of juvenile and adult leaf tissue of *Hedera helix*. Ash weight of the juvenile tissue exceeded the one of the adult tissue by 10 per cent. Nearly 50 per cent of this increased weight was due to the presence of greater amounts of K and

Ca. Juvenile tissue also contained greater amounts of trace elements. Greatest difference was found in the Mn content since the juvenile form had more than twice of the amount of the adult tissue. Only slightly more protein was found in the juvenile tissue when total protein was considered. When the proteins were fractionated, juvenile tissue was found to contain twice the microsomal protein as the adult tissue. However, no explanations for these differences are given.

Contrasting results have been obtained in two *Picea* species. In *Picea rubens* (MACLEAN & ROBERTSON, 1981), tree age has a significant influence on the trace element content of the foliage. However, in this case, the older the tree the higher the trace element content.

VON ARNOLD & ROOMANS (1983) analysed the mineral element contents of vegetative buds and needles collected from young and old trees of *Picea abies*. It was found that the contents of Cl, K and Na were significantly higher in buds from old trees than in buds from young plants. Since the Na concentration increased more with age than the K concentration, the K/Na ratio decreased with age.

Working with *Sequoia sempervirens*, VERSCHOORE-MARTOUZET (1985) tried to define biochemical criteria of juvenility or rejuvenation. She found that the K/Ca and peroxydase activity/total protein ratios of buds can be used to evaluate juvenility. Investigating different material of a mature 80-year-old and 25 m high tree, she found that the youngest material has the highest K/Ca ratio (Tab. 2).

The peroxydase activity/total protein ratio was investigated with in vitro material during the rooting phase. Different explants from a mature tree were cultured for several subcultures on multiplication medium containing cytokinin. With an increasing number of subcultures of mature explants, the rooting percentage increased as well as the maximum value of the ratio peroxydase activity/total protein which was achieved after 3 to 6 days on rooting medium. The values were almost as high as those of explants of seedlings and root-suckers which are juvenile. Since regaining a high rooting capacity - a sign for rejuvenation - was correlated with the peroxydase activity/total protein ratio, this criterion may serve as a marker for juvenility of woody plants.

In order to find biochemical markers for juvenility, JAY-ALLEMAND et al. (1988) analysed 21 polyphenols of hybrid walnut (*Juglans nigra x J. regia*) by HPLC. Annual shoot extracts of 1-year-old seedlings, mature trees and stump sprouts were analysed during growth and dormancy periods. Three ratios between 5 different polyphenols extracted during shoot growth appeared to be significantly related to the expression of juvenility. The structure of this phenolic compounds has still to be identified.

Tab. 2. K/Ca ratio of buds of different material of an 80-year-old Sequoia sempervirens tree (from VERSCHOORE-MARTOUZET, 1985)

Investigated material	ratio K/Ca
Seedling	4,54
root - sucker	3,30
grafted basal shoot	3,35
grafted shoot from the crown (15 m high)	1,95
basal shoot	0,97
shoot from the crown (15 m high)	0,75

CONCLUSION

Vegetative and generative propagation (flowering) of woody plants is highly affected by the physiological and ontogenetical age of the plant material. In general, cutting or in vitro propagation require juvenile shoots or explants. With increasing age there is a gradual loss of vegetative regeneration capacity. In breeding programs it is not possible to select individuals with desirable qualities before the trees have reached a certain age which is in general too old to allow vegetative propagation. Several methods like pruning, hedging, repeated grafting or serial cutting propagation, hormonal or nutritional treatments or in vitro culture are described to obtain rejuvenated material. Since the biochemical, physiological and genetical knowledge of aging and rejuvenation is still very poor, it is not possible to determine the degree of juvenility of plant material obtained by different rejuvenation techniques.

This would be of great importance because the classical rejuvenation methods probably cause not more than a partial rejuvenation which may not be sufficient in many cases. A better understanding of the molecular basis of differences between the juvenile and mature phase may allow the development of specific treatments to induce an artificial rejuvenation of all kinds of woody plants. On the other hand, our empirical knowledge of rejuvenation techniques has increased considerably during the recent years which may be a good precondition for such a research. Especially the work with somatic embryogenesis will offer interesting approaches to juvenility.

In tree breeding programs, however, it is very desirable
- to induce flowers on grafted plants of adult trees which usually do not flower for a long period under natural conditions
- to shorten the juvenile phase of seedlings in order to obtain early flowering or
- just to induce flowers on young plants which remain juvenile in other characteristics.

Some methods making use of a combination of gibberellin treatments with temperature and water stress, are already in practical application in several conifers. Like the rejuvenation techniques, the flower induction methods have mainly been developed by empirical experiments. In order to overcome still existing problems and to improve the effect of flower induction treatments, it is necessary to get a better basic knowledge of the biochemical, physiological, and genetic processes in the plant during flower induction.

LITERATURE

AITKEN J. and REILLY, K.J., 1978, Tissue culture of Radiata pine - Streamlining techniques for plantlet production, in: "Proc. 4th congress of plant tissue and cell culture," Univ. of Calgary, Alberta, Canada: 515.

ARNOLD, S. von, 1982, Factors influencing formation, development and rooting of adventitious shoots from embryos of *Picea abies* (L.) Karst., Plant Science Letters, 27: 275-287

ARNOLD, S. von, and ROOMANS, G.M., 1983, Analyses of mineral elements in vegetative buds and needles from young and old trees of *Picea abies*, Can. J. For. Res., 13: 669-693.

BLEYMÜLLER, H., 1976, Investigations on the dependence of flowering in Norway Spruce (*Picea abies* (L.) Karst.) upon age, Acta Horticult., 56: 169-172.

BONGA, J.M., 1982, Vegetative propagation in relation to juvenility, maturity, and rejuvenation, in:" Tissue culture in forestry," J.M. BONGA and D.J. DURZAN, eds., Nijhoff/Junk, The Hague: 387-412.

BONGA, J.M., 1984, Adventitious shoot formation in cultures of immature fe-male strobili of *Larix decidua*,. Physiol. Plant., 62 (3): 416-422.

BORCHERT, R., 1965, Gibberellic acid and rejuvenation of apical meristems in *Acacia melanoxylon*, Naturwissenschaften, 52 (3): 6-7.

BOULAY, M., 1979, Propagation in vitro du Douglas par micropropagation de germination aseptique et culture de bourgeons dormants, in: "Micro-propagation d'arbres forestiers," Etudes et Recherches, 12, AFOCEL, ed.,: 67-75.

BOULAY, M., 1987, In vitro propagation of tree species, in: "Plant Tissue and Cell Culture," C.E. GREEN, D.A. SOMERS, W.P. HACKETT and D.D. BIESBOER, eds., Alan R.Liss: 367-382.

CLARK, J.R., 1983, Age related changes in trees, J. Arboriculture, 9: 201-205.

DAMERI, R.M., CAFFARO, L., GASTALDO, P., and PROFUMO, P., 1986, Callus formation and embryogenesis with leaf explants of *Aesculus hippocastanum* L., Journal of Plant Physiology, 126 (1): 93-97.

DERMEN, H., 1948, Chimeral apple sports, and their propagation through adventitious buds, J. Hered. 39: 235-242.

DIELS, L., 1906, "Jugendformen und Blütenreife im Pflanzenreich," Berlin.

DOORENBOS, J., 1965, Juvenile and adult phases in woody plants, Handbuch der Pflanzenphysiologie XV (1): 1222-1235.

EICHHOLTZ, D.A., 1979, Adventive embryony in apple,. HortScience, 14 (6): 699-700.

FORTANIER, E.J., and JONKERS, H., 1976, Juvenility and maturity of plants as influenced by their ontogenetical and physiological ageing, Acta Horticulturae, 56: 37-44.

FRAMPTON, L.J., MOTT, R.L., and AMERSON, H.V., 1985, Field performance of loblolly Pine tissue culture plantlets, in: Proc. 18th So. For. Tree Improv. Conf., Long Beach, M.S., USA, p. 136-144.

FRANCLET, A., 1983, Rejuvenation: Theory and practical experiences in clonal silviculture, in: Proc. 19th meeting of the Canadian tree improvement association, Part 2: Symposium on clonal forestry: Its Impact on tree improvement and our future forests, Toronto, Ontario, 22.-26.8.1983, L. ZSUFFA, R.M. RAUTER, and C.W. YEATMAN, eds.,: 96-134.

FRTISCHE, R., 1948,: Untersuchungen über die Jugendformen des Apfel- und Birnbaumes und die Konsequenzen für die Unterlagen- und Sorten-züchtung, Ber. schweiz. bot. Ges., 58: 207.

FRYDMAN, V.M., and WAREING, P.F., 1973, Phase change in *Hedera helix* L. II. The possible role of roots as a source of shoot gibberellin-like substances, J. Expt. Bot., 24: 1139-1148.

GONZALEZ, M.L., VIEITEZ, A.M., and VIEITEZ, E., 1985, Somatic embryo-genesis from chestnut cotyledon tissue cultured in vitro, Scientia Horticulturae, 27 (1-2): 97-10.

GREENWOOD, M.S., 1984, Phase change in loblolly pine-shoot development as a function of age, Physiologia Plantarum, 61 (3): 518-523.

GREENWOOD, M.S., 1987, Rejuvenation of forest trees, Plant Growth Regulation, 6: 1-12.

HACKETT, W.P., 1985, Juvenility, maturation, and rejuvenation in woody plants, Horticultural Reviews, 7: 109-155.

HACKETT, W.P., 1987, Juvenility and maturity. in: "Cell and tissue culture in forestry," Vol. 1, J.M. BONGA, and D.J.DURZAN, eds., Nijhoff, Dordrecht: 216-231.

HAKMAN, I., and FOWKE, L.C., 1987, Somatic embryogenesis in *Picea glauca* (white spruce) and *Picea mariana* (black spruce), Canadian Journal of Botany, 65 (4): 655-659.

HANSCHE, P.E., 1986, Heritability of juvenility in peach, HortScience, 21 (5): 1197-1199.

HASHIZUME, H., 1962, Initiation and development of flower buds in *Cryptomeria japonica*, J. Japan. Forest. Soc., 44: 312-319.

HU, C.Y., and SUSSEX, I.M., 1971, In vitro development of embryoids on cotyledons of *Ilex aquifolium*, Phytomorphology, 21: 103 - 107.

JAMES, D.J., PASEY, A.J., and DEEMING, D.C., 1984, Adventitious embryogenesis and the in vitro culture of apple seed parts, J. Plant. Physiol., 115: 217 - 229.

JAY-ALLEMAND, C., CORNU, D., and MACHEIX, J.J., 1988, Biochemical attributes associated with rejuvenation of walnut tree, Plant Physiology and Biochemistry, 26 (2): 139-144.

JOERGENSEN, J., 1989, Somatic embryogenesis in *Aesculus hippocastanum* L. by culture of filament callus, J. Plant Physiol., (submitted).

LIBBY, W.J., 1976, Introductory Remarks, IUFRO, Joint meeting on advanced generation breeding. Bordeaux, June 1976, 8p.

LONGMAN, K.A., 1970, Initiation of flowering on first year cuttings of *Metasequoia glyptostroboides*, Nature, 227: 299-300.

LYRENE, P., 1981, Juvenility and production of fast-rooting cuttings from blueberry shoot cultures, J. Amer. Soc. Hort. Sci., 106: 396-398.

MACLEAN, K.S., and ROBERTSON, R.G., 1981, Trace element levels in red spruce and the effects of age, crown and seasonal changes, Commun. in Soil Sci. Plant Anal., 12 (5): 483-493.

MAPES, M.D., YOUNG, P.M., and ZAERR, J.B., 1981, Multiplication in vitro du Douglas (Pseudotsuga menziesii) par induction précoce d'un bourgeonnement, adventif et axillaire, in: "Colloque International sur le Culture 'In vitro' des Essences Forestières, IUFRO Section S2 01 5, Fontainebleau, France, AFOCEL, ed., Nangis: 109-114.

MCKEAND, S.E., 1985, Expression of mature characteristics by tissue culture plantlets derived from embryos of loblolly pine, J. Amer. Soc. Hort. Sci., 110 (5): 619-623.

MEIER-DINKEL, A., 1989, Micropropagation of birches. in: "Biotechnology in Agriculture and Forestry," Trees III, Y.P.S. BAJAJ, ed., Springer, Berlin, (submitted).

MERKLE, S.A., and SOMMER, H.E., 1986, Somatic embryogenesis in tissue cultures of *Liriodendron tulipifera*, Can. J. For. Res., 16: 420-422.

MEYER, M.M. Jr., 1983, Clonal propagation of perennial plants from flowers by tissue culture, Combined proceedings, International Plant Propagators' Society, 33: 402-407;

MILLIKAN, D.F., and JANKIEWICZ, L.S., 1966, Mineral and protein changes associated with juvenile and adult forms of *Hedera helix*, Bull. Acad. Pol. Sci. Ser. Sci. Biol. XIV, 11-12: 801-803.

MULLINS, M.G., NAIR, Y., and SAMPET, P., 1979, Rejuvenation in vitro: Induction of juvenile characters in an adult clone of *Vitis vinifera* L., Ann. Bot., 44: 623-627.

MULLINS, M.G., and SRINIVASAN, C., 1976, Somatic embryos and plantlets from an ancient clone of the grapevine (cv. Cabernet-Sauvignon) by apomixis in vitro, J. Exp. Bot., 27 (100): 1022-1030.

PASSECKER, F., 1943, Jugend- und Altersformen bei Obstbäumen und anderen Gehölzen, Biologia generalis, 17: 183.

PATON, D.M., WILLING, R.R., and PRYOR, L.D., 1981, Root-shoot gradients in *Eucalyptus* ontogeny, Ann. Bot., 47: 835-838.

PHARIS, R.P. and MORF, W. (1967): Experiments on the precocious flowering of western Red Cedar and four species of *Cupressus* with Gibberellin A$_3$ and A$_4$/A$_7$ mixture, Can. J. Bot., 45: 1519-1524.

PHARIS, R.P., ROSS, S.D., WAMPLE, R.L., and OWENS, J.N., 1976, Promotion of flowering in conifers of the Pinaceae by certain of the gibberellins, Acta Horticult., 56: 155-162.

RADOJEVIC, L., 1979, Somatic embryogenesis and plantlets from callus cultures of *Paulownia tomentosa* Stued., Z. Pflanzenphysiol., 91: 57 -62.

RADOJEVIC, L., VAJICIC, R., and NESOVIC, M., 1975, Embryogenesis in tissue culture of *Corylus avellana* L., Z. Pflanzenphysiol., 77: 33-41.

RANCILLAC, M., 1979, Mise au point d'une méthode de multiplication végétative in vitro du pin maritime (*Pinus pinaster* Sol.) pour la constitution de clones à partir de semences, in: "Micropropagation d'arbres forestiers," AFOCEL, ed.,: 41-48.

ROBBINS, W.J., 1957, Giberrellic acid and the reversal of adult *Hedera* to a juvenile state, American Journal of Botany, 11 (44): 743-746.

ROBINSON, L.W., and WAREING, P.F., 1969, Experiments on the juvenile-adult phase change in some woody species, New Phytol., 68: 67-78.

SCHAFFALITZKY DE MUCKADELL, M., 1954, Juvenile Stages in Woody Plants, Physiologia Plantarum, 7: 782-796.

SCHWABE, W.W., and AL-DOORI, A.H., 1973, Analysis of a juvenile-like condition affecting flowering in the black currant (*Ribes nigrum*), J. Expt. Bot., 24: 969-981.

SOMMER, H.E., and BROWN, C.L., 1980, Embryogenesis in tissue cultures of sweetgum, For. Sci., 26: 257 - 260.

SPETHMANN, W., 1986, Stecklingsvermehrung von Stiel- und Traubeneiche (*Quercus robur* L. and *Quercus petraea* (Matt.) Liebl.), Schriften aus der Forstlichen Fakultät der Universität Göttingen und der Niedersächsischen Forstlichen Versuchsanstalt, Band 86, 99 pp.

STERN, K., 1961, Über den Erfolg einer über drei Generationen geführten Auslese auf frühes Blühen bei *Betula verrucosa*, Silvae Genetica, 10: 48-51.

STRUVE, D.K., and LINEBERGER, R.D., 1988, Restoration of high adventitious root regeneration potential in mature *Betula papyrifera* Marsh. softwood stem cuttings, Can. J. For. Res., 18: 265-269.

SUSSEX, I., 1976, Phase change: Physiological and genetic aspects, Acta Horticulturae, 56: 275 - 280.

TULECKE, W., 1985, Somatic embryogenesis and plant regeneration from coty-ledons of walnut, Juglans regia L., Plant Science, 40 (1): 57-63.

TULECKE, W., 1987, Somatic embryogenesis in woody perennials, in: "Cell and tissue culture in forestry," Vol. 2, J.M. BONGA, and D.J. DURZAN, eds., Nijhoff,Dordrecht: 61-71.

VERSCHOORE-MARTOUZET, B., 1985, Etude de la variation topophysique au cours du clonage de *Sequoia sempervirens*, Thèse, Université Pierre et Marie Curie - Paris 6: 146 pp.

VIEITEZ, A.M., BALLESTER, A., and KLEINSCHMIT, J., 1977, Einfluβ von Alter und Wachstumsbedingungen auf die Entwicklung und Form von Douglasienstecklingen, Forstarchiv, 48 (4): 74 - 81.

VISSER, T., 1965, On the inheritance of the juvenile period in apple, Euphytica, 14: 125 - 134.

WAREING, P.F., and FRYDMAN, V.M., 1976, General aspects of phase change, with special reference to *Hedera helix* L., Acta Horticulturae, 56: 57-69.

WELLENSIEK, S.J., 1952, Rejuvenation of woody plants by formation of sphaeroblasts, Proceedings of the Koninklijke Nederlandse Akademie van Wetenschappen. Serie C. Biological and medical sciences 55: 567-573.

ZIMMERMAN, R.H., ed., 1976, Symposium on juvenility in woody perennials. College Park/ Maryland and Berlin, 1975, Acta Horticulturae, 56: 317 pp.

SECTION II

VEGETATIVE PROPAGATION: ADVANTAGES AND LIMITATIONS

SPECIAL PROBLEMS AND PROSPECTS IN THE PROPAGATION OF WOODY SPECIES

Trevor A. Thorpe and Indra S. Harry

Plant Physiology Research Group, Department of Biological Sciences,
University of Calgary, Calgary, Alberta, Canada   T2N 1N4

INTRODUCTION

Woody plants represent a vast array of types relative to their taxonomy
or use and include both angiosperms and gymnosperms. In general, these
are more difficult to propagate asexually than herbaceous species, which in
part is related to the phase change from juvenility to maturation that most
of them undergo. Thus, with few exceptions, methods for the large-scale
regeneration of true-to-type clones are limited. This is particularly true
for forest trees, the woody plants discussed in this article.

The potential benefits of using clonal planting stock in afforestation
or reforestation have been recognized for a long time. At least a 10%
increase in gain can be expected from planting selected clonal propagules
rather than selected seed families (Kleinschmidt, 1974). However, to achieve
the maximum possible genetic gain, both sexual reproduction and vegetative
multiplication should be used (Hasnain and Cheliak, 1986). Sexual reproduction is important for both introducing new genes to prevent inbreeding,
and for achieving genetic gain for those characteristics controlled by
additive gene effects. Asexual reproduction allows the multiplication of
elite full-sib families or individuals which exhibit significant gain due to
non-additive gene effects.

The traditional methods for vegetative propagation are rooted cuttings
or rooted needle fascicles (for pine species) and grafting; less frequently,
air layering has also been used. Although widely used with fruit trees and
ornamentals, these methods have been less successful with forest trees, in
part due to the rapid loss of rooting ability with age of trees and the
limited number of propagules that can be obtained in a reasonable time
(Thorpe and Biondi, 1984). As a result, considerable efforts have gone into
the use of tissue culture approaches for clonal propagation.

CLONAL PROPAGATION IN VITRO

Clonal multiplication of plants via tissue culture methods can be
carried out by three methods, namely (a) enhancing axillary bud breaking, (b)
production of adventitious buds, and (c) somatic embryogenesis (Murashige,
1978; Dunstan and Thorpe, 1986). The first method utilizes shoot tips,
lateral buds and small nodal cuttings. It produces the smallest number of

plants, as the number of shoots produced is limited by the number of axillary buds placed in culture. Regeneration of plantlets by either axillary bud breaking or adventitious budding involves the process of organogenesis, which can be achieved directly or indirectly (Dunstan and Thorpe, 1986). Direct pathways involve either the continued development of meristematic activity in preformed buds or the de novo induction of adventitious buds from loci that would not otherwise give rise to them. The indirect pathway often relies on an intermediate callus step, in which the induction of meristemoids and eventually shoot buds occur. Either pathway culminates with a rooting phase. Somatic, asexual or adventive embryogenesis is the development of embryos from cells that are not the product of gametic fusion. A bipolar structure is formed directly on the explant or more usually from callus. This approach has the greatest potential for forming large numbers of plantlets, but unfortunately it is the method which has been induced with the fewest number of species. While all three approaches to plantlet formation are commonly achieved in hardwoods, to date, only the latter two methods have been useful in softwoods.

The earliest reports of successful regeneration of forest tree plantlets via organogenesis in hardwoods was with Populus tremuloides (Winton, 1970) and in softwoods with Pinus palustris (Sommer et al., 1975). Regeneration via somatic embryogenesis was achieved with a hardwood, Ulmus americana in the mid 1970s (Durzan and Lopushanki, 1975), and with a softwood, Picea abies only very recently (Hakman et al., 1985). Today, lab-scale protocols exist for about 70 hardwoods and 30 softwoods (Kumar and Thorpe, unpublished).

JUVENILITY-MATURATION PROBLEMS

One of the major problems preventing a wide use of micropropagation is the inability to manipulate easily in vitro explants taken from mature trees. Most of the success reported involves the use of excised embryos and seedling parts as explants. The reason for this choice is that by the time trees are old enough for evaluation, they are often recalcitrant in culture. Thus, propagation is carried out with unproven material; and even if half-sib or better full-sib seed arising from controlled pollination is used, tree species are outbreeders and thus much genetic variation exists in any seed population. Some species show reasonably good juvenile-adult correlations, e.g., loblolly pine (Waxler and van Buijtenen, 1981), and thus micropropagation using juvenile explants should be acceptable. However, most tree breeders/silviculturists generally prefer to make selection from sexually mature trees at about half the rotation age. Nevertheless, many researchers feel that selection can be made at an earlier age (Thorpe and Hasnain, 1988).

Since it is easier to propagate juvenile than mature material, two options are available, namely (a) select the most juvenile tissues within a tree and (b) rejuvenate parts of the donor tree by special treatments prior to excision (Bonga, 1987). Cuttings from lower branches, especially from positions close to the trunk are more juvenile than other branches. Orthotrophic shoots that originate from the base of the trunk, or shoots that develop from sphaeroblasts are more juvenile than branches from other parts. Thus, the use of stump sprouts and the practice of coppicing are common. In some conifers, hedging or repeated pruning, which fixes the material at the "physiological age" of the tree at the time of initial pruning, is useful, e.g., with Pinus radiata (Libby et al., 1972).

In many species, especially conifers, juvenile sprouts are not readily available, and thus special rejuvenating or reactivating treatments have been applied before or during cloning (Bonga, 1987). Repeated spraying of branches intended for excision with cytokinin (N^6-benzyladenine being the most commonly used) has been patented (Abo El-Nil, 1982). Reactivation can also be achieved by serial grafting (Franclet et al., 1987) and repeated grafting of a scion from a mature tree onto seedling root stocks was found to accentuate and prolong juvenile behaviour in Hevea, Eucalyptus and Douglas fir. This approach has even allowed a 1000-year-old Tassili cyprus to be vegetatively propagated on a large scale. Similarly, serial rooting of mature cuttings have produced scions with more juvenile characteristics in both softwoods and hardwoods. Repeated subculture of shoot apices in cytokinin-containing medium has resulted in reactivated meristems in many species, including Prunus, Eucalyptus, Pinus pinaster and Sequoia. The degree of rejuvenation also increased with the number of subcultures.

Since it is not always possible to effect a level of rejuvenation suitable for clonal propagation, Bonga (1987) recommended the most juvenile somatic tissues available from the tree should be collected and the smallest possible explant should be used in order to disrupt any correlative controls that are present. One should also avoid tissues that have a history of a strongly fixed pattern of development, e.g., cambium. Lastly, one must adjust the culture conditions to accommodate certain explants; for example, the addition of NH_4NO_3 to the medium for the culture of vegetative buds from 17-20 year-old Douglas fir (Dunstan et al., 1986).

OTHER CULTURE PROBLEMS

Woody plants, particularly angiosperms, often secrete substances into the medium in response to wounding on excision. These brown/black pigments, oxidized polyphenols and tannins, often inhibit growth and development of the explants, and can even cause necrosis. Several approaches have been developed to deal with this problem (Pierik, 1987). These include the addition of activated charcoal, PVPP, antioxidants (citric acid, ascorbic acid, thiourea or L-cystine), and amino acids (glutamine, arginine or asparagine) to the medium. Diethyl-dithiocarbonate has also been added to the rinses after sterilization. The use of liquid medium allows for an easier and quicker dilution of toxic substances. Where browning is caused by photo-oxidation of the explant base, reducing the wounded surface of the explant, culture in the dark, or covering or painting of the basal parts of the culture vessels have reduced the problem. In some cases, a lowering of the salt concentration in the medium or the omission of the growth regulators have helped, but for some others this approach is counterproductive since growth and differentiation is drastically reduced. Soaking of the explants in water or solutions of antioxidants has also been advocated. Finally, in some cases, the only successful method is the very labour intensive process of frequent transfer of the explants onto fresh medium.

Tissues in culture also produce volatile substances which may accumulate in the culture vessels and become auto-intoxicating. These include carbon dioxide, ethylene, ethane, acetaldehyde and ethanol; the latter two are characteristic of callus cultures and woody nodal explants (Thomas and Murashige, 1979). However, the importance of some of these gases in de novo organogenesis in excised cotyledons of radiata pine has been observed (Kumar et al., 1987). If ethylene and CO_2 were allowed to build up in the culture flasks early in culture, morphogenesis was promoted. Excessive accumulation after bud initiation caused some dedifferentiation. Elimination of these gases inhibited organogenesis. These results suggest that manipulation of the gaseous environment in culture vessels might be valuable in optimizing large-scale micropropagation.

While it is relatively easy to surface sterilize explant material with bleach, $HgCl_2$, H_2O_2 etc., and eradicate potential bacterial and fungal contamination, not all contaminants are of surface origin. Infection also occurs from organisms trapped in areas not reached by the sterilants or they can be systemic. In the former, bark or bud scales, for example, can be removed before sterilization. However, systemic infection is extremely difficult to eliminate. Zimmerman (1986) isolated several types of bacteria from cultured apple shoots, which appeared sterile for several transfers before the contamination became apparent. Also, attempts to eliminate the contaminant with antibiotics resulted in severe damage to the plant tissue. This problem has also appeared in cultures of black and English walnut (Boulay, 1985), and although the bacteria did not seem to affect the growth of shoots during multiplication they caused severe problems during rooting. Internal infestations can be partially overcome by the use of antibiotics (Young et al., 1984; Boulay, 1985) but more research on types, concentration, duration and timing of applications is needed. To date, the most successful approach has been to elongate infected material rapidly in culture, and then subculture meristem-tips to fresh medium (Zimmerman, 1986).

Another problem often encountered in cultures of woody species is vitrification or hyperhydric malformation. This problem seems to be a physiological disorder where lignification is reduced, and where there is an increase in cell size due to the diffusion of water into these tissues (Gaspar et al., 1987). The leaves of vitrified shoots are often dark green, translucent, thick, curled, elongated and in conifers, the needles are often stuck together; the stems of such shoots are also thick, watery and brittle (Boulay, 1985; Gaspar et al., 1987). These types of shoots are difficult to multiply and root (Boulay, 1985) and when they are transferred to soil, they wilt quickly and are very susceptible to infection (Gaspar et al., 1987). Vitrification seems to be a multifaceted problem, and factors such as growth regulators, agar concentration (Debergh et al., 1981; Bornman and Vogelmann, 1984), ethylene (Kevers and Gaspar, 1985) and ammonium N (Daguin and Letouzé, 1986), have all been implicated. An increase in the agar concentration and the lowering of cytokinin and ammonia nitrogen levels have normalized vitrified tissues in some species (Gaspar et al., 1987).

FIELD PERFORMANCE

The evaluation of plantlet growth relative to seedling growth in conifers has only begun during the last five years. Parameters like field survival, growth rate, plagiotropism, and disease susceptibility have been monitored, and in general, the quality of the in vitro generated shoot had a direct correlation with its subsequent greenhouse and field performance. A crucial problem has been the difference in quality of roots between plantlets and seedlings. McKeand and Allen (1984) found that plantlets had thick, unbranched roots while seedlings had thinner and more lateral root development. After a 20-week period of evaluation, they found that seedlings were larger than plantlets although the relative growth rates were about the same. They also found that nutrient uptake (nitrogen and phosphorus) were again related to the root morphology: seedlings were more efficient based on per gram dry weight of root, but there was no difference when uptake was calculated as a function of root area. To generate better root quality in eucalyptus, an acclimatization period where the substrate was heated so that the roots were at a higher temperature relative to the shoots, promoted the development of the root system before planting out (see Boulay, 1985).

Field trials by the Weyerhaeuser Company, USA, showed a survival rate of 91% for plantlets, 99% for seedlings and 49% for rooted cuttings of Douglas fir. Since it was difficult to match plantlet and seedling sizes, there was a starting differential in size and by the end of five years, plantlets were still 11% shorter than seedlings. However, shoot growth increments per year were similar. Some plantlets were plagiotrophic at the planting out time, but after the third season, they had regained an orthotropic growth habit (Ritchie and Long, 1986). Four years of field evaluation showed that plantlets of loblolly pine lagged behind seedlings in growth (Amerson et al., 1988). Here again, this difference could be explained by the height differential at planting, and the ensuing first year lag in the growth of plantlets. However, by the fourth year, growth increments were similar for both types of material tested. Another very important finding of this study was that 5-year old plantlets seem to be less affected by the fusiform rust, a devastating fungal disease of loblolly pine. Genetic differences cannot account for this resistance since both test materials were derived from the same half-sib families.

In terms of overall appearance, subtle differences in plantlet and seedling shoot morphology were recognizable. McKeand (1985) noted that after two years, plantlets had numerous growth and morphological characteristics of mature trees, and Ritchie and Long (1986) observed that some plantlets had shorter and fewer branches, darker foliage and coarser needles. However, longer term monitoring data from these field plantings are necessary before any definitive conclusion can be reached.

COST OF PLANTLET PRODUCTION

The cost of generating plantlets vary with species and on the techniques used, but in general, in vitro multiplication has a higher unit cost than "macropropagation". The most advanced micropropagation technology is based on the multi-staged organogenic process, which is extremely labour and cost intensive (Thorpe and Hasnain, 1988). Therefore, this process can only be profitable if species cannot be multiplied conventionally, where the cost of macropropagation is high and special techniques (e.g. grafting) are necessary, and a high return is expected from genetic gain, yields, disease-free stock, etc. (George and Sherrington, 1984).

Studies done on production costs in New Zealand (Smith, 1986), France (Boulay, 1985) and Canada (Hasnain et al., 1986) indicated that plantlets cost three to ten times as much as seedlings, and labour costs represented 60-80% of the total. Clearly for in vitro techniques to be cost effective, there has to be certain improvements and the process has to be partially or fully automated. In the former, research aimed at increasing shoot multiplication rates by axillary or adventitious budding, and research on somatic embryogenesis and the production of artificial seed is being pursued (Thorpe and Hasnain, 1988). At Plant Biotech Industries in Israel, an automatic tissue culture system has been designed to integrate a bioreactor, a bioprocessor and an automated transplanting machine (Levin et al., 1988). This system can support both organogenic and embryogenic cultures. Prototype robots for use in various stages of the micropropagation process are also being developed (Fugita, 1989; Miles, 1989).

PROSPECTS FOR APPLICATIONS OF MICROPROPAGATION TECHNOLOGY

Micropropagation or clonal propagation is another tool for multiplying plant material, and as such will be incorporated wherever possible into existing crop and forestry systems. There are both short- and long-term prospects and applications seem to be dependent on species, costs, techniques used and to some extent on the players researching, promoting, adopting and diffusing this technology. In the short-term, micropropagation in forest trees will be involved with the multiplication of superior genotypes. A large number of plants can be produced from a few seeds, from an outstanding tree or from the limited material available from rare and endangered plant species. Also, characteristics like wood quality, disease and pollution resistance, drought and frost tolerance, morphology etc., can be selected for and produced on a commercial scale (Thorpe and Hasnain, 1988). Other important aspects are that certain ecotypes can be propagated for specific sites, genetic diversity (from different clones) can be maintained in plantations, hedged orchards are cheaper to manage than seed orchards, etc. (Carson, 1986). Longer term goals include increasing genetic gain from subsequent generations, producing commercial stock for reforestation, plantations, and orchards, and the utilization of techniques such as callus cultures and somatic embryogenesis once they are optimized. In each case, there has to be collaborative interaction between established breeding programs and tissue culture scientists for the most benefits.

One of the major advantages is the ability to exploit total genetic variance rather than just the additive portions (Timmis et al., 1987). Forestry has lagged behind conventional crops in terms of breeding and improvement. In North America, for example, most tree breeding programs were established in the early 1950s and most were based on the recurrent selection system. In this program, selection was made after each generation with interbreeding of selected material to provide for genetic recombination. While conventional breeding allows genetic gain in subsequent generations, tissue culture can allow exploitation of the genetic variation within a given generation through the cloning of outstanding individuals for testing and mass propagation.

Several woody species like poplars, wild cherry, redwood, eucalyptus and radiata pine are now commercially propagated in laboratories, and other species like sandalwood, birch, teak and loblolly pine show promise (see Haissig et al., 1987). To date, commercial exploitation methods are most advanced with the oil palm. Unilever (UK) started in vitro studies in 1968, and by 1975, a clonal unit was started in Malaysia. Production is planned at 200,000 plantlets per year with the ultimate goal being one million per year, using high quality germplasm. Other clonal centres are being established by multinational corporations or through government funding in parts of the world where oil palm can be grown (Brackpool et al., 1986). Tasman Forestry Ltd., New Zealand is currently producing large numbers of micropropagated plantlets of radiate pine for reforestation (Aitken–Christie, pers. com.). They hope to have over 2 million plantlets in the field by 1992. With time micropropagation will become more routine, and will be integrated with forest nursery operations (Haissig et al., 1987).

However, at the present time, there are several limitations associated with the use of this technology in forestry, as indicated in this article. A further concern is the criteria for judging a lab protocol as being optimum. In conifer propagation, for example, some in vitro treatments can lead to early maturation of needles on adventitious shoots, while others may lead to fewer but more vigorous shoots, or to shoots with plagiotropic growth habits. Clearly, culture conditions will have to be manipulated to produce vigorous, orthotropic juvenile plantlets for field comparisons with seedlings before the scale-up phase, which is required for large-scale clonal propagation.

Throughout the world, the markets for wood and wood products are escalating. Hardwoods such as oak, walnut, cherry etc. are viable economic commodities, while the markets for pulp, newsprint, paper and lumber from softwoods are increasing rapidly. Undoubtedly, the application of clonal propagation and other aspects of biotechnology will play a major role in producing the commercial planting stock necessary for the future.

REFERENCES

Abo El-Nil, M. M., 1982, Method for asexual reproduction of coniferous trees, U.S. Patent No. 4,353,184.

Amerson, H. V., Frampton Jr., L. J., Mott, R. L., and Spaine, P. C., 1988, Tissue culture of conifers using loblolly pine as a model, in: "Genetic Manipulation of Woody Plants," J. W. Hanover and D. E. Keathley, eds., Plenum Publishing, New York, pp. 117-137.

Bonga, J. M., 1987, Clonal propagation of mature trees: Problems and possible solutions, in: "Cell and Tissue Culture in Forestry," J. M. Bonga and D. J. Durzan, eds., Martinus Nijhoff Publ., Dordrecht, pp. 249-271.

Bornman, C. H., and Vogelmann, T. C., 1984, Effect of rigidity of gel medium on benzyladenine-induced adventitious bud formation and vitrification in Picea abies, Physiol. Plant., 61:505-512.

Boulay, M., 1985, Some practical aspects and applications of the micropropagation of forest trees, in: "In vitro Propagation of Forest Tree Species," Proceedings of the International Symposium, Bologna, Italy, May 1984, pp. 51-81.

Brackpool, A. L., Branton, R. L., and Blake, J., 1986, Regeneration in palms, in: "Cell Culture and Somatic Cell Genetics of Plants," I. K. Vasil, ed., Academic Press, New York, pp. 207-222.

Carson, M. J., 1986, Advantages of clonal forestry for Pinus radiata - real or imagined?, New Zealand J. For. Sci., 16:403-415.

Daguin, F., and Letouzé, 1986, Ammonium-induced vitrification in cultured tissues, Physiol. Plant., 66:94-98.

DeBergh, P., Harbaoui, Y., and Lemeur, R., 1981, Mass propagation of globe artichoke (Cynara scolymus): Evaluation of different hypotheses to overcome vitrification with special reference to water potential, Physiol. Plant., 53:181-187.

Dunstan, D. I., Mohammed, G. H., and Thorpe, T. A., 1986, Shoot production and elongation on explants from vegetative buds excised from 17 to 20 year-old Douglas fir (Pseudotsuga menziesii (Mirb.) Franco, New Zealand J. For. Sci., 16: 269-282.

Dunstan, D. I., and Thorpe, T. A., 1986, Regeneration in forest trees, in: "Cell Culture and Somatic Cell Genetics in Plants," Vol. 3, I. K. Vasil, ed., Academic Press, New York, pp. 223-241.

Durzan, D. J., and Lopushanski, S. M., 1975, Propagation of American elm via cell suspension cultures, Can. J. For. Res., 5:273-277.

Franclet, A., Boulay, M., Bekkaoui, F., Foret, Y., Verschoore-Martouzet, B., and Walker, N., 1987, in: "Cell and Tissue Culture in Forestry," Vol. 1, J. M. Bonga and D. J. Durzan, eds., Martinus Nijhoff Publ., Dordrecht, pp. 232-248.

Fugita, N. 1989. Application of robotics to micropropagation system. In Vitro 25: Abstr. 47, p. 22A.

Gaspar, Th., Kevers, C., DeBergh, P., Maene, L., Paques, M., and Boxus, Ph., 1987, Vitrification: Morphological, physiological and ecological aspects, in: "Cell and Tissue Culture in Forestry," Vol. 1, J. M. Bonga and D. J. Durzan, eds., Martinus Nijhoff Publ., Dordrecht, pp. 152-167.

George, E. F., and Sherrington, P. D., 1984, "Plant Propagation by Tissue Culture," Exegetics Ltd., England.

Haissig, B. E., Nelson, N. D., and Kidd, G. H., 1987, Trends in the use of tissue culture in forest improvement, Bio/Tech., 5:52-57.

Hakman, R., Fowke, L. C., von Arnold, S., and Eriksson, T., 1985, The development of somatic embryos in tissues initiated from immature embryos of Picea abies (Norway spruce), Plant Sci. Lett., 38:53-63.

Hasnain, S., and Cheliak, W., 1986, Tissue culture in forestry: Economic and genetic potential, For. Chron., 82:219-225.

Hasnain, S., Pigeon, R., and Overend, P. P., 1986, Economic analysis of the use of tissue culture for rapid forest improvement, For. Chron., 82: 240-245.

Kevers, C., and Gaspar, Th., 1985, Vitrification of carnation in vitro: changes in ethylene production, ACC level and capacity to convert ACC to ethylene, Plant Cell Tissue Org. Cult., 4:215-223.

Kleinschmidt, J., 1974, A programme for large-scale cutting propagation of Norway spruce, New Zealand J. For. Sci., 4:359-366.

Kumar, P. P., Reid, D. M., and Thorpe, T. A., 1987, The role of ethylene and carbon dioxide in differentiation of shoot buds in excised cotyledons of Pinus radiata in vitro, Physiol. Plant., 69: 244-252.

Levin, R., Gaba, V., Tal, B., Hirsch, S., De-Nola, D., and Vasil, I. K., 1988, Automated plant tissue culture for mass propagation, Bio/Tech., 6:1035-1040.

Libby, W. J., Brown, A. G., and Fielding, J. M., 1972, Effects of hedging radiata pine on production, rooting and early growth of cuttings, New Zealand J. For. Sci., 2:263-283.

McKeand, S. E., 1985, Expression of mature characteristics by tissue culture plantlets derived from embryos of loblolly pine, J. Am. Soc. Hort. Sci., 110:619-623.

McKeand, S. E., and Allen, H. L., 1984, Nutritional and root development factors affecting growth of tissue culture plantlets of loblolly pine, Physiol. Plant., 61:523-528.

Miles, G.E. 1989. Robotic transplanting for tissue culture. In Vitro 25: Abstr. 46, p. 22A.

Murashige, T., 1978, The impact of tissue culture on agriculture, in: "Frontiers of Plant Tissue Culture," T. A. Thorpe, ed., Univ. of Calgary Printing Services, Calgary, pp. 15-26.

Pierik, R. L. M., 1987, "In vitro Culture of Higher Plants," Martinus Nijhoff Publ., Dordrecht.

Ritchie, G. A., and Long, A. J., 1986, Field performance of micropropagated Douglas fir, New Zealand J. For. Sci., 16:343-356.

Smith, D.R., 1986, Forest and nut trees. I. Radiata pine (Pinus radiata D. Don). in: "Biotechnology in Agriculture and Forestry", Vol. 1: Trees, Y.P.S. Bajaj, ed., Springer-Verlag, Berlin, pp. 274-291.

Sommer, H.E., Brown, C.L., and Kormanik, P.D., 1975, Differentiation of plantlets in longleaf pine (Pinus palustris Mill.) tissue cultured in vitro. Bot. Gaz., 136: 196-200.

Thomas, D., and Murashige, T., 1979, Volatile emissions of plant tissue cultures. I. Identification of the major components, In Vitro, 15:654-658.

Thorpe, T. A., and Biondi, S., 1984, Conifers, in: "Handbook of Plant Cell Culture," Vol. 2, W. R. Sharp, D. A. Evans, P. V. Ammirato and Y. Yamada, eds., Macmillan, New York, pp. 435-470.

Thorpe, T. A., and Hasnain, S., 1988, Micropropagation of conifers: Methods, opportunities and costs, in: "Tree Improvement - Progressing Together," Proceedings 21st meeting of the Canadian Tree Improvement Association, Truro, NS, Aug. 1987, E. K. Morgenstern and J. B. Boyle, eds., Can. For. Serv., Ontario, pp. 68-84.

Timmis, R., Abo El-Nil, M. M., and Stonecypher, R. W., 1987, Potential genetic gain through tissue culture, in: "Tissue Culture in Forestry," J. M. Bonga and D. J. Durzan, eds., Martinus Nijhoff/Dr. W. Junk, The Hague, The Netherlands, pp. 198-215.

Waxler, M. S., and van Buijtenen, J. P., 1981, Early genetic evaluation of loblolly pine, Can. J. For. Res., 11:351-355.

Winton, L. L., 1970, Shoot and tree production from aspen tissue cultures, Am. J. Bot., 57:904-909.

Young, P. M., Hutchins, A. S., and Canfield, M. L., 1984, Use of antibiotics to control bacteria in shoot cultures of woody plants, Plant Sci. Lett., 34:203-209.

Zimmerman, R. H., 1986, Regeneration in woody ornamentals and fruit trees, in: "Cell Culture and Somatic Cell Genetics of Plants," Vol. 3, I. K. Vasil, ed., Academic Press, New York, pp. 243-258.

FACTORS AFFECTING TISSUE CULTURE SUCCESS IN MASS PROPAGATION

Pier-Luigi Pasqualetto

Azienda Agricola Meristema SRL, Laboratorio
di micropropagazione, Via Martiri della liberta'
n.13, 56030 Cascine di Buti (Pisa), Italy

INTRODUCTION

Tissue culture techniques are a wonderful and powerful tool in the hands of plant propagators. Micropropagation is probably the most widely applied tissue culture technique and its direct impact on commercial plant production is considerable (Debergh, 1987). Since Morel and Martin obtained virus-free dahlias by meristem culture in 1952, tissue culture techniques have progressed significantly. The list of plants vegetatively propagated _in vitro_ has increased in recent years, though many difficulties and problems still remain to be solved. In this paper I deal with the major factors affecting tissue culture success in a mass propagation system, discussing its _in vitro_ and _in vivo_ aspects.

THE CONCEPT OF REPEATABILITY

A good researcher knows that experiments in tissue culture should be repeated several times to obtain a reliable information on how to tackle the next step. Variability is the most frequent response of vegetative material in culture. When reading reports on tissue culture, one is struck by the fact that the results presented often show a high standar deviation. This happens because there is a considerable number of interacting factors and it is almost impossible to control all of them. Repeatability is the key word. A micropropagation system is only valid and reliable when the results obtained are constant in time, and this is not easy to achive.

CHOICE AND PRETREATMENT OF THE MOTHER PLANT (Stage 0)

Selected mother plants should be taken as starting material. Knowing the physiological and sanitary status is imperative for good results, since this has a great influence on the growth of the explant have once in culture. It is better to use virus-free plant as starting material. If one is obliged to work with plant material of unknown sanitary status, it is advisable to use small explants, for example meristem tips, to facilitate the separation of meristematic tissue from surrounding, more differentiated tissue and thus guarantee the success

Plant Aging: Basic and Applied Approaches
Edited by R. Rodríguez _et al._
Plenum Press, New York, 1990

of the following stages. Shoot tips can safely be used when mother plants are under sanitary control. Some problems of stage 1 cultures can be overcome by preparing the mother plants and/or explants during stage 0. The following is a list of the different types of intervention effected to obtain a mor hygienic or physiologically better adapted explants in stage 1 (Debergh, 1987): a) irrigation and nutrition via a trickle irrigation system or any other type of sub-irrigation: the upper parts of the plant, thus develop in a drier atmosphere and can yield cleaner explants (Debergh and Maene, 1984); b) control of the photoperiod under which the mother plants are grown helps to control the vegetative or generative state; c) appropriate cold treatments help to overcome dormancy problems; d) preliminary treatments with fungicides make for a more hygienic explants; e) appropriate pruning yields healthier, better sized and more active explants; f) use of forcing solutions for the source of explants or a pulse treatment with hormone solutions; g) growing the mother plants under etiolating conditions to yield elongated meristems or shoot tips. The problem of Juvenility-Maturity and Rejuvenation deserves a mention, particular with reference to forest trees. However, it is usually more difficult to establish shoot cultures from mature trees than from juvenile plants (Bonga 1987). Explants from the lower branches of the crown are normally easier to establish than explants from more mature tissues in the upper part of the crown. The following techniques may be adopted in stage 0: a) stump sprouts are a favourable juvenile alternative where they exist; b)spraying of buds with cytokinin before incubation in vitro ; c) repeated grafting of mature buds on juvenile rootstocks before excision for in vitro culture.

ESTABLISHMENT OF EXPLANTS (Stage 1)

With in vitro techniques to vegetatively propagate plants on a large scale, one has to choices: meristem tips or shoot tips. I have already pointed out that the sanitary status of the donor plant is a key factor. As regards stage 1 I would like to add another key word: time. Shoot tips generally take less time to adapt to life in culture than meristem tips and their subsequent growth will be more rapid, at least for 3 or 4 cycles.

Time of exicision is often a key factor for establishment of explants of woody species with their episodic growth. The best seasons for bud initiation of most trees are spring, coinciding with budbreak, and late summer. However, one should also consider the risk of contamination which is also season dependent (Bonga, 1987).

Microbial contamination of cultures . By microbial contamination I mean not only bacteria but also bacteria-like organisms, viruses, fungi and mycoplasma. Microbial contamination can be exogenous or endogenous. The problem of producing sterile explants is not an easy one to solve, particularly when endogenous bacteria are involved (Debergh, 1988). All interventions should be aimed at reducing the microbial population. An adequate sterilization procedure is very important. Many sterilizing agents are used: sodium hypoclorite, calcium hypoclorite and mercury chloride. The last one seems to be very effective against exogenous bacteria (precautions should be taken to recover the sterilizing solution to avoid environmental pollution). Some authors propose the use of clear agents, such as Gelrite, to make detection of contamination much easier. I have found that Gelrite produces vitrification of the explants and many types of culture respond unfavourably to the chemical properties of this gelling agent. Quite often, cultures can be contaminated with bacteria without any visible sign in the medium, and that is why it is so important to have good detection media at one's disposal (Debergh and Vanderschaeghe, 1987). Microbial contaminations frequently emerg after several cycles of culture: this is the case when endogenous bacteria were not eliminated during the initial sterilization and the cultures were only apparently aseptic, carrying microbes that developed very slowly. It often happens that microbial contamination do not impair the growth of cultures, and it would be interesting to discover which bacteria are harmful for the cultures and which are not. In my experience some contaminated cultures of gypsophila gave good quality plants in the field. Different classical media for early bacteria detection are available, but none of them allow certain definition of the culture as "bacteria-free." Many endogenous bacteria have probably not yet been identified; moreover contamination usually involves not a single bacteria but a population. The use of chemicals or antibiotics in the medium has not proved to be a very reliable remedy against bacteria. It has been reported that by taking meristems from contaminated cultures transferred to an elongation medium supplemented with antibiotics, it has been possible to recover cultures. I must remerber that _in vitro_ shoot tip cultures do not always yield pathogen-free plants (Wang, 1985). Indexing programs should be incorporated into various steps of _in vitro_ mass cloning, and should not be only limited to virus indexing but also extended to bacteria-like organisms, fungi or even mycoplasma (Debergh, 1987).

   Formation of phenols. Another problem connected with bud excision and initiation _in vitro_ is formation of phenols, which are extremely abundant in woody plants and often make growth and development impossible (Von Arnold, 1988). The methods to avoid this problem are reviewed by several authors (Anonymous, 1978; Compton and Preece, 1986). However, not all phenols are necessarily damaging. For example, phenol exudates originating from seedlings have stimulated _in vitro_ growth of birch buds (Thorpe, 1988).

SHOOT MULTIPLICATION (Stage 2)

   When the cleaned explant adapts itself to _in vitro_ life, one can start with the propagation stage. Micropropagation through axillary shoots is based on the break of apical dominance and the consequent growth of axillary buds. To date micropropagation using adventitious shoots or somatic embryogenesis have been very limited, since genetic stability is not always guaranteed with this technique. This aspect will not be discussed here. The success of the multiplication stage depends on the choice of suitable culture medium, culture atmosphere and propagation ratio. Avoiding the spread of contamination becomes strictly necessary at this stage.

   Culture medium. Any tissue culturist knows that an appropriate culture medium is the base of a good propagation system. Several formulations of media for the culture of woody plants can be found in the literature. Many plants will grow on a wide range of formulations, others appear specific. According to McCown and Sellmer (1987) general consideration on formulations of media are the following: a) high ionic strength has an inhibitory effect on the growth of many woody species; b) when low salt formulations do not support adequate growth of shoots, a marked improvement can be obtained by supplementing with NH_4NO_3; c) apex or shoot tip necrosis is often observed in actively growing cultures. It is most likely the result of Ca-deficiency in the tissues, due to (i) high humidity in the container (slowing transportation and consequently water/nutrient flow from the medium to the tissues), and (ii) the fact that Ca is not remobilized in plant tissues; d) high chloride concentrations can be responsible for unhealthy looking tissues and even their death. The use of an appropriate gelling agent at the correct concentration has proved to be an important factor in controlling vitrification in conifers and in broad-leaf trees.

    Culture atmosphere . Attention should be paid to the ambient temperature. Due to the greenhouse effect, the temperature in the container is higher than that in the culture room. This creates humidity inside the vessels and can have serious consequences on the quality of the plant material produced.

    Propagation ratio . _In vitro_ propagation via axyllary buds gives greater guarantees of genetic stability than other techniques (adventitious shoots, somatic embryogenesis). It is based on the use of cytokinins to block apical dominance, allowing the growth of lateral shoots. Generally the propagation ratio is a direct consequence of the amount of cytokinin: increasing BAP, 2iP or Kinetin etc, causes the propagation ratio to rise. As reported by Debergh (1987), such interventions can sometimes have detrimental effects in a later stages: a) a system which was originally designed for axillary bud development becomes adventitious; b) the rootability of the shoots is lowered; c) physiological disorders _in vitro_ , such as vitrification, are evoked; d) poor survival of _in vitro_ rooted plants upon transfer ex vitro; e) abnormal physiology of micropropagated plants in the field.

    Avoiding the spread of contamination . Contaminations can invade a culture due to lack of precautions during manipulations (Debergh and Vanderschaeghe, 1987). It has been clearly demonstrated that some bacteria can develop in alcohol, used to flame the scalpel and forceps. It is also important to check each jar carefully for any visible sign of contamination in the medium, not to mix plant material from one jar with that from other jars. In this way the undesirable effect of contaminations can be limited.

ELONGATION (Stage 4)

    In this stage cultures are induced to form strong, apically-dominant shoots. The medium is quite often different from that of stage 2. The amount of cytokinin is reduced to encourage apical dominance, and the ionic strength is generally lower. The use of liquid medium to be added on top of the exhausted stage medium has been proposed (Maene and Debergh, 1985). However a prerequisite for this type of intervention is that very homogeneous propagula are produced in stage 2.

ROOTING (Stage 4)

 Rooting can be accomplished either _in vitro_ or _in vivo_ . For economic reasons, the current tendency is to switch the rooting stage from an _in vitro_ step to an _ex vitro_ one wherever possible. _In vitro_ rooting includes root induction and root elongation in the culture jar. Root initiation is effected by treating the shoots coming from the elongation stage with exogenous auxins IBA, NAA or IAA. Pulse treatment of excised shoots in a medium with a high concentration of auxin followed by root growth in a hormone-free medium has proved to be more efficient than a longer treatment in a low concentration for conifers (Bornman, 1983). Enhanced rooting has been reported with auxins combined with other compounds having growth regulating effects such as coumarin in Pinus sylvestris (Bornman and Janson, 1980) and fusicoccin in Betula pendula

(Dembny et al., 1988). As it became clear that simplifying the rooting stage would be economically advantageous, various techniques for direct _in vivo_ rooting of micropropagated shoots were developed. There are two-step and one-step processes. In the former, shoots are pretreated to induce rooting with a solution containing a high level of auxin and then planted for rooting under high humidity. Rooting induction is performed in non-sterile conditions or using aseptic techniques. The latter (one-step processes) simply involve treating the cut basal ends of micropropagated shoots with an auxin carried on talc powder before inserting them into a rooting medium or plug and placing them under mist or high humidity conditions.

ACCLIMATIZATION

Acclimatization of micropropageted plants is complex and different for each crop. Success in micropropagation requires that shoots or plantlets that have been growing heterotrophically under conditions of very high humidity (90-100%) undergo adaptation to become autotrophic and to grow under conditions of moderate to low humidity (Zimmerman, 1988). Three main factors are important during weaning: humidity, light intensity and substrate. Humidity is a key word during acclimatization. Generally a double tunnel system is used for fruit plants. After approssimately one week, depending on the crop, the first tunnel is gradually removed until the plant is completely adapted to grow under _in vivo_ conditions. Light intensity is a very important factor for the success of weaning. High light intensity during the first days may be detrimental for the plants. On a sunny day light intensity is very high, up to 100.000 lux, while laboratory conditions it never exceeds 6.000 lux. For the first few days plants should be grown under shade and gradually adapted to a higher intensity.
Peat composition can influence the growth of the young plantlets. Several kinds of peat are on the market. Attention should be paid to the pH value which greatly influences the rate of growth. The substrate should also allow good draining to avoid fungal attacks.

FIELD PERFORMANCE

Micropropagation is a valid technique when the plants produced are a copy of the mother plant. Evaluating the genetic stability of micropropagated plants is difficult. Chromosome counts can determine the ploidy level but do not establish the phenotype of the micropropagated plant. For woody plants, phenotypic determinations require long-term evaluations in the field (Zimmerman, 1988).

CONCLUDING REMARKS

I have considered the main factors which influence the success of mass propagation, bearing in mind that there are many differences in cultural requirements not only between species but even within species. The researcher also plays an important role in the success of mass-propagating a crop. While it is imortant to standardize procedures for a more efficient production planning, creative research is needed when a new crop is studied for the first time. The cost of producing plants by micropropagation is still a limiting factor.

REFERENCES

Anonymous, 1978. In vitro multiplication of woody species. Round Table Conf. , Gembloux, Belgium, Centre Rech. Agron. Etat, 1-295.

Bonga, J. H. M., 1987. Clonal propagation of mature trees: problems and possible solutions. In : "Cell and tissue culture in forestry". Vol. 1:249-271. Bonga, J. H. M., and Durzan, D. J., eds., Martinus Nijhoff Publishers.

Bornman, C. H., 1983. Possibilities and constraints in the regeneration of the trees from cotyledonary needles of Picea abies in vitro . Physiol. Plant. , 57:5-16.

Bornman, C. H., and Janson E., 1980. Organogenesis in cultured Pinus sylvestris tissue. Z. Pflanzenphysiol. , 96:1-6.

Compton, M. E., and Preece, J. E., 1986. Exudation and explant establishment. IAPTC Newsletter , 50:9-37.

Debergh, P. C., 1987. Improving micropropagation. IAPTC Newsletter , 51:2-9.

Debergh, P. C., 1988. Micropropagation of woody species. State of the art on in vitro aspects. Acta Hort. , 227:287-295.

Debergh, P. C., and Maene, L. J., 1984. A scheme for commercial propagation of ornamental plants by tissue culture. Scientia Hort. , 14:335-345.

Debergh, P. C., and Vanderschaeghe, A. M., 1987. Some symptoms indicating the presence of bacterial contaminants in plant tissue cultures. Acta Hort. , 225:77-82.

Dembny, H., Zoglauer, K., Muromtsev, G. S., and Goring, H., 1988. Effect of fusicoccin in adventitious root formation of birch shoot tips Betula pendula (Roth) cultured in vitro . Plant Cell Physiol. , 29(2):237-242.

Maene, L. J., and Debergh, P. C., 1985. Liquid media additious to established tissue cultures to improve elongation and rooting in vivo . PCTOC. , 5:23-33.

McCown, B. H., and Sellmer, J. C., 1987. General media and vessels suitable for woody plant culture. In : "Cell and tissue culture in forestry". Vol. 1:4-16. Bonga, J. H. M., and Durzan, D. J., eds., Martinus Nijhoff Publishers.

Thorpe, T. A., 1988. Physiology of bud induction in conifers in vitro . In : "Genetic manipulation of woody plants". Vol. 44. Hanover, J. W., and Keathley, D. E., eds. Basic life sciences. Plenum Press, New York and London.

Von Arnold, S., 1988. Tissue culture methods for clonal propagation of forest trees. IAPTC Newsletter , 56:2-13.

Wang, P. J., 1985. Producing pathogen-free plants using tissue culture. IAPTC Newsletter , 46:2-8.

Zimmerman, R. H., 1988. Micropropagation of woody plants: post tissue culture aspects. Acta Hort. , 227:489-499.

EFFECTIVE HANDLING OF PLANT TISSUE CULTURE

R. Sánchez Tamés, B. Fernandez Muñiz, J. P. Majada

Lab. Fisiología Vegetal
Departamento de Biología de Organismos y Sistemas
Universidad de Oviedo. Oviedo, Spain

Introduction

Since in vitro cell, organ and tissue culture have become widespread tools in the study of plant differentiation, and in its practical application to the production of great numbers of plants, it is convenient to consider the possibilities of these techniques for the improvement of the procedures amenable to these endings.

When we talk about effectiveness in plant tissue culture, we are thinking in obtaining results that surpass those obtained by other methods. When we try to evaluate in vitro results we need to have another way to reach the same endings. From all the techniques and manipulations we can submit cells, organs or plants, the most extended and amenable for comparison with other well stablished techniques is micropropagation, where the results obtained in vitro can be compaired with the traditional methods of plant propagation.

Commercial micropropagation

Micropropagation techniques currently in use require very labor intense operations and are consequently expensive and difficult to manage in large scale applications.

Several factors influence the decision   to commercially micropropagate a plant through tissue culture: Rapid multiplication rates, production of disease free plants, uniquenes as reproduction method and economics of alternate propagation systems.

Moreover, micropagation is probably the tissue culture technique with the largest economical turnover, in a way that its direct impact on commercial plant production is considerable. Notwithstanding the commercial success there are still a lot of problems which hamper full economical

The process of micropropagation

As defined by Murashige (1974), micropropagation can be divided in five steps :

      Stage 0   Mother plant selection and preparation.
      Stage I   Aseptic culture stablishment.
      Stage II   Production of suitable propagules.
      Stage III Preparation for growth ex vitro.
      Stage IV   Transfer to the outside environment.

The aim of any program to make plant micropropagation competitive can be directed against any of the five stages. Keeping in mind that this can be achived through one or several ways:
      Reducing hand labour, (60-90% of the production costs)
      Enhancing operations flexibility.
      Reducing loses. (Contamination, weaning).
      Increasing multiplication rate.

Several approaches can be taken:
      Optimization of already existing techniques.
      Partial robotization.
      New technologies.

The increasing of in vitro produced plants and the higher and higher number of species and cultivars multiplied in one laboratory makes production management more and more difficult. Recording all kinds of data flowing from production takes a lot of time. The difficulties of handling all this information can be overcome by the use of PC programmed for the needs of one particular laboratory. Software designed at satisfying the demands of an in vitro plant producer, exists already and it contributes a lot at cutting time and expenses in trials when a new material is introduced at the production line.

Where automatize

Stage 0.- Mother plant selection should be based on data collected from records on plant physiology and sanitary status, criterions should be the product of weighing pros and cons and the final decission is not prone to automatic treatment, although computers can be involved in record keeping, this stage does not have any meaningful weight in cost production but is at the roots of a reliable and succesful micropropagation program.

Stage I.- Choosing the right explant for a micropropagation program is not an easy and simple task, of the several factors that can exert some influence on explant behaviour, none of them (except age of mother plant in some cases) can be predicted a priori. A system of trial and error, with ample screening of the several possible explants in different media and conditions, will help to decide which is the best. The state of the art is not so well developed as to predict through an analysis of endogenous hormone content which is the best explant, and the right media and proportions in which hormones shoul be added, the technique, if feassible, should be budgeted in economic terms.

However at this stage there are previous preparative processes where optimization of already existing techniques and automatization reduces hand labour (Cleaning and sterilization of glassware, media preparation, filling of the vessels, etc.) and progress on this lines is already a fact.

Efficiency of the operator at explant preparation with some help for automatization in, for example, explant cutting or slicing. All the sterilization shakings and rinsings with sterile distilled water can be made also automatically, gaining in cost of labor and in some cases being more effective than the manual method (Maurice et al. 1985). Even in the sterile cabinet some degree of automatization could be achieved, for example by conveyor belts which would allow more time for subculturing or transferying. Opening and closing of containers can also be made automatically.

Stage II.- This stage is where more changes can be introduced and should be considered in order to abridge the gap between what is done and what we think should be done. Once the explant is introduced in the vessel with the adequate amount of the right medium and put in the right environmental conditions, any change in the outside of the vessel can be easily accomplished (lighting, photoperiod, temperature) but if we want to modifie something inside the vessel hand labor is at first stake. At the lab this is done by hand, changing explants to a new flask, with fresh or new medium, or adding new liquid medium to the flasks. However here

Automatization can play a very important role in several ways:
        Use of vessels with greater capacity.
        Introduction and extrusion of media.
        Control of gasses in the vessel atmosphere (Carbon dioxide, oxygen and ethylene).

Increasing the size of containers affects economics by improving labour efficency. Larger containers, with higher numbers of explants, increase operator productivity per unit time. However, losses due to vessel contamination are magnified as vessel size is increased.

During culture time there is a depletion of nutrients and growth regulators in the medium, with this system the amount of nutrients can be regulated, avoiding problems of toxicity and growth regulators pulsed at the proper times to direct differentiation and growth in the desired pattern. Moreover, a pumping system will contribute to keep in control medium pH along the culture period (Skirvin et al. 1986), also it can help to the removal of unwanted plant exudates that somehow impair growth and differentiation, (and to know the nature of such exudates). Last, but not least, provides for contamination control through prophylaxis applications (Tisserat y Vandercook, 1985).

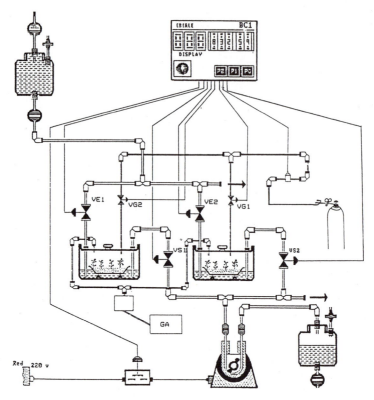

Figure 1. A simple PC-controlled system designed to allow gas (VG1,VG2) and liquid input (VE1,VE2) and output (VS1,VS2) in sterile conditions, avoiding handling of the explants for the whole culture period. GA is for gas analyzer.

Nutrient solution can be incorporated sterile, inside the vessel, in the amount needed and at the times required for a good development of the explants, but what is more important it can be introduced at the right temperature, and kept cooler than the inside atmosphere (4 °C difference will suffice), in order to have lower RH and avoid vitrification responses (Koch et al. 1987). This lowering in RH also contributes to a better cutticle development what will be crucial in Stage III.

Vessel atmosphere can also be controlled in order to have a better growth and multiplication rate. Normally, at laboratory stage, the test tubes or glass vessels, do not allow a rapid interchange of carbon dioxide between the outside and inside atmosphere (Jacksson et al., 1987), so that carbon dioxide is used up in photosynthesis at the begining of the light period, coming down to values close to the compensation point and making photoautotrophy impossible for most of the day (Fujiwara et al. 1988). Carbon dioxide concentration can be kept high inside the vessels, or forcing interchange through permable caps or pumping it inside the vessels, if its size is big enough, through sterile filters. If automatization is feassible, pumping will be selected, because at the same time a purging of the systeme can be achieved, getting rid of gases that can impair growth and differentiation (Mahon & Dennis, 1985). If photosynthetic photon flux is high enough to keep its pace with carbon dioxide concentration, autotrophic growth can be achieved (Bassi & Spencer, 1979), allowing a decrease of sugar concentration in the nutritive medium, even getting rid of them altogether if the explant has enough chlorophyll, as a consequence the growth of contaminating bacteria and fungi can be prevented (Kosai et al., 1987).

Stage III.- There are several reasons which make the weaning of plants a very important stage in a succesful commercial program of in vitro micropropagation:

        In vitro plantlets show low photosynthetic rate

        They are not fully autotrophic

        High transpiration rate due to poor cuticle development and abnormal stomatal functioning.

        Abnormal root development

        Physiological disorders.

Some of these pitfalls can be averted with the proposed system, in the same box or vessel in which the culture has been carried out, the atmosphere can be controlled and managed to avoid some of the misbehaviour of the plants. Not always agar is the best substrate for micropropagation (Henderson & Kinnerley, 1988), if the right substrate is used to support root growth of the explants and a careful managemant of the timing and amount of liquid that reaches te vessel is kept, the possibilitis of having strong and healthy plants increase considerably so a reason for having loses is avoided. At the end of the process, with an adequate program, plants will be adapted to the same conditions they are going to find in the outside. Moreover there is no need of handling the plants, so that expenses can be reduced.

Stage IV.- The last step in the process of producing a plant through in vitro culture, is the most labour demanding and so the one most challenging for automatization (Fujiwara et al., 1988). Once the plants are aclimitized, they should be potted or taken to the definitive place one by one. If clumps of plants were obtained, they should be individually separated, by hand or through some mechanical device, and one by one they should be transplanted to a definitive growth medium or one where individual shoots can root and produce a whole plant. A system of continous hedging has been developed (Aitken-Christie and Jones, 1987) cutting the shoots at a predetermined height regardless of their size, this produce a crop in which 50% of the shoots were of an unnacceptable size for rooting. The shoots with the right length were potted for rooting.

Robotization

Robotics are going to play a very important role in commercial micropropagation, as we have seen the production of multiple shoots or plantlets, can be achieved with no very sophisticated equipement, however all the procedures can be robotized in a more complex way, avoiding the use of hand labour.

We may pose the question under two perspectives: the technological one (Could robotics be applied to in vitroplant cultures?) or the economical one (Should robotics be applied to in vitro plant micropropagation?).

Man is always trying to implement new ways and tools to free himself from tedious and/or hard work, increase production and gain productivity. Something of the type is occuring now in plant micropropagation.

We saw that with a PC and slight modifications of the traditional methods improvements can be achived. Technological innovation led to the mating of a computer with a mechanical arm and a sensor hence to a large number of industrial applications. Fast developments are underway to overcome the three main bottelenecks: sensory information, processing and users education in robotics.

The field where most promising seems robotization is in the production of somatic embryos or atificial seeds (levis ent al., 1988). Through image analysys, and some of the developments alredy existent in the pharmaceutical industry for sorting and encapsulation the way is more straightforward tha it looks for micropropagation. Prototypes of robots exist already, but the problem is now an economic one (Miwa, 1987).

From an economical point of view, plant culture is a serie of relatively low added value operations. Dreams, hopes and R&D projects for their respective robotizaion seem to evolve and even to anticipate the lowering of robot cost. As synthesis of historical and economical aspects tend therefore to show that the application of robotics to plant culture is only limited by one's imagination.

Conclusion

In many cases current industrial micropropagation processes are very similar to the laboratory technique from which they have been derived. They are labour intensive and relatively unmechanised, the plants produced are expensive compared to other methods of propagation and are often variable in quality. If we applied robotics to the process, we can speed up the whole operation but due to implicit shortcomings, still some drawbacks will skeep, hence the need to think about devising new technologies which could avoid some of them, (quality variability, lack of uniformity) and which make difficult and expensive the devising of sensors able to cope with all the circumstances.

References

Aitken-Christie, J. & Jones, C. 1987. Towards automation:Radiata pine shoot hedges in vitro.- Plant

Cell Tissue and Organ Culture 8: 185-196.

Bassi, P. K. & Spencer, M. S. 1979. A cuvette design for measurement of ethylene production and carbon dioxide exchange by intact shoots under controlled environmental.- Plant Physiology 64: 488-490.

Fujiwara, F., Kozai, T. & Watanabe, I. 1988. Development of a Photoautotrophic tissue culture system for shoots and/or plantlets at rooting and acclimatization stages.- Acta Horticulturae 230: 153-158.

Henderson, W. E. & Kinnersley, A. M. 1988. Corn starch as an alternative gelling agent for plant tissue culture.- Plant Cell, Organ Tissue Culture 15(1): 17-22.

Jackson, M. B., Abbott, A. J., Belcher, A. R. & Hall, K. C. 1987. Gas exchange in plant tissue cultures.- In Advances in the chemical manipulation of plant tissue cultures Proc. Meet. British Plant Growth Regulator Group (M. B. Jackson, S. H. Mantell and J. Blake, edss), pppp. 57-71., London.

Koch, G. W., Winner, W. E., Nardone, A. & Mooney, H. A. 1987. A system for controlling the root and shoot environment for plant growth studies.- Environmental and Experimental Botany 27,n.4: 365-377.

Kozai, T., Oki, H. & Fujiwara, K. 1987. Effects of CO2 enrichment and sucrose concentration under high photosynthetic photon fluxes on growth of tissue-cultured Cymbidum plantlets during the preparation stage.- Symposium Florizel, Arlon - Belgium. Plant Micropropagation in Horticultural Industries : 135-141.

Levin, R., Gaba, V., Tal, B., Hirsch, S. & De-Nola, D. 1988. Automated plant tissue culture for mass propagation.- Bio/Technology 6: 1035-1040.

Mahon, J. D. & Dennis, M. F. 1985. An automated computer-controlled gas-exchange system for continual monitoring of physiological activities in plants.- Can.J.Bot. 63: 2213-2220.

Maurice, V., Vandercook, C. E. & Tisserat, B. 1985. Automated plant surface sterilization system.- Physiol.Veg. 23(1): 127-133.

Miwa, Y. 1987. Plant tissue culture robot oerated by a shape memory alloy actuator and a new type of sensing.- Colloque Electronique et Pilotage des Plantes. Monaco September 14,15 and 16, Moet.Hennessy : 7.

Murashige, T. 1974. Plant propagation through tissue cultures. Ann. Rev. Plant Physiol. 25: 135-166.

Skirvin, R. M., Chu, M. C., Mann, M. L., Young, H., Sullivan, J.& Fermanian, T. 1986. Stability of tissue culture medium pH as a function of autoclaving time, and cultured plant material.- Plant Cell Reports 5: 292-294.

Tisserat, B. & Vandercook, C. 1985. Development of an automated plant culture system.- Plant Cell Tissue Organ Culture 5: 107-117.

ACKNOWLEDGEMENT

This work was supported by grants from the FICYT (Fundación para la Investigación Científica y Técnica) of the Principado de Asturias (Spain).

IN VITRO MICROMULTIPLICATION OF GRAPEVINE:

Effect of age, genotype and culture conditions on

induction of callus in *Vitis spp.* leaf segments

K.A. Roubelakis-Angelakis and K. C. Katsirdakis

Department of Biology. University of Crete, and
Institite of Molecular Biology and Biotechnology
P.O. Box 1470, 711 10 Heraklion, GREECE

INTRODUCTION

Stable callus with embryogenic or organogenic potentiality leading
to plant regeneration is a prerequisite for *in vitro* selection and
genetic transformation of plants. Krul and Worley (1977) obtained
callogenesis from internodes, petioles, flower parts and leaf veins of 2
Vitis vinifera cultivars and 2 interspecific hybrids using Murashige and
Skoog (MS, 1962) medium supplemented with 1 mg/l 2,4-D and 0.1 mg/l
6-BAP. Also, Hawker *et al.* (1973) and Jona and Webb (1978) induced
callogenesis in immature grape tissue. Recently, Katsirdakis and
Roubelakis-Angelakis (1990) studied the quantitative and qualitative
callogenic response of leaf segments in 2 *Vitis vinifera* cultivars and
6 *Vitis spp.* interspesific hybrids to 2 culture media and 48
combinations of plant growth regulators. Best results in callus quality
were found at 2 or 4 μM 6-BAP in combination with 5 μM NAA. In contrary,
5 or 10 μM 2,4-D and 2 or 4 μM 6-BAP with 5 μM NAA gave callus of poor
quality. NAA alone caused development of root radicles. There was a
varying response among genotypes (Katsirdakis and Roubelakis-Angelakis,
1990). Maximum quantity of callus from internode explants of an
interspecific hybrid was obtained at 1/5/5 μM, BA/2,4-D/NOA,
respectively (Rajasekaran and Mullins, 1981).

Among the factors which have been reported to affect tissue culture
success are the physiological or ontogenic age of the donor organ, the
environment of the mother plant and the light regime under which it is
grown (Ketel *et al.*, 1985; Monette, 1983; Yu and Meredith, 1986).
Photoperiod and season are also known to influence callus production
from stem explants (Alleweldt and Radler, 1962; Brezeanu *et al.*, 1980;
Wang *et al.*, 1984).

In this report we present results on the interaction effects on
quantitative callus genesis from grapevine leaf explants of the
following parameters: environment of growth of mother plants,
developmental stage of donor tissues, composition of culture media and
genotype at 2/5 μM 6-BAP/NAA, respectively, which were found to favor
good quality callus genesis (Katsirdakis and Roubelakis-Angelakis,
1990). In addition a brief presentation of our recent results on
micromultiplication techniques is included (Roubelakis-Angelakis and
Zivanovitch, 1990; Katsirdakis and Roubelakis-Angelakis, 1990).

TISSUE CULTURE TECHNIQUES

The *in vitro* methods which could be used for regeneration of grapevine appear in Fig. 1. All of them have been studied and sufficient information has been accumulated in the literature (for review see Krul and Mowbray, 1984; and Krul, 1988). The methods, which require further investigation is the plant regeneration from callus via organogenesis and from protoplasts, via either somatic embryogenesis or organogenesis.

The techniques which seem more promising for grapevine micromultiplication *in vitro* on a commercial scale are the shoot proliferation followed by root genesis and the rooting of one-node green shoot explants.

IN VITRO SHOOT MULTIPLICATION

The shoot proliferative capacity of apical shoot segments of 12 *Vitis* genotypes (4 *vinifera* and 8 interspecific hybrids and clones) was very low in the absence of exogenous cytokinin. Full strength MS medium gave superior results compared to half strength MS, Nitsch and Nitsch (Nitsch and Nitsch, 1967) and Roubelakis medium (Roubelakis-Angelakis and Zivanovitch, 1990).

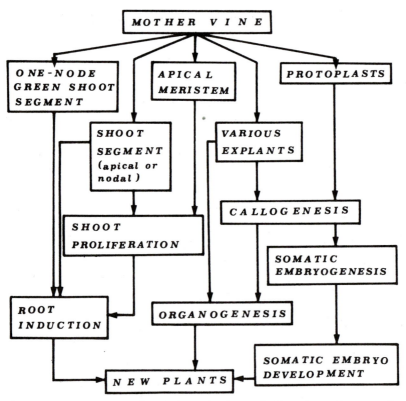

Fig. 1. Methods for *in vitro* micromultiplication of grapevine

Interaction of cytokinin and genotype and to a lesser extend of culture medium seemed to affect more the proliferating capacity of explants. Six-BAP was the most effective cytokinin. The optimum concentration ranged from 2 to 8 µM and varied among genotypes in accordance with previous results (Goussard, 1981; Reisch, 1986). Concentration of 6-BAP higher than the optimum exhibited an inhibitory effect. Work with other genotypes showed that half stength MS supplemented with higher BA concentration improved growth compared to full stength MS plus half BA concentration (Reisch, 1986). Kinetin alone had no effect in our tested genotypes and in hybrids of *V. labrusca* (Reisch, 1986). In combination with other cytokinin, kinetin was reported to be effective in inducing shoot multiplication in grapevine genotypes (Sasahara *et al.*, 1981).

ROOT INDUCTION IN GREEN SHOOT SEGMENTS

Micropropagation of grapevine by *in vitro* rooting of one-node green shoot segment is an efficient method for commercial use. Systematic study of the rooting behavior of 15 *Vitis* genotypes (8 *vinifera* cvs and 7 interspecific hybrids and clones which are used as rootstocks), revealed that there was a strong genotype-dependent morphogenetic response (Roubelakis-Angelakis and Zivanovitch, 1990). In MS medium and in the absence of exogenous auxin most genotypes did not or developed poor root system. However, in a modified culture medium (which we named Roubelakis, R) without IBA, all genotypes developed strong root system. The cv. Liatiko and the hybrid SO_4 exhibited the lowest rhizogenic response in the absence of auxin. By increasing IBA concentration, rhizogenic response was higher in most cases in R medium compared to MS medium. The optimum IBA concentration ranged from 3 to 8 µM. Other authors have proposed the use of MS or Nitsch and Nitsch medium for rhizogenesis of grapevine shoots, supplemented with 0.5 µM NAA (Barlass and Skene, 1980); or 0.1 µM IBA (Novak and Junova, 1983); or 5 µM 2,4-D and NOA (Rajasekaran and Mullins, 1981). The observation that R medium can induce rhizogenesis in one-node green shoot explant from grapevines without exogenous auxin is of significant importance because, presence of plant growth regulator(s) may increase the possibility for genetic instability during *in vitro* multiplication.

CALLOGENESIS: EFFECT OF AGE, GENOTYPE AND CULTURE CONDITIONS

Donor plants were two-years-old vines of 5 *Vitis spp.* genotypes which were grown in a controlled temperature glasshouse. The plants were meristem derived from heat-treated vines. Their freedom of the more important grapevine viruses was tested by ELISA or by using indicator plants. The genotypes used were the *Vitis vinifera L. cvs* Soultanina (syn. Thomson Seedless) and Italia, the *Vitis rupestris* clone Rupestris du Lot, and the interspecific hybrids ARG No 1 (Aramon X *Rup.* Ganzin No 1) and Richter 110 (*Vitis berlandieri* X *Vitis rupestris*). In addition, 6-weeks old plants were aseptically *in vitro* grown. They were derived from one-node green shoot segments on solidified R medium (Roubelakis-Angelakis and Zivanovitch, 1990). Culture conditions were temperature of 25 ± 1 °C, photoperiod 16/8 h, and total energy of 1500 $\mu W \cdot cm^{-2}$ provided by cool-white fluorescent lamps. The genotypes were the same as above.

Leaves from greenhouse grown vines were collected from the 10th (top), 6th and 1st (basal) node of green shoots and transferred to the laboratory on crushed ice as soon as possible. Then they were rinsed with tap water and surface sterilized as described previously

(Roubelakis- Angelakis and Zivanovitch, 1990). Also, leaves from *in vitro* plants were sampled from the 7th (top) 4th and 1st (basal) node of developing shoots and aseptically used.

The tested culture media were full strength and half strength MS, (Murashige and Skoog, 1962), and R media (Roubelakis-Angelakis and Zivanovitch, 1990). All culture media were supplemented with 2/5 μM, 6-BAP/NAA, respectively. The pH of the media was adjusted to 5.7 and 6.4 for MS and R media, respectively prior to autoclaving at 120 ^{o}C for 20 min.

Leaf segments were aseptically positioned on solidified medium in 60x15mm plastic Petri dishes. Each segment had an area of approximately 64 mm^2 and an average weight of 20 mg. The dishes were sealed with parafilm and incubated at 26 ^{o}C in the dark for 4 weeks. There was a factorial experimental design with 5 replicates per treatment. At the end of the 4-week incubation period, fresh weight of calluses was recorded. Data were subjected to factorial analysis of variance and multiple rank analysis by using the Statgraphics PC program.

Induction of callogenesis in grapevine leaf segments is a multifactor-dependent process. In the absence of exogenously supplied growth regulators callogenesis in grapevine leaf explants was very limited. Previous results indicated that full strength MS medium was superior compared to half strength MS at all tested cytokinin and auxin concentrations for all genotypes. Addition of cytokinin and auxin gave superior quantitative callogenic results in most genotypes and optimum concentrations varied among genotypes. Furthermore, all studied genotypes gave good qualitative callogenic response in either full or half-strength MS medium supplemented with 6-BAP and NAA at 2 or 4 and 5 μM, respectively (Katsirdakis and Roubelakis-Angelakis, 1990).

In Fig. 2 the effect of developmental stage of donor leaves, genotype and culture media on quantitative callogenesis is presented under optimum auxin and cytokinin concentrations for qualitative callogenic response, for *in vitro* and for greenhouse derived explants.

The overall callogenic response in all treatments was significantly superior in full strength MS medium compared to half strength MS and R media in agreement with previous results (Katsirdakis and Roubelakis-Angelakis, 1990). It seems that higher concentrations of mineral elements and carbon source are necessary for maximum callogenic response.

Interaction effects among developmental stage (age) of donor leaves, genotype and culture medium appear in Table 1, for both conditions of growth of donor plants. Only the interaction between culture medium and genotype was significant at P=0.05 in both cases.

The overall quantitative and qualitative callogenic response of leaf explants from *in vitro* grown plants was superior than from greenhouse grown plants (Fig. 2). Similar quantitative callogenic response of leaf explants from same genotypes cultured on 2/5/5 μM 6-BAP/2,4D/NAA, respectively was found (unpublished data).

The developmental stage of the donor leaf significantly affected the callogenic response of leaf explants from greenhouse grown donor vines, whereas in explants from *in vitro* grown plants there was not a significant difference (P=0.05) among the 3 developmental stages. As dedifferentiation of cells is a prerequisite before cell division and callus development occurs, it is rather apparent why rapidly growing

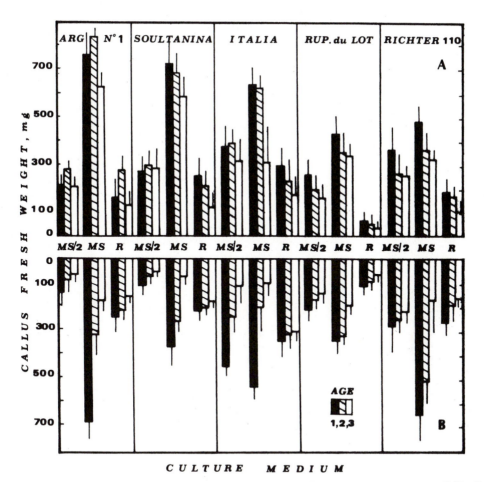

Fig. 2. Quantitative callogenesis from grapevine leaf explants at 2/5 μM 6-BAP/NAA, respectively. Interaction effect between genotype, culture medium and developmental stage of donor leaf. Culture media were full (MS) and half strength (MS/2) Murashige and Skoog and Roubelakis (R). Developmental stages are designated as 1, 2 and 3 for youngest-top (7th or 10th), middle (4th or 6th) and oldest-basal (1st) leaf of *in vitro* (A) or greenhouse (B) grown plants, respectively.

plant organs, consisting of cells with higher totipotency show higher callogenic response. In addition, Yu and Meredith (1986) found a strong negative correlation between survival *in vitro* and preexisting phenolic content of the explants. Thus, the greater callogenic response of younger leaves, may be due at least partially to their lower concentration in phenolics. Furthermore, the surface sterilization of leaves from greenhouse grown plants may have resulted in damage of the young leaves, which in turn could affect their callogenic behavior.

There was a significant difference in the callogenic response among genotypes (Fig. 2). Multiple rank analysis revealed that the tested genotypes were classified as follows, from the best to worst; Soultanina, ARG No 1, Italia, Rupestris du Lot and R 110, for *in vitro* explants. For greenhouse explants, the classification was R 110, Italia, ARG No 1, Rupestris du Lot and Soultanina. However, it is worth noting

Table 1. F-test for interaction effect between genotype, culture medium and developmental stage (age) of donor leaf on quantitative callogenic response (mg produced callus).

Interaction	In vitro grown donor plants	Greenhouse grown donor plants
Age X Genotype	NS	NS
Culture medium X Age	NS	NS
Culture medium X Genotype	5 %	5 %

that surface sterilization could have resulted in the lower callogenic response of the latter explants and thus the above classification does not reflect the callogenic potentiality of those genotypes.

The results presented in this report suggest that genotype and ontogenic status and to a lesser extend culture medium and conditions of growth of mother vines determine the degree of callogenic response in grapevine leaf explants. *In vitro* grown mother plants seem to offer better starting material for callus development because surface sterilization is ommited which may have detrimental effect; also donor organs show anatomical, biochemical and developmental characteristics which favor expession of cellular totipotency.

SUMMARY

Callus induction and growth is a multifactor-dependent process. Callogenesis in grapevine (*Vitis spp.*) interspecific hybrids, cultivars and clones was studied in the presence of 6-BAP and NAA at concentrations optimum for qualitative callogenic response (Katsirdakis and Roubelakis-Angelakis, 1990). A factorial experimental design was applied to study the interaction effect of genotype, culture medium, developmental stage of donor organ and conditions of growth of mother plants. Results revealed that explants from *in vitro* grown vines and from actively developing leaves showed best callogenic response. In most cases, better results were obtained in MS culture medium whereas there was a strong genotype dependent response. There was an interaction effect at 5 % level between culture medium and genotype.

REFERENCES

Alleweldt, G., and Radler, F., 1962, Inter-relationship between photoperiodic behavior of grapes and growth of plant tissue cultures. Plant Physiol., 37(3): 376-379.

Barlass, M., and Skene, K.G.M., 1980, Studies on the fragmented shoot apex of grapevine. II. Factors affecting growth and differentiation. J. Exp. Bot., 31 (121): 489-495.

Brezeanu, A., Iordan, M., and Rosu, A., 1980, The micropropagation of callus from tissue of somatic origin. Rev. Roum. Biol. Ser. Biol. Vegetale, 25: 135-142.

Goussard, P.G., 1981, Effects of cytokinins on elongation, proliferation and total mass of shoots derived from shoot apices of grapevine cultured *in vitro*. Vitis 20: 228-234.

Hawker, J.S., Downton, W.J.S., Wiskich, D., and Mullins, M.G., 1973, Callus and cell culture from grape berries. Hort. Science, 8(5): 398-399.

Jona, R., and Webb, K.J., 1978,   Callus and axillary bud culture of *Vitis vinifera* Sylvaner Riesling. Sci. Hort., 9(1): 55-60.

Katsirdakis, K.C., and Roubelakis-Angelakis, 1990, Organo- and callogenic potentiality of leaf segments from *Vitis spp.* genotypes. J. Amer. Soc. Hort. Sci., accepted.

Ketel, D.H., Bretelez, H., and de Groot, B., 1985,   Effect of explant origin on growth and differentiation of calli from *Tagetes* species. J. Plant Physiol., 118: 327-333.

Krul, W.R., and Worley, J.F., 1977,   Formation of adventitious embryos in callus cultures of Seyval french hybrid grape.   J. Amer. Soc. Hort. Sci., 102(3): 360-363.

Krul, W.R., and Mowbray, G.H., 1984 , Grapes, In: Hand Book of Plant Cell Culture, W.R. Sharp, D.A. Evans, P.V. Ammirato, and Y. Yamada, eds, Vol. 2, pp. 396-434, Acad. Press, London.

Krul, W.R., 1988, Recent advances in protoplast culture of horticultural crops: small fruits. Sci. Hortic., 37: 231-246.

Monette, P.L., 1983,   Influence of size of culture vessel on *in vitro* proliferation of grape in a liquid medium. Plant Cell Tissue Organ Cult., 2: 327-332.

Murashige, T., and Skoog, F., 1962,   A revised medium for rapid growth and bioassay with tobacco tissue culture.   Physiol. Plant., 15: 473-497.

Nitsch, C., and Nitsch, J.P., 1967,   The induction of flowering *in vitro* in stem segments of *Plumbage indica* L.I. The production of vegetative buds. Planta 72:355-370.

Novak, F.J., and Junova, Z., 1983,   Clonal propagation of grapevine through *in vitro* axillary bud culture. Sci. Hort., 18:321-340.

Rajasekaran, K., and Mullins, M.G., 1981,   Organogenesis in internode explants of grapevines. Vitis 20:218-227.

Reisch, B.I., 1986, Influence of genotype and cytokinins on *in vitro* shoot proliferation of grapes. J. Amer. Soc. Hort. Sci., 111(1): 138-141.

Roubelakis-Angelakis, K.A., and S.A. Zivanovitch, 1990, Morphogenetic responses of grapevine (*Vitis spp.*) genotypes to plant growth regulators and culture media. J. Amer. Soc. Hort. Sci., accepted.

Sasahara, H., Tada, K., Irs, M., Takazawa, T., and Tazaki, M., 1981, Regeneration of plantlets by meristem tip culture for virus-free grapevine. J. Japan. Soc. Hort. Sci., 50:169-175.

Wang, Y., Ge., K., Zou, G., Yang, J., and Yel, M., 1984, Callus induction and plantlet regeneration in grapevines. Int. Symp. on Genetic Manipulation in Crops, Beijing, People's Republic of China, Abstr., 22-26 Oct. 1984.

Yu, Dan-hua, and Meredith, C.P., 1986,   The influence of explant origin on tissue browning and shoot production in shoot tip cultures of grapevine. J. Amer. Soc. Hort. Sci., 111 (6): 972-975.

ACKNOWLEDGMENTS

This work was partially supported by the Greek Ministry   of Research,Energy and Technology and the Science for Stability   Programme.

PLANT CELLS AND PROTOPLAST IMMOBILIZATION AS TOOLS FOR STUDIES

ON CELL FUNCTION, METABOLISM AND DIFFERENTIATION

M. Salomé S. Pais*   and Joaquim M.S. Cabral**

* Departamento de Biologia Vegetal
– Faculdade de Ciências de Lisboa, Bloco C2,
Campo Grande 1700 Lisboa – Portugal
** Laboratório de Engenharia Bioquímica
– Instituto Superior Técnico, Av. Rovisco Pais,
1000 Lisboa – Portugal

INTRODUCTION

Immobilization methods have been successfully used for the immobilization of enzymes for industrial purposes. More recently, immobilization of microorganisms whole cells has been developed in many laboratories and also in industrial processes. According to Mattiasson (1983) immobilized cell preparations offer some advantages over the conventional preparation technology that can be summarised as follows:1) higher reaction rates due to increased cell densities; 2)higher specific product yield; 3)possibilities for continuous operation; 4)high dilution rates without the problems of wash out; 5)better and easier control; 6)reduced demands of costly fermentors; 7)easier product up-grading.

Immobilization of microbial cells may even be more advantageous than immobilization of the corresponding enzymes, since immobilized viable cell preparations can maintain the catalytic activities of labile enzymes or can be of great interest when sequential reactions involving several enzymes are concerned. Immobilized viable cells can regenerate and retain the coenzymes responsible for catlytic reactions. They appeared as the only methodology to solve problems with coenzyme regeneration.

Co-immobilization of microorganisms with enzymes, of microorganisms with organelles and enzymes with organelles have been performed (for a revision see Hahn-Hagerdal,1983).

The immobilization of eukariotic mammalian and plant cells appeared in the sequence of animal and plant cell cultures and some methods have been developed following adaptations from those used for microbial cells.

The recognition that there exists, in some plant cultured cells, a link between differentiation and expression of secondary metabolism leading to the production and accumulation of industrially interesting compounds, explains some of the failures of conventional plant cell cultures to produce the secondary metabolites characteristic from "in vivo" plants.

Plant Aging: Basic and Applied Approaches
Edited by R. Rodríguez *et al.*
Plenum Press, New York, 1990

Immobilization of plant cells has been developed in a way to use differentiated cells for the increase of production of phytochemicals or for the use of plant cells as an enzymatic system.

Recently, immobilization methods have been used for protoplast culture, some advantages of immobilization having been recognised when compared with culture in liquid medium.

In this paper we will present the immobilization methods for plant cells and protoplasts, giving an approache for their use in studies on cell function, metabolism and differentiation.

METHODS FOR PLANT CELLS AND PROTOPLAST IMOBILIZATION

From the methods generally used for immobilization of whole cells, those commonly used for plant cells and protoplasts are: entrapment in gels, entrapment in nets or foams, immobilization in hollow-fiber membranes and adsorption to microcarriers.

Entrapment in gels

The gel entrapment method involves entrapping of the cell within the interstitial spaces of water-insoluble polymer gels. Cells can be entrapped either by polymerization, by ionic network formation, or by precipitation by pH, temperature, or solvent changes.

Entrapment by polymerization

The procedure for the preparation of insoluble-gel networks is identical to that employed for the preparation of gel commonly used for disk electrophoresis. This method is based on the free- radical polymerization of acrylamide in an aqueous solution. Because of the solubility in water of the linear polymers, and in order to use them as matrices, they have to be insolubilized by cross-linking with bifunctional compounds, usually N',N'-methylenebisacrylamide (BIS).

In the gel-entrapment method of whole cells, the free-radical polymerization of acrylamide is conducted in an aqueous solution containing the whole cell and a cross-linking agent. Polymerization is commonly carried out in the absence of oxygen and at lower temperatures (10 C) in order not to damage the cell during the operation. The polymerization reaction is initiated by potassium persulfate ($K_2S_2O_8$) or riboflavin and catalyzed by dimethylaminopropionitrile (DMAPN) or N,N,N',N'-tetramethylene- diamine (TEMED). The resulting gel block can be mecanically dispersed into particles of defined size. However, these gels are quite weak in a mechanical sense and have an open network with a broad distribution of pore sizes. These disadvantages are overcome by optimizing the degree of cross-linking. However, this entrapping method has a major disadvantage: the toxicity of the acrylamide monomer, the cross-linking agent (BIS) and the initiator (TEMED). Thus the cell exposure to free-radical polymerization results in a decrease not only in the enzymic activity, but also in the viability of the cell. Fukui et al., 1978 had developed a method to entrap enzymes and whole cells with photo-cross-linkable resin prepolymers and urethane resin prepolymers of either hydrophilic or hydrophobic character. These oligomers are derivatives of poly(ethylene glycol) (ENT), hydrophilic; poly(propylene glycol) (ENTP), Hydrophobic and polybutadiene (ENTB), hydrophobic, of different chain lenghts.

A typical preparation procedure for these prepolymers consists of reacting equimolar amounts of hydroxyethylacrylate and isophorone diisocyanate at 70 C in the presence of a suitable catalyst, such as organic tin compounds or ternary amines. After 2h a half-molar ratio of the poly(ethylene glycol) is added – the reaction proceeds during 5h at 70 C. The resulting product is the corresponding photo-cross-linkable resin prepolymer (ENT). In this process one part of the water-soluble prepolymer is mixed with 0.01 part of an initiator (photosensitizer), usually benzoin ethylether, and the mixture is melted by warming at 60 C. To this molten mixture is added a whole-cell suspension and the resultant mixture is illuminated at 360 nm for 3 min, with a 2-kW high pressure mercury lamp to initiate a free-radical polymerization reaction that entraps the cell in a nonionic hydrophobic resin. The resin gel thus formed is cut into small pieces and used. In the case of water-insoluble prepolymers, these and the initiator are dissolved in a solvent (benzene:heptane = 1:1 by volume), and the cells are added as suspension in the same solvent. The mixture is photo-cross-linked in the same way as the water-soluble prepolymers

Entrapment by ionic -network formation

This technique of entrapment is an example of coarctation, the polymerization of polyelectrolytes by multivalent ions. One of the most popular methods is the immobilization of whole cells by entrapment in calcium alginate. In this very versatile procedure, only nontoxic compounds are used and so it has been preferentially chosen for the immobilization of living cells and very sensitive cells such as plant cells (Brodelius et al., 1979, 1980) and protoplasts (Scheurich et al., 1980).

The process of immobilization includes the preparation of a solution of sodium alginate, the addition of the cell mass, and the dispersion of the mixture into a counterion solution which results in the formation of a uniform, spherical, highly microporous structure that retains cells, organelles and enzyme molecules larger than the pores. The pores are large and continuous such that substrate molecules may diffuse throughout the beads. The mechanical properties of alginate gels are related to the distribution of the D-mannuronic and D-glucuronic residues and to the molecular weight and degree of dispersion within the pellets

However, this mild, simple, rapid and excellent method has some disadvantages. One is the use of calcium alginate beads in a medium containing calcium-chelating agents, such as phosphates and certain cations such as Mg^{2+} or K^{2+}, which results in the disruption of the gel by solubilizing the bound Ca^{2+}. Recently, Birnbaum et al. (1981) circumvented the instability of alginate in phosphate-containing media by treating the gel with polyamines and cross-linking. This treatment provides a stabilized peripheral layer that prevents the release of cells. Cell leakage from the matrices has also occurred when cell division within the pellets takes place and when the pellets are used in stirred vessels.

Entrapment by precipitation

Gels may be formed by precipitation of some natural and synthetic polymers by changing one or more parameters in the solution, such as temperature, salinity, pH, or solvent.

Collagen has been widely used as an enzyme-, cell-, and organelle- immobilization matrix . Collagen is the most abundant protein constituent of higher vertebrates. It can be readily isolated from a number of biological sources and reconstituted into various forms without losing its native structure. This, in conjunction with its ready availability from a large number of biological species-from fish to cattle- makes it an inexpensive matrix.

To increase the mechanical strenght of the collagen-whole-cell membrane and to maximize the amount of whole cells retained, mainly in conjunction with the complexation method, a tanning step is performed by addition of suitable bifunctional reagents, usually glutaraldehyde, to the dried collagen-whole-cell complex membrane.

Gel entrapment of whole cells is also possible in gelatin. This material is dissolved in an aqueous medium at 40-50 C and is mixed with the cell suspension and cooled at around 10 C. This gel-entrapped whole cell may be obtained in a particulate form, by stirring the aqueous cell-gelatin suspension into an organic liquid poorly miscible or immiscible in water at 50 C, rapidly cooled to 10 C whereby cell- and gelatin-containing particles are formed. However, because this gel does not possess sufficient physical stability to be used alone, it is necessary to incorporate a cross-linking agent-formaldehyde or glutaraldehyde- to increase its mechanical stability.

A very promising matrix, K-carrageenan, for immobilization of living cells was reported by Wada et al.(1979). K-carrageenan is a polysaccharide from seaweeds, used as nontoxic food aditive. Gel entrapment of whole cells based on this matrix has been accomplished by mixing a cell suspension at 45-50 C with a solution of K-carrageenan at the same tamperature. The gel is formed by cooling, as in the case of agar, or by contact with metal ions. ammonium ion, amines, or water-miscible organic solvent such as methanol or acetone. The gel formed can be granulated into particles with a suitable size and shape, and in this case where the gel strenght of the particles was not satisfactory, the immobilized cells can be treated with hardening agents, such as glutaraldehyde and 1,6-diaminohexane. Alternatively the gel can be further strenghtened by incorporation of locust bean gum, which is a D-galacto-D-mannan extracted from locust bean and used for increasing gel strenght of jelly foods .

Entrapment of cells in agar gell is an obvious method of cell immobilization but it has been little used, presumably because of the poor mechanical strenght of the gel and the characteristics of the gel being associated with oxygen- and product-diffusion limitations.

The major disadvantages of the entrapment-by-precipitation method is the unavoidable phase-parameter change which is attributable to the precipitation of the gel, because heat or organic solvent application can damage the cells.

Fiber-entrapment method

A method of immobilizing enzymes and whole cells by entrapment within microcavities of synthetic fibers has been developed by Dinelli (1972). This method is a variant procedure of entrapment by precipitation with solvents. Biocatalysts can be entrapped in fibers and continuously produced by the conventional wet-spinning techniques for the manufacture of man-made fibers using apparatus very similar to that used in the textile industry.

Adsorption method

Adsorption of cells to supports is dependent on the characteristics of the environment. Because the adsorption or adhesion phenomenon is mainly based on electrostatic interactions – that is, van der Waals forces, and ionic and hydrogen bonds – between the cell surface and the support material, the actual zeta potential on both of them plays an important role in cell-support interaction. From the zeta potential it is possible to obtain an approximate value for the surface-charge density (Abramson et al., 1942). Clearly cells will be attracted to surfaces of opposite zeta potential. If cells have the same zeta potential as a surface, attachment is still possible provided the electrostatic barrier can be penetrated by small surface projections (Van Oss et al., 1975; Grinnell, 1978). Zeta potential and the adhesin of microorganisms to surfaces have been reviewed by several authors (Daniels, 1971; Marshall, 1976; Martin, 1978).

Another factor that influences the adsorption and is strictly dependent on the properties of the cell is the cell wall composition. The charged nature of the cell wall is chiefly determined by the distribution of carboxyl and amino groups of the peptide amino acid, diaminopimelic acid and 2-amino-2-deoxyhexose residues of the cell wall surfaces, which may directly interact with the solid surface.

Carrier properties other than zeta potential will also influence the adsorption of the cells to solid supports. One of the most important properties is the carrier composition. So all glasses and ceramics consist of varying proportions of aluminium, silicon, magnesium, zirconium, titanium and other oxides that in solution act as ion-exchange materials. The corresponding hydroxides (hydrous oxides) can be formed; this allows the replacing of the hydroxyl groups on the carrier surface by suitable amino or carboxyl groups on the cell surface by a chelation mechanism.

IMMOBILIZATION OF PLANT CELLS FOR THE BIOSYNTHESIS AND BIOTRANSFORMATION OF VALUABLE SECONDARY METABOLITES

Plant cells have been immobilized for the production and biotransformation of useful fine chemicals (Brodelius, 1979; Alfermann et al., 1974). Other immobilized plant cell systems have been tested for continuous milk clotting using plant proteases (Cabral et al., 1984).

Viability of immobilized plant cells

It has been assumed that, when cells are immobilized for secondary metabolites production, only the metabolism concerned with the product formation has to be intact. However, in some processes, multienzymes are concerned which implies that immobilized cells maintain their viability and, consequently, retain their production capacity.

Viability of immobilized cells can be estimated by substrate utilization, product formation or assessement of cell integrity (Rosevear, Kennedy and Cabral, 1987).

The viability of immobilized cells can be demonstrated by standard procedures such as plasmolysis, respiration, cell growth and cell division (Brodelius, 1983, 1986). According to Brodelius (1983) Catharanthus cells entrapped in agarose or agar grow better than those entrapped in alginate this growth being slightly higher than for K-carrageenan entrapped cells. These results show that cells immobilized in alginate are inhibited in their growth which may be due to the gel structure and charge (Brodelius ,1983). Agarose is essentially uncharged while alginate is strongly negatively charged . The greater porosity of agarose may be responsible for the easier diffusion of nutrients leading to higher growth rates.

From the results obtained for immobilization of Silybum marianum suspended cells, it can be observed that from different gel entrapment supports, alginate enables the best retention and clotting activities when compared with free cells. Immobilization by metal linking (hydrous titanium) allowed retention activities similar to those obtained for Ca-alginate. Entrapment in gelatin or polyacrylamide, where retention of activity was zero, appeared inapropriate for immobilization of suspended cells of this species (Cabral et al., 1984). If the clotting activity of plant cells entrapped in Ca alginate, K carrageenan and agar is compared with that of free cells, Ca alginate presents activities similar to those of free cells and higher than those of immobilized in K-carrageenan or agar (Fonseca et al, 1987). In this case, it can indicate that the entrapment support that enables better retention of activity of immobilized cells is Ca alginate. Brodelius (1983) has also compared different entrapment supports in the production of ajmalicine isomers having verified that Ca alginate and agarose are the most appropriate for immobilization of these plant cells.

Why to immobilize plant cells for secondary metabolites production?

According to Yeoman (1987) three main characteristics from immobilized plant cells have to be examined to understand the advantages of immobilization of plant cells over the use of free cells for secondary metabolites production.

(1) The aggregate cell masses present higher cell-to cell contact and greater degree of cell diversity and differentiation when compared with cell suspension cultures. For some entrapped plant cells and, particularly for pepper cells immobilized in polyurethane foams, degrees of differentiation similar to those obtained for "calli" were observed (Lindsey et al., 1983). Collinge and Yeoman (1986) achieved the formation of roots by immobilized cells of Hyocyamus muticus. Other signs of organization in immobilized cells such as shoot and embryoid formation where also reported by Yeoman (1987).

(2) One of the main conditionants of growth of suspended cells as uniform nondifferentiated clumps, is the reduction of gradients (Yeoman, 1987) which influences the expression of biosynthetic pathways essential to secondary metabolite production and accumulation. Immobilization of cells accounts

for the creation of gradients leading to higher production rates.

(3)   It must be mentioned the advantage of immobilized cells in cultures manipulation. Immobilized whole plant cell systems can be used for performing different tests such as medium composition alterations, introduction of different precursors for the biosynthesis of the desired secondary metabolites, influence of different growth regulators without loss of biomass. The ability to change cells from a growth state to a nongrowth state offers also possibilities for studies of plant cell growth.

The fact that plant cells take long time to produce the appropriate level of biomass that is much longer (several weeks) than that in which product accumulates (few days, 1 week) is responsible for the high costs of the product produced (Fowler, 1983). The possibility of immobilization of plant cells enabling the alternation of long periods of synthesis and accumulation with small periods of rejuvenation can contribute to minimise problems with reuse of plant cells.

Biosynthesis of secondary metabolites

The biosynthetic capacity of viable preparations of Catharanthus roseus has been first demonstrated by Brodelius et al. (1979). When the entrapped cells are compared for the production of ajmalicine in a medium devoid of growth regulators, it was verified that alginate entrapped cells present the highest production yields. These results can be interpreted taking into account the comparison of Brodelius (1983) on the effect of alginate in reduction of cell growth. This having been verified, it is in agreement with the higher ability of alginate entrapped cells to produce secondary metabolites. This result is in concordance with those obtained for freely suspended cells according which secondary metabolites are produced in the stationary phase. Studies on the novo synthesis of anthraquinones by immobilized cells, revealed that, in the same conditions, alginate entrapped cells produce as much as 10 times of anthraquinones per cell as freely suspended cells (Brodelius et al., 1980).

Immobilized plant cells can also be used as systems for studies on secondary metabolites precursors. Freely suspended cells have been widely used to study the biosynthetic pathways of the indole alkaloid ajmalicine. The results reported by Brodelius and Nilsson (1980) show that immobilization of Catharanthus roseus cells can use tryptamine and secologanin for the synthesis of ajmalicine, being alginate the most appropriate entrapment agent. In fact, in this case, alginate entrapped cells synthesize, from the added distant precursors, after 5 days of incubation, approximately 12 times as much ajmalicine as it is formed, 2 weeks after incubation, by freely suspended cells. Wichers et al. (1983) have reported that alginate entrapped cells of Mucuma pruriens are able to use exogenously supplied tyrosine in the production of L-Dopa which is released into the culture medium at about 90%.

Cells of Capsicum frutescens immobilized in a matrix of polyurethane foams produce levels of capsaicin 2 to 3 folds the levels produced by freely suspended cells (Lindsey and Yeoman, 1984). Moreover, it was verified that supplementing the medium with phenylalanine and isocapric acid (precursors of capsaicin) the capsaicin production by immobilized plant cells is increased. The use of 14C phenylalanine enabled these authors to recognise the existence of a reciprocal relationship between protein synthesis and capsaicin production (Lindsey and Yeomann, 1984). Lindsey (1985) has also reported that starvation and the absence of growth regulators in the culture medium stimulates capsaicin production by immobilized cells of C. frutescens. These immobilized cells could also be used for studies on the effect of oxygen stress in secondary metabolite production. According to Wilkinson et al. (1988), dissolved oxygen concentration affects the appearence of capsaicin in the medium. Using phenylalanine or vallilylamine as precursors, when dissolved oxygen is found at 60% saturation, the capsaicin yield is less than 1mg/l. However, if dissolved oxygen falls to 0.5%, the concentration of capsaicin in the medium rises to about 17mg/l. This effect of oxygen concentration can be interpreted following Maurel and Pareilleux (1985) according whom during air sparing some volatile compounds, such as ethylene (growth regulator) and dissolved gases (carbon dioxide) may be stripped from the medium. These compounds, kept at critical concentrations, may stimulate secondary metabolism.

Biotransformation of secondary metabolites

During the last decades, great number of biotransformations have been studied using plant cell cultures (Reinhard, 1974; Alfermann et al., 1974; Furuya, 1978; Alfermann et al., 1980; Jones and Velicky, 1981). Different substrates such as steroids, alkaloids, terpenoids, phenols, coumarins and cardenolides have been used. (for revisions see Reinhard and Alfermann, 1980; Kurz and Constabel, 1985). The reactions involved in biotransformation are oxidations, reductions, hydroxylations, demethylations, glycosylations, esterifications, epoxidations and isomerizations, being most of them stereospecific. The stereospecificity is the main advantage of using plant cells or enzymes when compared with chemical synthesis (Suga et al, 1983). The ability of plant cell cultures (v.g. tobacco cell cultures) to descrimine between the different enantiomers account for the advantage of using plant cells or enzymes for the synthesis of important medicinal compounds by chemical and biological approaches. Endo et al. (1988) have recently proposed a shema for a combined chemical and biological approche for the production of vinblastine.

The biotransformation of cardiac glycosides by Digitalis lanata has been insistently studied due to the high hydroxylating capacity of a number of D.lanata cell lines that are able to transform β-methyldigitoxin to β-methyldigoxin at a high tranformation rate (Heins, 1978; Alfermann et al., 1980). Alfermann et al.(1980) immobilized this cell line by entrapment in calcium alginate and verified that immobilized cells expressed aproximatly 50% of the activity of freely suspended cells. This reduction of activity, when compared with that of freely suspended cells, is well compensated by the constant conversion (over 200 days) by immobilized cells. Moreover, since β-methyldigoxin is found in the culture medium, immobilized Digitalis cells appear highly suitable for biotransformation. Recently, it has been reported by Peterson et al.(1988) the characteristics of digitoxin-12-B-hydroxylase from D. lanata cell cultures and the immobilization of this enzyme by entrapment of microsomes in Ca-alginate.

Alginate entrapped cells of Daucus carotta have been used for 5-β-hydroxylation of the aglycones digitoxigenin and gitoxigenin (Jones and Velicky, 1982). These authors could demonstrate that the growth of these cells is limited by decreasing concentration of nitrogen source.

Immobilization of whole plant cells, organelles or enzymes may be very important to perform the combined approches for biotransformation proposed by Endo et al. (1988).

Permeabilization of immobilized plant cells

Secondary metabolites produced by freely suspended cells are often stored within the vacuoles. Very few examples (v.g shiconin and berberine) of released secondary products are known.
The transport mechanisms that are responsible for translocation of compounds over the tonoplast are not fully understood. If, in some cases, it is known the anastomose of vesicles containing secondary products, in others actif transport (requiring ATP) as well as ion trap mechanisms are possible. In other cases, such as flavonoid biosynthesis, part of the biosynthetic pathway takes place at the level of the tonoplast. One of the main problems for the use of cell suspension cultures in secondary metabolites production is the intracelular accumulation and the difficulties of their release without damage of the cells. Immobilized cells have been used to study the permeabilization of both tonoplast and plasma membrane for product release. The reversible permeabilization of various plant cells entrapped in agarose has been achieved by treatment with a medium containing an appropriate concentration of DMSO (100 mg/ml). This treatment allowed the release of stored products without loss of cells viability. Thus, the immobilized permeabilized cells can be used over a cyclic process including a growth phase (when necessary), a production phase and a permeabilization procedure. For revisions see Brodelius and Nilsson (1983) and Brodelius in Fonseca et al. (1988). Inspite of some results on permeabilization of immobilized cells, it is yet very difficult, if not impossible, for many cases, to release secondary metabolites without damage of plant cells. Immobilized plant cells can, however, be used as models for studies of membrane stability and permeabilization probably enabling approaches for the release of intracellular products.

PROTOPLAST IMMOBILIZATION

Since the first studies of protoplast isolation, different procedures have been reported for protoplast culture using liquid or solidified media. Among the solidified media, agar was the first gelling agent to be used (Nagata and Takebe, 1971). However, agar preparations present toxicity for a number of protoplasts. Only some robuste ones, used for cloning purposes, could be cultured on this gelling agent. Calcium alginate was also used for plant protoplasts culture giving platting efficiencies similar to those obtained with agar (Adaoha-Mbanaso and Roscoe, 1982). Platting efficiency was markedly improved in a number of species by culturing protoplasts in media solidified with agarose (Shillito et al., 1983). These authors, comparing the platting efficiency of agarose with that of agar concluded that platting in agarose gives improved colony formation over platting in cleaned agar or liquid medium, the effect being more evident at low initial concentrations but occurring also for high initial densities. The toxic effects of agar are diffusible as demonstrated for protoplasts cultured in agarose layer over agar or cleaned agar. Inspite of the high platting efficiency obtained for protoplasts cultured in agarose, it appeared for some cases, inadequate for full development of colonies probably due to defficient diffusion of nutrients. Taking this result into account, Shillito et al. (1983) proposed the bead type culture in which protoplasts are first immobilized in agarose and, after a number of days, protoplasts immobilized in agarose are introduced in liquid medium. This procedure will facilitate nutrient diffusion. Some examples can be reported demonstrating the advantages of protoplast immobilization in agarose. For Petunia hybrida protoplasts it was observed that in the bead-type system, the % of protoplasts forming colonies (after 2 weeks) was higher than in cultures not transferred to liquid medium (Shillito et al., 1983). Results from Kreuger-Lebus et al. (1983) also revealed that platting efficiency of colonies recovery from Lycopersicon esculentum increased about 10 fold when bead-type culture is used. Following these results, Shillito el al., 1983) point out the following main advantages of immobilization in agarose over culture in liquid medium: (1) agarose appears to stabilize fragile protoplasts; (2) the development of individual protoplasts can be followed which facilitates cloning; (3) microinjection of protoplasts can be carried out in agarose immobilized protoplasts; (4) in bead-type system, the medium in which beads are suspended \ can be changed without disturbing the cells which may be useful for removing the toxic products of growth, maintaining selection pressures following experiments in genetic engineering or fusion and carrying out of pulse treatments on protoplasts to induce or synchronise division.

The effect of immobilization of protoplasts in polymer gels has been studied in terms of 5-B-hydroxylation of digitoxigenin to periplogenin which is dependent on the viability of Daucus carotta cells (Jones and Velicky, 1982). The protoplasts immobilized in K-carrageenan or agarose present full viability and are stabilized against mechanical stress. Otherwise, immobilized protoplasts showed increased tolerance to osmotic shock when compared with free protoplasts (Linse and Brodelius, 1983). The method of adsorption to microcarriers used for immobilization of animal cells, was also tested for plant protoplasts. Bornman et al. (1983) showed that the attachement of protoplasts to microcarries is promoted by concanavalin A and soybean agglutinin and that they maintain the viability unnafected. The same authors, following studies on attachement and electron microspical evidences of anchorage considered that immobilized plant protoplasts could be helpful in studies of: (1) identification and possibly fusion; (2) properties of binding, coalescence and aggregation; (3) cell wall biogenesis; (4) effects of herbicides, toxins and inhibitors bound to microcarriers coating; (5) co-cultivation with transformed bacteria; (6) studies on photosynthesis and respiration among others.

Protoplast immobilization and their use in physiological studies

From the most widespread uses of plant protoplast culture, plant regeneration and production of hybrids and cybrids by fusion and transformation have to be mentioned.

Culturable protoplast systems offer a lot of advantages in selection on cellular level, require small space for large number of individuals, are suitable for direct gene transfer, for membrane studies, plastids and vacuole isolation. Protoplasts are a pre-requesite for somatic hybridization and for chemically mediated DNA uptake as well as for induced DNA uptake by electroporation (Eriksson, 1988).

Protoplasts and studies of plasma membrane function

In plant cells, many important functions such as regulation of ion and metabolite transport and action, host/endophyte interactions, recognition phenomena and cell wall biosynthesis and accumulation are controlled by plasma membrane. Studies of soybean protoplasts plasma membrane fluidity after photobleaching analysis, confirmed the electron microscopic results reported for these protoplasts by Williamson et al. (1976). From these results it appears that carbohydrates, lipids and proteins are able to diffuse in the plane of plant plasma membrane and that there exists two domains within the plasma membrane that influence the rate of mouvement (Metcalf et al. (1986).

Endocytosis by plant protoplasts

Protoplasts have revealed particularly suitable for studies of endocytosis via coated and smooth vesicles. Recent studies of Tanchak et al. (1984), Thanchak and Fowke (1987) have demonstrated that soybean protoplasts can bind cationized ferritin that is uptake via coated pits to coated vesicles until delivery of ferritin in the partially coated reticulum, dictyosomes and multivesicular bodies to the vacuole. The recognition of the existence of this pathway may contribute for the understanding of host/endophyte interactions namely those dealing with introduction in host cells of phytoregulators produced by the host or of elicitors to target sites in plant cells or even toxins produced by the endophyte (Barroso and Pais, 1987). Further research using plant protoplasts may be important to clearly elucidate this role of coated membranes.

Smooth vesicles, generally larger than coated vesicles can introduce macromolecules into plant protoplasts (Saleem and Cutler, 1986). Bacteria, such as <u>Agrobacterium</u> sp. or <u>Escherichia colli</u> were introduced into plant protoplasts by Hasezawa et al (1983) and by Matsui et al. (1983), this uptake being impossible in whole plant cells due to the organized cell wall. This ability of protoplasts for bacteria uptake may be responsible for their use in DNA transfer via <u>Agrobacterium</u>.

Culturable protoplasts may also be of interest for studies on the mechanisms by which nuclei or organelles enter protoplasts. (For a revision see Fowke and Gamborg, 1980; Fowke et al., 1981).

Protoplasts may also provide a useful tool for studies on cytoskeleton and organelles associated with plasma membrane particularly those involved in endocytosis. The use of protoplasts in cell wall regeneration are abundant. Some of them are concerned with the role of cell organelles in the synthesis of cell wall precursors, with transport mechanism involving the cytoskeleton and with the role of plasma membrane in micrifibril formation. Tarchevsky (1983), using protoplasts cultured in light, have shown that exogenous glucose incorporation into structural polyssacharides is activated by light and that photophosphorylation plays an important role in providing high energy phosphates to the synthesizing structural polyssaccharides.

Herbicide absorption

Few reports are known on herbicide absorption by protoplasts. Studies of Darmstadt et al. (1983) have demonstrated that atrazine absorption by protoplasts is completed after 10 sec. and that no further absorption occurred until 5 min. Atrazine accumulates in ranges of 1 to 4 times the external concentration while 2,4-D accumulates between 2 and 16 times depending on the pH is 6.5 and 4.5 respectively. These results can be atributed to the fact that 2,4-D is a weak acid and that ionized 2,4-D is more membrane permeable and thus a higher absorption by protoplasts occurs at pH 4.5.
Being known the chemical structure and physical properties of different herbicides, protoplasts may help to predict the efficiency of absorption of these molecules by plant cells as well as the appropriate pH for application.

Studies on ion contents in protoplasts

From the ions known to be absorbed by plant cells, calcium has been considered as a regulator of a wide range of development responses in plants such as mitosis, protein secretion, photosynthesis, pollen tube growth and root gravitropism (Hepler and Wayne, 1985). Ca 2+ appears as playing a central role as regulator of cell function in plant cells. However Ca 2+ is a cytotoxic element for animal cells appearing in the cytoplasm in a range of 10-100 nM (Tsien, 1983). Little is known concerning Ca 2+ concentration in plant cells cytoplasm probably due to difficulties inherent to the methods to determine CA 2+ levels in whole cells. Protoplasts appear as a good material to study Ca 2+ levels in plant cell cytoplasm. Douglas et al. (1988), working with barley aleurone protoplasts showed that, when a protoplast incubation medium is used at pH 4 and 4.5, indo-1 (a Ca 2+ sensitive dye) is accumulated in the cytoplasm and not in the vacuole. At the contrary, if the incubation medium is at pH 7, there is no indo-1 accumulation. According to these authors, the pH dependent loading of indo-1 into barley aleurone protoplasts can follow the model proposed by Goldsmith (1977) for the accumulation of IAA and other weak acids by plant cells.
Plant protoplasts have been recently used to study the sensitivity of plant cells to auxins (Ephritikhine et al.,1987).

Action of plant growth regulators

There is a generalized use of growth regulators in cells and protoplast cultures. There are few data concerning the levels of growth regulators in the cells as well as how they act in the diverse physiological roles they play. Once again, protoplasts constitute a good material for experiments on the contents and metabolism of phytoregulators in plant cells.
Sandberg and Crosier (1985) detected the presence of IAA metabolites in protoplasts of different species. These protoplasts seemed to be unable to metabolize ABA (Loveys and Robinson,1987). Jolles and Pillet (1988) working with maize

root protoplasts verified that the highest levels of IAA and ABA were detected in protoplasts obtained from the root cap and apex. These authors could also demonstrate that the different IAA and ABA concentrations along the different zones of the root influences their reaction to applied hormones. They also verified that when protoplasts are cultured in the presence of D-L cis trans (3H) ABA, phaseic acid (PA), dihydropaseic acid (DAP) and a glycoside of ABA could be detected in the incubation medium at about 76% of the total amounst of PA and DPA detected.

The advantages of using protoplasts for studies on growth regulators effects has been emphasized by Batchelor and Elliot (1983) who demonstrated that in Amaranthus tricolor protoplasts, benzyladenine stimulates betacyanin production by a Ca 2+ dependent protein modulator. The effect of anti-calmodulin drugs on H+ extrusion by protoplasts enabled Elliot (1983) to suggest the effect of calmodulin or of a phospholipid-protein interaction in stimulating a membrane bound ATPase concerned in K+ mouvement. It can be postulated that by controling the Ca 2+-flux, Ca 2+ modulation of K+ permeability can be achieved by growth regulators (Batchelor and Elliot, 1983).

Studies on enzymes from secondary metabolism

Plant cell cultures have been largely used for studies of the enzymes concerned in secondary metabolites production. (For a revision see Stockigt and Schubel,1988).

Protoplasts from anthocyanin-containing cell suspension cultures of Daucus carotta were used for the obtention of vacuoles in order to understand the transport of anthocyanidin or their glycosides, taking into account the activity of glucosyl-transferase in cell cultures of this species (Hopp et al.,1983).
Following the experiments of Linse and Brodelius (1983) on the activity of 5-β-hydroxylase in K-carragenean immobilized protoplasts we can consider that immobilized protoplasts can be used for studies on the enzymatic regulation of secondary metabolites production.

CONCLUDING REMARKS

From this brief and non exaust if revision on the uses of immobilized plant cells and protoplasts, we can consider that although the immobilizing technologies applied to plant cells and protoplasts are very recent, advantages and perspectives of their use appeared very quickly. We can now expect that, in a very next future, immobilized plant cells and protoplasts will constitute good tools for studies on cellular physiology giving approaches for the understanding of cell function and differentiation important for developments of many plant biotechnology domains.

REFERENCES

Abramson, H. A., Mayer, L. S. and Goren, M. H., 1942, Electrophoresis of proteins and the chemistry of the cell surfaces, Van Nostrand-Reinhold. Princeton, New Jersey.
Adaoha-Mbanaso,E.N. and Roscoe,D.H.,1982, Plant Sci. Lett. 25:61-66.

Alfermann, A. W., Bergmann, W., Figur, C., Helmbold, U., Schwantag, D., Schuller, I. and Reinhard, E., 1974, In Plant Biotechnology (Mantell, S.H. and Smith, E. Eds.)pp. 67-74. Cambridge Univ. Press.

Alfermann, A. W., Schuller, I. and Reinhard, E., 1980, Planta Med. 40:218-223.

Barroso, J. and Pais, M. M. S., 1987, The New Phytol.,105:67-70.

Batchelor, S.M. and Elliot, D.C., 1983, Experientia Suppl.45:178-179.

Bornman, C.H., Olesen, P. and Zachrisson, A.,1983, Experientia Suppl. 45:270-271.

Brodelius, P., 1983, In Immobilized cells and organelles, vol.I (Mattiasson, B. Ed.) pp. 28-54, CRC Press.

Brodelius, P., 1986, InHandbook of plant cell culture, vol.4 — Techniques and applications (Evans, D.A., Sharp, W.R. and Ammirato, P.V. Eds.)pp. 287-315. MacMillan.

Brodelius, P., Deus, B., Mosbach, K. and Zenk, M.H., 1980, In: Enzyme Engineering vol.5 (Weetall, H. H. and Royer, G.P. Eds.): pp. 373-381. Plenum Press, New York.

Brodelius, P., Deus, B., Mosbach, K. and Zenk, M. H., 1979, FEBS Lett.,103:93-97.

Brodelius, P. and Nilsson, K., 1980, FEBS Lett. 122:312- 316.

Brodelius, P. and Nilsson, K., 1983,Eur. J. Appl. Microbiol. Biotechnol. 17: 275-280.

Cabral, J. M. S., Fevereiro, P., Novais, J. M. and Pais, M. S. S., 1984, Annals New York Acad. Sci., 434: 501- 503.

Chetham, P.S.J., 1979, Enzyme Microb. Technol.,1:183-186.

Collinge, M. A. and Yeoman, M. M., 1986, In Secondary metabolism in plant cell cultures (Morris, P., Scragg, A. H., Stafford, A. and Fowler, M. W. Eds.). Cambridge Univ. Press.

Daniels, S. L., 1971, Dev. Ind. Microbiol,13:211.

Darmstadt, G.L., Balke, N.Z. and Schrader, L.E., 1983, Experientia Suppl. 45:202-203.

Dinelli, D.,1972, Process Biochem,7:9-13.

Douglas, S.B., Asok, K.B. and Russel, L.J., 1988, In. Progress in Plant Protoplast Research (Puite,K.J., Dons,J.J.M.,Huizing, H.J., Kool, A.J.,Koornneef,M. and Krens, F.A. Edit.) pp. 139-140, Kluwer Acad. Publish.

Elliot,D.C., 1983, Plant hysiol.72:215-218.

Ephritikhine,G.,Barbier-Brygoo,H., Muller, J. F. and Guern, J. ,1987, Plant Physiol 83:801-804.

Endo, T., Goodbody, A., Vukovic, J. and Misawa, M., 1988, Phytochemistry, 27 2147-2149.

Eriksson, T.R., 1988,In (Puite,K.J., Dons, J.J.M., Huizing, H.J., Kool, A.J., Koornneef,M. and Krens, F.A. Edit.) pp. 7-14, Kluwer Acad. Publish.

Fonseca, M. M. R., Cabral, J. M. S., Fevereiro, P. and Pais, M. S. S. , 1987, Annals New York Acad. Sci., 501: 358- 361

Fonseca, M. M. R., Mavituna, F. and Brodelius, P., 1988, In Plant Cell Biotechnology NATO ASI Ser. H (Pais, M. S. S., Mavituna, F. and Novais, J. M. Eds.) pp. 389-402. Springer Verlag.

Fowler, M. W., 1983, In Plant Biotechnology (Mantell, S. H. and Smith, H. Eds.) pp. 3-38. Cambridge Univ. Press.

Fowke, L. C. and Gamborg, G. L., 1980, Int. Rev. Cytol., 68: 9-51.

Fowke, L.C., Marchant, H.J. and Gresshoff, P.M., 1981,Can. J. Bot.,59:1021-1025.

Fukui, S, Tanaka, A. and Gellf, G., 1978, _Enzyme Eng._, **4**:299.

Furuya, T., 1978, _In_ Frontiers of plant tissue culture (Thorpe, T. Ed.) Int. Assoc. Plant Tissue Culture, Calgary (Canada)pp. 191–200.

Grinnell, F., 1978, _Int. Rev. Cytol._, **53**:65–69.

Goldsmith,M.H.M.,1977,_Ann. Rev. Plant Physiol._**28**:439–478.

Hahn-Hagerdal, B., 1983, _In_ Immobilized cells and organelles, vol.II (Mattiason, B., Ed.) pp. 79–94. CRC Press.

Hasezawa,S., Matsui, C., Nagata, T. and Syono, K., 1983, _Can.J. Bot._,**61**1052–1057.

Hepler,P. K. and Wayne,R. O., 1985, _Ann. Rev. Plant Physiol._,**36**: 397–439.

Heins, M., 1978, Doct. Thesis Univ. Tubingen.

Hopp, W., Mock, H.P. and Seitz, H. V., 1983,_Experientia_ Suppl.**45**:188–189.

Jolles, Ch. and Pillet,P.E., 1988, _In_ Progress in plant protoplasts research (Puite,K.J., Dons,J.J.M., Huizing, H.J., Kool, A.J.,Koornneef,M. and Krens,F.A. Eds.) pp.147–148, Kluwer Acad. Publ.

Jones, A. and Velicky, I. A., 1981, _Planta Med._, **42**: 160–166.

Jones, A. and Velicky, I. A., 1982, _Biotechnol. Lett._, **3**: 551–555.

Kreuger-Lebus, S.,Potrykus, I. and Imamura, J. ,1983, _Experientia_ Suppl.,**45**46–47.

Kurz, W. G. W. and Constabel, F., 1985, CRC Critical Reviews in Biotechnology **2**: 105–118.

Lindsey, K., 1985, _Planta_, **165**:126–133.

Lindsey, K. and Yeoman, M. M., 1984, _Planta_: **162**:495–501.

Lindsey, K., Yeoman, M. M., Black, G. M. and Mavituna, F., 1983, _FEBS Lett._ , **155**: 143–149.

Linse,L. and Brodelius,P., 1983, _Experientia_ Suppl.**45**:260–261.

Loveys,B.R. and Robinson,S.,1987,_Plant Science_,**49**:23–30.

Marshall, K.C., 1976, Interfaces in Microbiol. Ecology. Harvard Univ. Press, Cambridge.

Martin, P., 1978, M.Sc. Thesis, Univ. Western Ontario.

Mattiasson, B., 1983, _In_ Immobilized cells and organelles, vol. II (Mattiasson, B. Ed.) CRC Press pp. 23–40.

Matsui,C.,Hasezawa,S.,Tanaka,N. and Syono, K.,1983, _Plant Cell Reports_, **2(1)**:30–32.

Maurel, B. and Pareilleux, A., 1985, _Biotechnol. Lett._,**7**:313–318.

Metcalf,T.N.,Wang,J.L. and Schindler, M., 1986, _Proc. Nat. Acad. Sci._ **83**:95–99

Nagata,T. and Takebe,I., 1971, _Planta_,**99**:12–20.

Peterson, M., Alfermann, A.W. and Seitz, H. U., 1988,_In_ Plant Cell Biotechnology, NATO ASI Ser.H Cell Biology, vol. 18 (Pais, M. S. S., Mavituna, F. and Novais, J. M.., Eds.)pp. 365–371. Spriger-Verlag.

Reinhard, E., 1974, _In_ Tissue Culture and Plant Science (Street, H. E. Ed.) pp. 433– 459, Academic Press.

Reinhard, E. and Alfermann, A. W., 1980, _Adv. Biochem. Engin._, **16**:49–83.

Rosevear, A., Kennedy, J. F. and Cabral, J. M. S., 1987, Immobilized enzymes and cells. IVP Publish. Ltd.

Saleem,M. and Cutler,A.,J., 1986, _J. Plant Physiol._,**124**11–21.

Sandberg,G. and Crosier,A., 1985, _In_ The physiological properties of plant protoplasts, (Pilet,P.E.Ed.) pp.209–218, Springer– Verlag.

Scheurich, P. Schnabl, H., Zimmermann, U. and Klein, J., 1980, _Biochem. Biophys. Acta_,**598**:645–649.

Shillito,R.D.,Paszkowski,J. and Potrykus,I., 1983, _Experientia_

Suppl. **45**:266-267.

Stockigt, J. and Schubel, H., 1988, <u>In</u> Plant Cell Biotechogy, NATO ASI Ser.H Cell Biology, vol. 18 (Pais, M. S. S., Mavituna, F. and Novais, J. M.m, Eds.)pp. 251-264. Springer Verlag.

Suga, T., Hirata, T., Hamada, H. and Futatsugi, M., 1983, <u>Plant Cell Reports</u>, **2**:186-188.

Tanchak, M. A., Griffing, L. R., Mersey, B. G. and Fowke, L. C., 1984, <u>Planta</u>,**162**:481-486.

Tanchak, M. A. and Fowke, L. C., 1987, <u>Protoplasma</u>, **138**:173-182.

Tarchevsky,I.A., 1983, <u>Experientia</u> Suppl. **45**:192-193.

Tsien,R.Y ,1983, <u>Ann. Rev. Biophys. Bioeng.</u>, **12**: 91-116.

Van Oss, C.J., Gilman,G.F. and Newman,A.W., 1975, Phagocytic engulfment and cell adhesiveness. "Decker", New York.

Wada, M., Kato,J. and Chibata, I., 1979, <u>Eur. J. Appl. Microbiol. Biotechnol</u>,**8**: 241-244.

Wichers, H.J., Malingrè, T. M. and Huizing, H. L., 1983, <u>Planta</u>, **158**:482-486.

Wilkinson, A. K., Williams, P. D. and Mavituna, F., 1988, <u>In</u>Plant Cell Biotechnology, NATO ASI Ser. H: Cell Biology, vol.18 (Pais, M. S. S., Mavituna, F. and Novais, J. M. Eds.)pp. 373-377. Springer-Verlag.

Williamson,F.A.,Fowke, L.C., Constabel, F.C. and Gamborg, O.L., 1976, <u>Protoplasma</u> **89**:305-316.

Yeoman, M. M., 1987, <u>In</u> Cell Culture and Somatic Cell Genetics of Plants, vol.4, Cell Culture in Phytochemistry (Constabel, F. and Vasil, I. K., Eds.)pp. 197- 215. Academic Press Inc.

RECOVERY OF SOMATIC VARIATION IN RESISTANCE

OF POPULUS TO SEPTORIA MUSIVA

M.E. Ostry, D.D. Skilling, O.Y. Lee-Stadelmann[1], and W.P. Hackett[1]

North Central Forest Experiment Station, U.S. Dept. of Agriculture
[1]Dept. of Horticultural Science, University of Minnesota
St. Paul, MN 55108

Tissue culture has been used primarily to clonally propagate plants. However, genetic variation has been observed among regenerated plants. Passage of plant cells through a tissue culture cycle can result in increased spontaneous phenotypic and genetic variation. Somaclonal variation is the term used to describe variation exhibited by plants obtained from aseptic culture (Larkin and Scowcroft, 1981). Somatic variation is now considered a general phenomenon and somaclonal variation in disease resistance has been observed in many agronomic crops (Larkin, 1987).

Although the cause of somaclonal variation is not yet completely understood, the phenotypic and genetic variation found indicates that many factors are involved. Somaclonal variation can be preexisting genetic variation that is expressed in regenerated plants, or it can be induced by the tissue culture process itself (Scowcroft and Larkin, 1983). Working with tomatoes (Lycopersicon esculentum), Evans and Sharp (1983) provided classical genetic proof that tissue culture can be mutagenic. The plant species, genotype, source of explant used for culture initiation, and perhaps most importantly, the duration of the culture cycle may influence the variability observed (Lorz, 1984). Meins (1983) suggested that tissue culture induces cellular destabilization that can result in diverse heritable changes. In addition to epigenetic variation, tissue culture can cause genetic changes ranging from single base pair changes to chromosome deletions, translocations, and changes in chromosome number (Evans, 1986).

Tissue and cell culture techniques can be valuable tools for developing trees with improved characteristics. Additive and nonadditive traits can be captured, and aseptic culture may provide a new source of genetic diversity. Because of the long generation times of trees and the possibility of introducing desired traits not possible through traditional breeding, somaclonal variation may offer an advantage to forest tree improvement. Somatic variation in a tree species has been documented. One growing season after planting, a wide variation in height, number of branches, and leaf traits was detected in trees from five Populus X euramericana clones regenerated from callus cultures (Lester and Berbee, 1977).

Based on the above findings, we decided to investigate the potential application of somaclonal variation for increasing resistance to Septoria musiva Peck in Populus. Populus was chosen because of its worldwide importance as a source of fiber and energy and its amenability to whole plant regeneration through a variety of cell and tissue culture systems. Additionally, biomass yields from hybrid poplar plantations are now limited by the foliar and canker diseases caused by the fungal pathogen S. musiva.

In order to investigate the potential of somaclonal variation for increasing disease resistance in Populus, a rapid, in vitro bioassay for resistance to S. musiva is needed. Field testing poplars for resistance to Septoria is time-consuming and expensive. In addition, results may vary because of inconsistent inoculum concentrations and widely fluctuating weather conditions critical for infection and disease development. Other biotic and abiotic diseases also

Plant Aging: Basic and Applied Approaches
Edited by R. Rodríguez et al.
Plenum Press, New York, 1990

113

may complicate field evaluations. An in vitro bioassay to assess resistance of poplars to a leafspot caused by Marssonina brunnea has been described by Spiers (1978).

The in vitro bioassay developed for assessment of resistance of Populus clones to S. musiva involved the inoculation of leaf discs on 2% water agar in petri dishes with conidial suspensions under continuously lighted conditions (Ostry et al, 1988). The area of necrosis on leaf discs was measured beginning the 4th day after inoculation using a dot grid. Measurements were made every 2 days and continued for 32 days or until the control leaf discs started to become necrotic.

This leaf disc bioassay effectively evaluated the relative resistance of poplar clones to infection by S. musiva. The test is easily repeatable and provides a rapid method to screen a large number of clones for disease resistance. The leaf disc bioassay was sufficiently sensitive to distinguish among clones with high, moderate, or low resistance. Clones highly resistant or highly susceptible were most easily identified. There is a continuous range in the degree of resistance to S. musiva among hybrid poplars (Ostry and McNabb, 1986) and this leaf disc bioassay can rapidly determine approximately where a clone ranks. The generally high correlation of the bioassay results with those from the field means that this technique can be used to rapidly identify poplar clones highly resistant to S. musiva.

This in vitro bioassay was used for screening for variation for resistance to Septoria in plants derived from tissue culture (Ostry and Skilling, 1988). In vitro culture of tissue from hybrid poplars known to be susceptible or resistant to S. musiva was used to generate plants to be tested. One-cm-long stem internode explants were excised from the shoot tips and placed horizontally into shell vials containing 10 ml of Woody Plant Medium (WPM) (Lloyd and McCown, 1980) plus 20 g/L sucrose, 6 g/L Difco Bacto agar, and 2.3 μM (2,4-Dichlorophenoxy) acetic acid (2,4-D). Cultures were maintained in the dark in an incubator at 25°C.

Three to four weeks after culture initiation, proliferating callus at the cut ends of the explants was excised, subcultured onto fresh WPM containing 0.45 μM 2,4-D, and returned to the incubator. Proliferating callus cultures were subsequently divided and subcultured onto fresh medium every 3-4 wk. Callus cultures ranged in age from 5-13 mo before shoot proliferation was induced. Adventitious bud formation and shoot proliferation were induced by transferring callus cultures to WPM containing 1.1 μM 6-benzylaminopurine (BA) and 0.27 μM 1-naphthaleneacetic acid (NAA). Cultures were incubated under light (3000-lux) with a 16-hr photoperiod at 25°C. Elongated shoots were excised and rooted in a 2:1 peat-perlite medium under 100% humidity in a continuously lighted growth room (20-25°C, 2000-lux). After acclimation to lower humidity, plants were transferred to the greenhouse (18-30°C, 18-hr photoperiod). All tissue-cultured-derived plants had been growing in the greenhouse for at least 1 mo. before being tested for resistance to Septoria using the leaf disc bioassay.

We found no obvious mutant phenotypes among the regenerated plants from any of the clones except for two plants with mutant leaves that did not root. Phenotypically, all plants that rooted resembled their parent source plants. Plants with increased resistance were recovered from two of the three Septoria-susceptible clones used in this study (NE 41, NE 299, and NE 319). Many regenerants from clones NE 41 and NE 299 were significantly more resistant to Septoria than their source plants.

Regression analysis showed the regenerants from clone NE 319, a moderately resistant clone, were as susceptible to Septoria as the parent plant. Most regenerants from the two resistant clones (NE 293 and NE 314) were as resistant as their parent plants. However, a range in disease resistance was exhibited by the regenerants from all the clones. The variation in disease severity exhibited among the regenerants depended on clone. Some of the regenerated plants were as susceptible or more susceptible than the parent clone and some were clearly more resistant. Some of the leaf discs never reached 50% necrosis, and several did not exhibit any disease symptoms.

We could not detect a clear trend in the rate of somaclonal variation in regenerants from callus cultures of different ages. Variants with increased levels of resistance were recovered from callus of all ages. However, somatic variation for Septoria resistance was not detected in plantlets derived in vitro by axillary shoot multiplication.

Somaclonal variants for disease resistance have been planted in the field. After 2 years in field tests where Septoria inoculum is pervasive, 44% of the callus-derived plants exhibiting high resistance in the leaf disc bioassay were free from Septoria canker and 61% of those were free of <u>Septoria</u> leaf spot. After 3 years of field observation, resistant somaclones have been propagated from hardwood cuttings in the greenhouse. When plants from these cuttings were tested for resistance using the leaf disc assay, 22% of the field resistant clones showed resistance as high as that obtained in the original assay by which they were selected. This indicates that the somaclonal variation for <u>Septoria</u> leaf spot resistance is stable through convention cuttage propagation procedures.

The origin of this somaclonal variation is being studied by comparing the incidence of variation in plantlets derived from protoplasts, cell suspensions, callus, excised root cultures, and stem micro-cross sections. The cytological basis of this somaclonal variation is also being studied.

For the results reported above, it took about 5 to 13 months depending on clone to develop the callus cultures from which adventitious shoots were induced. Adventitious shoot induction took another 2 to 4 months and the number of shoots formed varied markedly from callus to callus within a clone and also between clones. Some clones were more recalcitrant than others under identical culture conditions. These observations led us to investigate more efficient methods for regenerating plantlets via adventitious bud formation.

In previous experiments, we had observed that either longitudinal or transverse incisions of 1 cm leaf midvein sections increased the number of adventitious shoots formed <u>in vitro</u>. This led us to devise tests of the influence of explant size on adventitious bud formation using hybrid poplar leaf midveins and stem internodes. In addition, we tested the influence of raising the $Ca(NO_3)_2$ concentration in WPM on adventitious bud formation.

A Lancer Vibratome was used to make micro-crossections (MCS) of midveins from hybrid poplar leaves (Lee-Stadelmann et al, 1989). MCS of 100, 200, 300, 400, and 500 μm in length were cultured on WPM containing 0.2 mg/l benzylaminopurine (BA). In the presence of NAA (0.01 mg/l), 400 and 500 μm sections had equal bud forming capacity and produced as many adventitious buds (average 4 shoots/MCS) as a 1 cm explant under the same culture conditions. This means that MCS of midveins are about 25 times more efficient in shoot regeneration than a 1 cm explant. The conditions used were suboptimal for the 200 μm and 100 μm sections. Shoot numbers from the 200 μm MCS were considerably reduced and the 100 μm MCS did not form buds at all. The size of lamina remnants was critical with 1 to 2 mm on each side of the midvein being optimal for bud formation.

With 400 μm midvein MCS, buds are macroscopically visible in 2 to 2.5 weeks, ready for rooting in 4 weeks, rooted and ready for transplanting to <u>in vivo</u> conditions after 8 weeks. MCS have been used with 3 clones of hybrid poplar. Midvein MCS work well with 2 clones but not with the third clone. MCS of young stem internodes work well with all 3 clones and for each clone more adventitious buds are produced on 400 μm stem MCS than on 400 μm midvein MCS.

Plantlets derived from midvein MCS of a <u>Septoria</u>-susceptible clone are currently being tested for resistance using the leaf disc bioassay.

The midvein MCS (300 and 400 μm) were not only an effective system for production of adventitious shoots, but also allowed direct microscopic observations of early morphogenetic events <u>in vitro</u>. These direct observations indicate that the callus and buds initiate from a group of small chlorophyllous cells located in areas of the midvein that are adjacent to the mesophyll cells.

Midvein and internode MCS are a rapid and efficient method for <u>in vitro</u> regeneration of adventitious plantlets. If derived plantlets display variation for <u>Septoria</u> resistance, it could be used as a more efficient way for recovery of somaclonal variants; if derived plantlets do not display variation for resistance, it may be an efficient method of clonal propagation.

LITERATURE

Evans, D.A., and Sharp, W.R., 1983, Single gene mutations in tomato plants regenerated from tissue culture. <u>Science</u>, 221:949-951.

Evans, D.A. and Sharp, W.R., 1986, Applications of somaclonal variation, <u>Bio/Technology</u>, 4:528-532.

Larkin, P.J., and Scowcroft, W.R., 1981, Somaclonal variation - a novel source of variability from cell cultures for plant improvement, Thoer. Appl. Genet., 60:197-214.

Larkin, P.J., 1987, Somaclonal variation: history, method, and meaning, Iowa State Journal of Research, 61:393-434.

Lee-Stadelmann, O.Y., Lee, S.W., Hackett, W.P., and Read, P.E., 1989, The formation of adventitious buds in vitro on micro-crossections of hybrid Populus midveins, Plant Sci. 61:263-272.

Lester, D.T., and Berbee, J.G., 1977, Within-clone variation among black poplar trees derived from callus culture, Forest Sci., 23:122-131.

Lloyd, G., and McCown, B., 1980, Commercially-feasible micropropagation of mountain laurel, Kalmia latifolia, by use of shoot tip culture, Proc. Inter. Soc. Plant Prop., 30:421-427.

Lorz, H., 1984, Variability in tissue culture derived plants, Pages 103-114, in: "Genetic Manipulation: Impact on Man and Society", W. Arber, ed., Cambridge Univ. Press, 250 pp.

Meins, F., Jr., 1983, Heritable variation in plant cell culture, Ann Rev. Physiol, 32:327-346.

Ostry, M.E., and Skilling, D.D., 1988, Somatic variation in resistance of Populus to Septoria musiva, Plant Disease, 72:724-727.

Ostry, M.E., and McNabb, H.S., Jr., 1986, Populus species and hybrid clones resistant to Melampsora, Marssonina, and Septoria, Res. Pap. NC-272, St. Paul, MN: U.S. Department of Agriculture, Forest Service, North Central Forest Experiment Station, 7 p.

Ostry, M.E., McRoberts, R.E., Ward, K.T., and Resendez, R., Screening hybrid poplars in vitro for resistance to leaf spot caused by S. musiva, Plant Disease, 72:497-499.

Scowcroft, W.R., and Larkin, P.J., 1983, Somaclonal variation, cell selection and genotype improvement, Pages 153-168. in: "Comprehensive Biotechnology", Vol. 3. C.W. Robinson, and H.J. Howell, eds., Pergamon Press, Oxford.

Spiers, A.G., 1978, An agar leaf-disc technique for screening poplars for resistance to Marssonina, Plant Dis. Rep., 62:144-147.

SHOOT-TIP GRAFTING IN VITRO OF WOODY SPECIES AND

ITS INFLUENCE ON PLANT AGE

L. Navarro

Instituto Valenciano de Investigaciones Agrarias (IVIA)
Moncada, Valencia, Spain

INTRODUCTION

The development of the technique of shoot-tip grafting in vitro
(STG) was a consequence of the high economic losses caused by citrus
virus and virus-like diseases. They produce decline of vigor, yield and
quality, they restrict the use of several rootstocks and in some cases
they kill the trees, or make them completely unproductive. With the
development of the modern Citrus Industry it became very evident that the
use of virus-free propagative material was the only means to obtain an
adequate benefit from the citrus orchards.

The most widely used method in the past to recover virus-free citrus
plants was the selection of nucellar seedlings of polyembryonic cultivars
(Roistacher, 1977). The effectiveness of this technique is based on the
fact that most citrus viruses are not transmitted through the process of
embryogenesis and that nucellar plants, that are produced by asexual
embryogenesis, are true-to-type. The limitation of this method is that
nucellar plants have juvenile characters. In citrus there are many disadvantages associated with the juvenile phase. These include plants with
excessive thorniness, upright habit of growth, excessive tree vigor, lack
of flowering during several years, tendency to alternate bearing, and
poor fruit quality of initial crops. A long period is required to loose
this adverse characters and to bring nucellar seedlings into production
and become commercially acceptable (Roistacher, 1977). With some
species, like sweet oranges, up to 20 years may be needed to completely
loose the juvenile characters.

Virus-free nucellar citrus plants of monoembryonic and polyembryonic
citrus cultivars can also be obtained by nucellus and ovule culture in
vitro through a process of somatic embryogenesis (Navarro and Juarez,
1977) but, in addition to juvenile characters, a high proportion of the
monoembryonic plants are not true-to-type (Navarro et al., 1985).

Thermotherapy has been used to recover virus-free citrus plants
without juvenile characters. However, this method has been ineffective
to eliminate some important and widespread diseases (Roistacher, 1977).

In this context, a new method to recover citrus plants free from all
virus and virus-like diseases, and without juvenile characters was
needed. The first attempts in this direction were done by shoot-tip

Plant Aging: Basic and Applied Approaches
Edited by R. Rodríguez *et al.*
Plenum Press, New York, 1990

culture in vitro, a technique extensively used to recover healthy herbaceous plants. However, all attempts to develop citrus plants from shoot tips failed. An alternative was to graft the shoot tips from infected plants on young rootstock seedlings growing in vitro. A few plants free of viruses and without juvenile characters were initially recovered (Murashige et al., 1972) and the method, named shoot-tip grafting in vitro, was further studied in detail, and a routine procedure established (Navarro et al., 1975).

THE TECHNIQUE OF STG FOR CITRUS

The STG technique comprises the following steps: rootstock preparation, scion preparation, grafting, culture in vitro of grafted plants, and transfer to soil.

Rootstock preparation. Rootstocks are obtained by seed germination in vitro. Seeds are peeled by removing both seed coats and surface sterilized. They are individually planted in 25 x 150 mm culture tubes containing 25 ml of the plant cell culture salt solution of Murashige and Skoog, solidified with 1% Bacto Agar. Culture tubes are incubated at constant 27°C in darkness for two weeks.

Most studies with STG have been done using Troyer citrange as rootstock. However, any rootstock which is graft-compatible with the shoot-tip scion variety could be used for STG (Navarro, 1988). Differences in grafting success have been observed among rootstocks, but detailed comparative studies have not been conducted. In our laboratory we routinely use Etrog citron to graft lemon shoot tips and Troyer citrange for the other citrus species.

Scion preparation. Shoot tips are obtained in routine work from growing vegetative flushes of field or greenhouse grown plants. The latter source has the advantage that flushes can be induced when necessary, avoiding the seasonal dependency of field trees; in addition, greenhouse plants allow a warm pre-treatment to enhance the elimination of some pathogens. Plants are completely defoliated by hand and placed in the greenhouse, or in a growth chamber at 32°C. In 8-15 days, many buds sprout and produce flushes.

Flushes 3 cm long or less are used to avoid abscissing or degenerating shoot tips. They are stripped of larger leaves, cut to about 1 cm long, and surface sterilized.

Shoot tips can also be isolated from dormant buds or from buds growing in vitro. In both cases the incidence of successful grafts and pathogen elimination is reduced. Flushes produced by budwood cultured in vitro are a very good source of shoot tips and the system is being used as a procedure for plant introduction in the Citrus Quarantine Station in Spain (Navarro et al., 1984).

STG has been successfully used to recover plants of more than 30 citrus species and citrus relatives, and many hybrids.

Grafting procedure. A two-week-old rootstock seedling is removed from the test tube, under aseptic conditions, and is decapitated, leaving about 1.5 cm of the epicotyl. The root is cut to a length of 4-6 cm and the cotyledons and their axillary buds removed.

Grafting can be done by different methods, but the best results are obtained by grafting at the top of the decapitated epicotyl, or in an inverted-T incision (Navarro et al., 1975). The incision is made by a 1 mm long vertical cut, starting at the point of decapitation, and a 1-2 mm long horizontal cut. The cuts are made through the cortex, to the cambium, and the flaps of the incision are slightly lifted to expose the cortex.

The remaining leaves of the flush, except for the youngest three leaf primordia, are removed with the aid of dissecting instruments and microscope. The shoot tip, composed of the apical meristem and three leaf primordia, and measuring 0.1 - 0.2 mm, is then excised with a razor blade sliver attached to a surgical handle. The shoot tip is placed inside the incision of the rootstock, with its cut surface in contact with the cortex exposed by the horizontal cut of the incision or at the top of the decapitated seedling, in contact with the vascular ring.

The frequency of successful grafts increases with the size of the shoot tip, buth the frequency of recovery of healthy plants decreases (Navarro et al., 1976). Consequently, it is necessary to choose a shoot-tip size that give a realistic degree of grafting success, and allow a reasonable number of pathogen-free plants. Shoot tips composed of the apical meristem and three leaf primordia are used routinely in our laboratory with good results. The size varies from 0.1 to 0.2 mm, depending on the citrus species.

Culture "in vitro" of grafted plants. Grafted plants are cultured in a liquid nutrient medium containing the plant cell culture salt solution of Murashige and Skoog, modified White's vitamins, and 75 g/l sucrose (Navarro et al., 1975). The nutrient medium is distributed into 25 x 150 mm test tubes in 25 ml aliquots. A folded filter paper platform, perforated in its center for insertion of the root portion of the rootstock, is placed in the nutrient solution. The cultures are kept at constant 27°C and exposed 16 hr daily to 50 u E m^{-2} sec^{-1} illumination.

Four to six weeks after grafting, the successfully grafted shoot tips produce two to four developed leaves, and they are ready to be transplanted to soil. Following the described procedure we routinely obtain 60-70% of successful grafts.

Transfer to soil. Micrografted plants are transferred to pots containing a steam sterilized soil mix. Pots are enclosed in polyethylene bags, which are closed with rubber bands, and placed in a shaded area of a temperature-controlled greenhouse set at 18-25°C. After 8 to 10 days the bags are opened, and after another 8 to 10 days they are removed, and the plants grown under regular greenhouse conditions. With this procedure we routinely obtain over 95% survival on transplanting with good subsequent growth.

THE TECHNIQUE OF SHOOT-TIP GRAFTING FOR OTHER WOODY SPECIES

The STG technique previously described for citrus species has also been used with some modifications to recover plants of the following woody species: almond (Prunus amigdalus Batsch) (Navarro et al., 1989), apple (Malus plumila Mill) (Alskieff and Villemour, 1978; Huang and Millican, 1980), apricot (Deogratias et al., 1979; Martinez et al., 1979) avocado (Persea americana Mill) (De Lange and Nell, 1985), camellia (Camellia japonica L.) (Crézé, 1985), cherry (Prunus avium L.) (Deogratias et al., 1986), European plum (Prunus domestica L.) (Navarro et al., 1989), grapevine (Vitis vinifera L.) (Engelbrecht and Schwerdtfeger, 1979), Japanese plum (Prunus salicina Lind) (Navarro et al., 1979), peach (Prunus persica Batsch) (Alskieff, 1977; Arregui et al., 1986; Deogratias et al., 1986; Martinez et al., 1979; Mosella et al.,1979; Navarro et al., 1982) Pine (Pinus pinaster Ait.) (Tranvan and David, 1985), Pollizo plum (Prunus insititia L.) (Navarro et al., 1989),and Sequoia (Sequoiadendron giganteum Buchholz) (Montieuuis, 1986, 1987). The percentage of grafting success has varied very much among species and laboratories. Seeds with dormancy are stratified either in vivo, or in vitro (Jonard, 1986) for various periods of time, depending on the species. In our laboratory we use the Nemaguard peach as a polyvalent rootstock for the different Prunus species. We dry store the seeds at 4°C, and we germinate them in vitro after removal of the integuments.

APPLICATIONS OF SHOOT-TIP GRAFTING IN VITRO

The main application of shoot-tip grafting has been the recovery of healthy plants (Navarro, 1988). In citrus, STG has been effective to eliminate the following citrus virus, and virus-like diseases: chachexia, concave gum, cristacortis, dweet mottle, exocortis, greening, impietratura, infectious variegation, psorosis A and B, ringspot, seedling yellows, stubborn, tatter leaf, tristeza, and yellow vein.

In other species, STG has been effective to eliminate the following viruses: chlorotic leaf spot, fan leaf, leaf roll, plum pox, prune dwarf, prunus ringspot, spy-decline-stem pitting, and stem grooving.

The pathogens eliminated by STG include viroids, closterovirus, ILAR virus, poty virus, nepovirus, Spiroplasma citri, fastidious bacteria, and several graft-transmissible diseases of unknown etiology. This suggests the effectiveness of the technique to eliminate a wide variety of pathogens.

STG is being used in citrus sanitation programs in the main producing countries (Navarro, 1988). In Spain we have the largest program (Navarro, 1976, Navarro et al., 1980, 1981, 1988), and about 20 million healthy plants, originally recovered by STG, have been already planted in the field.

Practical application of STG in sanitation programs of other woody species is not as advanced as in citrus. In Spain there is a project to recover healthy stone fruit plants by STG (Arregui et al., 1986). The first release to the nurseries of healthy material from thirteen peach clones has already been made.

Quarantine is another practical application of STG. A tissue culture procedure has been developed for citrus that basically consists of culturing in vitro the imported budwood to induce the formation of flushes, that are used as source of shoot tips for grafting in vitro (Navarro et al., 1984). The resulting plants are cultured in a quarantine glasshouse for indexing. The only material actually introduced into the country by this method is a 0.1-0.2 mm long shoot tip that is free of pests and diseases. This method has been successfully used in Spain to introduce sixty citrus varieties from different countries.

Separation of viruses in mixed infections is another application of STG with important implications in plant pathology studies. As mentioned above, STG has a better efficiency to eliminate some pathogens than others and, consequently, it is possible to recover plants infected with one pathogen from original plants infected with several pathogens (Navarro, 1981; Navarro et al., 1982). STG has been used with this purpose to establish the pure collections of citrus psorosis isolates and citrus viroids of our Institute. STG can also be used for studies on graft incompatibilities, on graft-union histology, and on physiology of grafting (Jonard, 1986; Navarro, 1988).

INFLUENCE OF SHOOT-TIP GRAFTING ON PLANT AGE

Citrus plants recovered by STG maitain the same ontogenic age as the shoot tip source plant. Micrografted plants normally flower and set fruit within two years from grafting. Among the several thousand plants recovered by this technique, we have never observed any reversion to the juvenile phase using mature plants as source of shoot-tips. When shoot-tips are excised from juvenile plants, the resulting micrografted plants also have juvenile characters. Only in the citrus sanitation program in Spain, about 20 million plants originated in STG have been planted in commercial orchards, performing as mature plants.

In many cases micrografted plants are more vigorous than the originally infected shoot-tip source plant. Particularly in the case of lemon, there has been an important increase in vigor, and the micrografted plants have more thorns than the original plant. Nevertheless, the more vigorous healthy plants produce equal, or most likely, earlier and higher crops. Since flowering and fruiting are the main characters associated with the mature phase of citrus, we understand that the increase in vigor of micrografted plants is produced by the elimination of virus and virus-like pathogens that are detrimental to plant growth.

In some cases, micrografted plants have been used as source of shoot-tips for STG. This successive micrografting did not induce the rejuvenation of the plants.

The gross morphology of citrus shoot tips is different in mature and in juvenile plants of the same variety. However, no differences have been found on grafting success among the two sources of shoot tips.

The available information indicates that other fruit trees rcovered by STG also maintain the same ontogenic age as the shoot-tip source plant. Most stone fruit plants recovered by STG in our laboratory, flower within one or two years from grafting.

Despite all the data indicating that the application of STG does not change the ontogenic age of the source plants, some authors are using this technique in an attempt to rejuvenate mature forest trees, based on the rejuvenation resulting from successive grafting in some species (Franclet, 1977).

In Sequoiadendron giganteum it was found that some micrografted plants showed a juvenile type of growth during in vitro culture, but they resumed the mature type of growth when plants were transplanted to soil (Monteuuis, 1986, 1987). In Pinus pinaster some micrografted plants showed and apparent reversion to the juvenile phase and they maintain the juvenile type of growth three months after transplanting to soil (Monteuuis, personal communication). More data are needed to clarify whether or not shoot tip grafting of conifers produce a true rejuvenation of adult plants, or just a reinvigoration of plants.

CONCLUSION

STG is used to recover pathogen-free plants of fruit trees without juvenile characters. Particularly in citrus, the application of STG in sanination and quarantine programs is having a very important economic impact on the citrus industry of several countries. There are some data that indicate the possibility that the application of STG to some conifers may produce a rejuvenation, or at least a reinvigoration of mature trees. If these data are confirmed, STG will play an important role in propagation of selected mature forest trees.

REFERENCES

Alskieff, J., 1977. Sur le greffage in vitro d'apex sur des plantules décapitées de Pêcher (Prunus persica Batsch). C.R. Acad. Sc. Paris 284:2499-2502.

Alskieff, J., and Villemur, P., 1978. Greffage in vitro d'apex sur des plantules décapitées de pommier (Malus pumila Mill). C.R. Acad.Sc. Paris 278:1115-1118.

Arregui, J.M., Cambra, M., Juarez, J., Llacer, G. Navarro, L., and Rodriguez, J., 1986. Obtención de melocotoneros libres de virus de variedades autóctonas de Murcia mediante microinjerto de ápices caulinares in vitro, p. 24-29, in: Actas II Cong. Soc. Española Ciencias Hort., Vol 1, Córdoba.

Crézé, J., 1985. Elimination des virus de Camelia par greffe d'apex in vitro. C.R. Acad. Agr. France 71:572-576.

De Lange, J.H., and Nel, M., 1985. Shoot tip grafting of avocado for virus and viroid elimination. S. African Avocado Grow. Ass. Yearbook 8:66-69.

Deogratias, J.M., Juarez, J., Arregui, J.M., Ortega, C., Castellani,-- Llacer, G., Dosba, F., Navarro, L., 1989. Study of growth parameters on apricot shoot-tip grafting in vitro (STG) Acta Horticulturae (In press).

Deogratias, J.M., Lutz, A., and Dosba, F., 1986. In vitro shoot tip micrografting from juvenile and adult Prunus avium L. and Prunus persica (L.) Batsch to produce virus-free plants. Acta Horticulturae 193:139-145, 1986.

Engelbrecht, D.J., and Schwerdtfeger, U., 1979. In vitro grafting of grapevine shoot apices as an aid to the recovery of virus-free clones. Phytolactica 11:183-185.

Franclet, A., 1977. Manipulation des pieds-mères et amelioration de la qualité des boutures. AFOCEL, Etudes et Recherches 8:3-20.

Huang, S.C., and Millican, D.F., 1980. In vitro micrografting of apple shoot tips. HortScience 15:741-743.

Jonard, R., 1986. Micrografting and its application to tree improvement, p. 31-48, in: Y.P.S. Bajaj (ed.), Biotechnology in Agriculture and Forestry, Vol 1, Springer-Verlag, Berlin.

Martinez, J. Hugard, J., and Jonard, R., 1979. Sur différents combina- tions de greffages des apex réalisés in vitro entre le Pêcher (Prunus persica Batsch), Abricotier (Prunus armeniaca L.) et Mirobolan (Prunus cesarifera Ehrh). C.R. Acad. Sc. Paris 288:759-762.

Monteuuis, O., 1986. Microgreffage de points végétatifs de Sequoiaden- dron giganteum Buchholz sur des jeunes semis cultivés in vitro C.R.Acad.Sc.Paris 302:223-225.

Monteuuis, O, 1987. Microgreffage du sequoia géant. Ann. AFOCEL 1986, 39-61.

Mosella, Ch.L, Riedel, M., and Jonard, R., 1979. Sur améliorations apportées aux techniques de microgreffage des apex in vitro chez les arbres fruitiers. Cas du Pêcher (Prunus persica Batsch). C.R. Acad.Sc. Paris 289:505-508.

Murashige, T., Bitters, W.P., Rangan, T.S., Nauer, E.M., Roistacher, C.N., Holliday, B.P., 1972. A technique of shoot apex grafting and its utilization towards recovering virus-free citrus clones. HortScience 7:118-119.

Navarro, L., 1976. The citrus variety improvement program in Spain, p. 198-202, in: E.C. Calavan, (ed.), Proc. 7th Conference Int. Organization Citrus Virol., IOCV, Riverside.

Navarro, L., 1981. Shoot-tip grafting in vitro and its applications: a review. Proc. Int. Soc. Citriculture 1:452-456.

Navarro, L., 1988. Applications of shoot tip grafting in vitro to woody species. Acta Horticulturae 227:43-55.

Navarro, L., Ballester, J.F., Juarez, J., Pina, J.A., Arregui, J.M., and Bono, R., 1981. Development of a program for disease-free citrus budwood in Spain. Proc. Int. Soc. Citriculture 1:452-456.

Navarro, L., Ballester, J.F., Juarez, J., Pina, J.A., Arregui, J.M., Bono, R., Fernández de Córdova, L., and Ortega, C., 1980. The citrus variety improvement program in Spain (CVIPS) after four years, p. 289-294, in: E.C. Calavan, S.M. Garnsey, and L.W. Timmer (eds.), Proc. 8th Conference Int. Organization Citrus Virol., IOCV, Riverside.

Navarro, L, Juarez, J., 1977. Elimination of citrus pathogens in propagative budwood. II In vitro propagation. Proc. Int. Soc. Citriculture 3:973-987.

Navarro, L., Juarez, J., Arregui, J.M., Llacer, G., and Cambra, M., 1989, Unpublished results.

Navarro, L., Juarez, J., Pina, J.A., and Ballester, J.F., 1984. The citrus quarantine station in Spain. p. 289-294, in: S.M. Garnsey, L.W. Timmer, and J.A. Dodds (eds.), Proc. 9th Conference Int. Organization Citrus Virol., IOCV Riverside.

Navarro, L., Juarez, J., Pina, J.A., Ballester, J.F. and Arregui, J.M., 1988. The citrus variety improvement program in Spain (CVIPS) after eleven years, p. 400-406, in: L.W. Timmer, S.M. Garnsey, and L. Navarro (eds.) Proc. 10th Conference Int. Organization Citrus Virol., IOCV, Riverside.

Navarro, L., Llacer, G., Cambra, M., Arregui, J.M., and Juarez, J., 1982. Shoot-tip grafting in vitro for elimination of viruses in peach plants (Prunus persica Batsch). Acta Horticulturae 130:185-192.

Navarro, L., Ortiz, J.M., Juarez, L., 1985. Aberrant citrus plants obtained by somatic embryogenesis of nucelli cultured in vitro. HortScience 20:214-215.

Navarro, L., Roistacher, C.N., and Murashige. T., 1976. Effect of size and source of shoot tips on psorosis-A and exocortis content of navel orange plants obtained by shoot-tip grafting in vitro, p. 194-197, in: E.C. Calavan (ed.), Proc. 7th Conference Int. Organization Citrus Virol., IOCV, Riverside.

Roistacher, C.N., 1977. Elimination of citrus pathogens in propagative budwood. I.. Budwood selection, indexing and thermotherapy. Proc. Int. Soc. Citriculture 3:965-972.

Tranvan, H., David, A., 1985. Greffage in vitro du pin maritime (Pinus pinaster). Can. J. Bot. 63:1017-1020.

THE CONTROL BY CRYOPRESERVATION OF AGE-RELATED

CHANGES IN PLANT TISSUE CULTURES

Erica E. Benson

Department of Agriculture and Horticulture
Nottingham University School of Agriculture
Sutton Bonington, Loughborough, Leics. UK. LE12 5RD

Keith Harding

Department of Genetics, School of Biological Sciences
Medical School, Queens Medical Centre
University of Nottingham, Nottingham, UK. NG7 2UH

INTRODUCTION

Cryopreservation is the process by which viable tissues are stored at ultra-low temperatures (-196°C) in liquid nitrogen. Metabolism is suspended and evidence suggests that genetic stability is maintained. The following provides a brief account of cryopreservation methodology (see also Withers, 1987; Benson and Withers, 1988).

1.   Pregrowth and cryoprotection

This can enhance the survival potential of the tissue in several ways. Thus, osmotically active compounds such as mannitol and sorbitol promote cellular dehydration, thereby reducing the amount of water which will be frozen. Dimethyl sulphoxide (DMSO) added during pregrowth and compounds associated with plant stress responses (proline, abscisic acid) or "conditioning" stress treatments such as cold hardening have also been shown to increase survival. The application of cryoprotectants is usually essential to survival and a wide range of compounds are effective (DMSO, sugars, glycerol and polyols, PEG, PVP). In the case of cell suspensions and callus tissue, cryoprotection usually involves the application of a "cocktail" comprising several cryoprotectants. However, for shoot-tip meristems it is more usual to use a single additive, which is often DMSO. Cryoprotectant efficacy is dependent on the application of the above compounds at relatively high concentrations (5-10% or 0.5-1.0 M) for up to one hour before freezing. The mode of action of cryoprotectants is thought to be largely colligative. The plant tissues are protected from damaging solution effects as cell water is removed during dehydration and freezing.

2.   Freezing

Tissues can be either directly immersed into liquid nitrogen or they can undergo controlled freezing in which intermediate temperature changes occur before they are exposed to the terminal temperature of -196°C (liquid nitrogen).

The latter procedure is most commonly used and it is important to optimise the rate of freezing, the terminal transfer temperature before liquid nitrogen storage, and the duration of the transfer temperature. It is important to consider the physiology of the tissues when choosing a freezing program. Highly meristematic tissues with low water contents are more susceptible to dehydration damage on freezing. Thus, a rapid or intermediate freezing protocol would prove more beneficial to survival. Under these conditions excessive dehydration will be reduced. In contrast highly vacuolated tissues require treatments which enable the removal of cellular water before cryopreservation can be successfully achieved. The success of cryopreservation is therefore largely dependent on maintaining a balance between the damaging effects of intracellular ice formation and dehydration.

3.   Thawing and recovery

Rapid thawing (40-45°C) is the most beneficial approach for plant tissues. In the case of cell suspensions, recovery may be enhanced if the cells are plated onto semi-solid media for a short period before transferring to liquid culture. The plant growth regulator content of the recovery media can greatly influence the survival and regeneration of frozen shoot-tips.

Increasingly, cryopreservation is being applied to the storage of plant tissue cultures. One of its most important applications concerns the control of age-related changes that occur when plants are maintained in vitro. Evidence suggests that prolonged growth in culture can lead to genetic and physiological deterioration, especially if tissues are maintained in a disorganised state (Ramulu et al., 1989). This can have far reaching consequences for the commercial application of plant biotechnology. The following sections review the application of cryopreservation in controlling the age-related changes that can occur in tissue cultures.

CHANGES IN MORPHOGENIC POTENTIAL

Many in vitro manipulations are dependent on the passage of cultures through a dedifferentiated state. The maintenance of morphogenic potential and regeneration capacity in cell and callus tissue is therefore crucial. Unfortunately, it has often been observed that the ability to regenerate plants from de-differentiated tissues declines with time. Chen et al. (1985) report aging phenomena in callus obtained from wheat, and they show the importance of cryopreservation in maintaining cultures which display a high regeneration potential. Cold hardening treatments and applications of abscisic acid substantially improved post-freeze survival. Similarly, Ulrich et al. (1982) were able to maintain post-freeze somatic embryogenesis and plant regeneration in date palm callus. Finkle et al. (1985) also regenerated normal plantlets from cryopreserved callus cultures of alfalfa. In both studies a cryoprotectant cocktail containing polyethylene glycol, dimethylsulphoxide and glucose was used in combination with a slow freezing regime (-1°C/min to -30°C/min) before transferring to liquid nitrogen). Shillito et al. (1989) have cryopreserved embryogenic cell lines of maize in order to ensure that a constant supply of responsive cells is available for protoplast isolation. Maize cells were thus frozen at a stage when they first yielded viable protoplasts. This was considered most important since the source cultures exhibited reduced regeneration with increase in age. It is also probable that freshly initiated, competent cultures are more amenable to cryopreservation. Calli obtained from the frozen maize lines formed cell suspensions suitable for protoplast isolation at an increased rate compared to the original starter callus. Cell suspensions from these post-freeze regenerants were successfully cryopreserved for a second time. This study exemplifies the importance of introducing cryopreservation methodology at the initial stages of culture before the adverse effects of culture instability occur.

Protoplasts have been isolated from freeze-recovered cells of rice. However, a recovery period is essential for successful protoplast preparation (Cella et al., 1982). Yields of rice protoplasts were much reduced immediately after thawing. On microscopic examination protoplast damage was observed after treatment with cell wall degrading enzymes. Under certain circumstances it may be desirable to cryopreserve protoplasts in preference to cell suspension cultures. Thus, Takeuchi et al. (1982) have demonstrated post-thaw cell division and callus formation in cryopreserved protoplasts of carrot, soybean and wheat. Optimum cooling rates were species dependent and a cryoprotectant cocktail of DMSO and glucose was used. Microscopy revealed cell wall regeneration in the protoplasts within several hours of thawing and totipotency was high.

CHANGES IN METABOLIC STABILITY

Alterations in secondary metabolism can occur if cell suspensions are maintained, without selection, for extended periods (Deus-Neumann and Zenk, 1984). Holden et al. (1986) suggest that progress in and exploitation of many tissue culture techniques, especially those concerned with secondary metabolite production, are vulnerable to culture instability. These workers show clonal instability for capsaicin production over a 20 month maintenance period in callus cultures of Capsicum frutescens. However, it was also noted that morphologically stable clones retained their capacity to produce the secondary metabolite compared to morphologically unstable clones. The latter were derived from initially fast growing cultures which later developed hard compact structures and accumulated phenolics in preference to capsaicin. The period of time that the clones were maintained in culture in the undifferentiated state greatly affected their regeneration potential. This phenomenon caused practical problems and confounded attempts to develop optimum tissue culture conditions for regeneration and product synthesis. Although it must be a priority to understand the basis of culture instability in cell lines, cryopreservation may provide an indirect means by which culture instability can be "controlled" in high metabolite producers. These may be selected and then cryopreserved thus providing a repository of stable cultures. In addition cryopreservation may be a useful adjunct to the more basic studies required to devise optimum growth parameters etc. for such labile cultures. The maintenance of post-freeze biosynthetic capacity has been demonstrated for a wide range of species, and in this specific application cryopreservation has been largely successful. Examples include the maintenance of post-freeze biosynthetic capacity for steroid production in Dioscorea deltoidea (Butenko et al., 1984; Volkova et al., 1986), post-freeze accumulation and biotransformation of alkaloids in Digitalis lanata (Diettrich et al., 1982, 1985; Seitz et al., 1983), and alkaloid production in Catharanthus roseus (Chen et al., 1984).

Cryopreservation protocols used in the freezing of secondary product producing cell lines are continually being improved. Thus, Reuff et al. (1988) have found that the duration of a sorbitol post-freeze treatment can markedly affect survival in cryopreserved cell cultures of Coleus blumei. The effects of three different thawing regimes were also investigated for this species. Cells benefited from short thawing periods at a relatively high temperature (60°C). This contrasts with the usual thawing range of 40-45°C. These workers found that the capacity of C. blumei to resume normal growth was reproducibly high when the cells were thawed in a microwave oven. They suggest that this positive effect is due to homogeneous rewarming under microwave irradiation. The accumulation of rosmarinic acid in the thawed cells was found to be similar to that of the non-cryopreserved controls. It may be possible within cryopreservation protocols to exploit a plant's natural adaptive capacity to withstand low temperature stress, and Butenko et al. (1984) have taken this approach in the cryopreservation of Panax ginseng cells. Ginseng grows in areas where winter temperatures can reach as low as -30°C. Dormant shoots and roots can withstand these temperatures. A cold hardening regime involving 2-10°C temperature treatments over several days in the presence of high sucrose concentrations proved to ensure successful post-cryopreservation recovery and secondary product formation in ginseng. The above researchers investigated the effects of various amino acids on recovery in D. deltoidea cells. Asparagine and alanine proved to be the most effective and secondary product synthesis was unchanged after the cryopreservation treatments.

CHANGES IN GENETIC STABILITY

Age-related genetic change in tissue cultures can be described as somaclonal variation (heritable mutations among the progeny of regenerated plants). Epigenetic (transient) changes also occur during prolonged in vitro culture, however these apparently are not transmitted to the sexual progeny. Whilst genetic instability may be considered a major problem in preventing uniformity in regenerants, it may provide a promising adjunct to conventional breeding programs (Evans, 1989). The manifestation of instability may occur at phenotypic, cytological, biochemical and molecular levels. Severe genetic damage can be observed in the phenotype of regenerated plants and their sexual or vegetative progeny, for which morphological characters can display quantitative variation. Moreover, although plants regenerated from tissue culture have been described as "normal-looking" (Bajaj, 1986; Ramulu et al., 1989), it can be discovered after close examination that some regenerated plants which appear normal are in fact variant for one or several characters (Harding, personal observations). The application of a large number of accepted plant descriptors in the screening of tissue culture regenerants can result in the detection of such less obvious variants. Phenotypic variation can often be associated with variation in the number of chromosomes in plants regenerated from tissue cultures (Ramulu et al., 1989; Feher et al., 1989). Instability of genetic processes during mitosis can lead to cells becoming polyploid and aneuploid. A number of other alterations, such as the formation of chromosomal bridges and fragments, have been observed (Kovacs, 1985; Pijnacker et al., 1986).

Minor changes in the biochemical fabric of the plant can occur without resulting in any detectable changes at the phenotypic and cytological levels. Variations in isozyme patterns have been demonstrated in a number of plant species. A variant in the electrophoretic mobility of alcohol dehydrogenase was detected in maize after screening many regenerated plants derived from tissue culture (Brettell et al., 1988). Variation at this locus was caused by a single base pair change in exon 6 of this gene sequence. Moreover, stability at phenotypic, cytological and biochemical levels does not necessarily imply integrity of DNA sequences. Indeed, genetic rearrangements of non-coding sequences such as satellite DNA may have very little effect on the whole plant. Although minor genetic changes of this type may go undetected, the significance of such alterations has yet to be determined. It would appear that variation in the number of multi-copy sequences can be tolerated. The 25s component of the ribosomal genes of potato was reduced by 70% in plants derived from tissue culture (Landsmann and Uhrig, 1985). Similarly, nuclear DNA has been shown to be amplified 75-fold in cultured cells of rice (Zheng et al., 1987).

The introduction of foreign DNA into plant cells (transformation) is of considerable importance to several areas of plant biotechnology. Genetic stability of transformed lines is essential for the maintenance and expression of these traits (Stiekema et al., 1988); the effects of somaclonal variation may interfere with the organisation and expression of foreign genes (Deroles and Gardner, 1988a and b).

A means by which the undesirable effects of somaclonal variation could be controlled is by the application of cryopreservation. This allows the storage of freshly initiated, genetically stable cultures, which may be reintroduced into tissue culture when required. Such storage technology can then be applied to the conservation of elite genotypes, transformed cultures and somatic hybrids. However, before liquid nitrogen storage can be used as a routine method for conserving important genotypes it is essential to determine the effects of freezing protocols on molecular stability.

To date there remains a paucity of information on this aspect of cryopreservation. However, a limited number of studies do indirectly suggest that tissues recovered from cryopreservation are genetically stable. Cytological studies of cells obtained from cryopreserved cultures of D. deltoidea showed post-freeze stability in chromosome number as compared to unfrozen cultures (Volkova et al., 1986). Enzyme polymorphism for alcohol dehydrogenase, esterase, peroxidase, phosphoglucomutase and phosphoglucoisomerase has been studied in cryopreserved date palm callus. Isoenzyme patterns were similar in frozen and unfrozen tissues (Ulrich et al., 1982). Mutant cell lines appear to retain their biochemical and physiological characteristics after cryopreservation. Thus, steroid production capacity was maintained after freezing in mutants of D. deltoidea (Volkova et al., 1986), and a salt tolerant line of alfalfa was stable on thawing (Finkle et al., 1985). Amino acid analog-resistance was unaltered in cell suspension cultures of carrot and tobacco after cryopreservation (Hauptman and Widholm, 1982). Similarly, Strauss et al. (1985) reported the maintenance of resistance to O-methylthreonine in frozen/thawed cell suspensions of resistant strains of rosa.

CONCLUSIONS

Clearly the age-related changes that take place in tissue cultures are detrimental. Cryopreservation may provide a useful tool in the control of aging phenomena. However, this technique has yet to be applied for all the plant species amenable to tissue culture, and the further development of cryopreservation protocols is needed. In the first instance, low temperature storage can be used to safeguard germplasm which may suffer physiological and genetic change, if maintained in culture for long periods of time. However, cryopreservation may also have an important role in aiding basic studies aimed at understanding in vitro aging.

REFERENCES

Bajaj, Y. P. S., 1986. Cryopreservation of potato somaclones, pp 244-250, in: "Somaclonal Variation and Crop Improvement". J. Semal, ed., Nijhoff Dordrecht, Lancaster.

Benson, E. E. and Withers, L. A., 1988. The application of germplasm storage in biotechnology, pp 430-433, in: "Plant Cell Biotechnology", NATO ASI Series H. Cell Biology Volume 18. M. S. S. Pais, F. Mavituna and J. M. Novais, eds, Springer-Verlag, Heidelberg.

Brettell, R. S., Dennis, E. S., Scowcroft, W. R. and Peacock, W. J., 1986. Molecular analysis of a somaclonal mutant of maize alcohol dehydrogenase. Mol. Gen. Genet., 202:235-239.

Butenko, R. G., Popov, A. S., Volkova, L. A., Chernyak, N. D. and Nosov, A. M., 1984. Recovery of cell cultures and their biosynthetic capacity after storage of Dioscorea deltoidea and Panax ginseng cells in liquid nitrogen. Pl. Sci. Letts, 33:285-292.

Cella, R., Colombo, R., Galli, M. G., Nielsen, E., Rollo, F. and Sula, F., 1982. Freeze-preservation of rice cells : a physiological study of freeze-thawed cells. Physiol. Plant., 55:274-284.

Chen, T. H. H., Kartha, K. K., Leung, N. L., Kurz, W. G. W., Chatson, K. B. and Constabel, F., 1984. Cryopreservation of alkaloid-producing cell cultures of periwinkle (Catharanthus roseus). Plant Physiol., 75:726-731.

Chen, T. H. H., Kartha, K. K. and Gusta, L. V., 1985. Cryopreservation of wheat suspension culture and regenerable callus. Plant Cell Tissue Organ Culture, 4:101-109.

Diettrich, B., Popov, A. S., Pfeiffer, B., Neumann, D., Butenko, R.. and Luckner, M., 1982. Cryopreservation of Digitalis lanata cells grown in vitro. Precultivation and recultivation. J. Plant Physiol., 126:63-73.

Diettrich, B., Haack, U., Popov, A. S., Butenko, R. G. and Luckner, M., 1985. Long-term storage in liquid nitrogen of an embryonic cell strain of Digitalis lanata. Biochem. Physiol. Pflanzen, 180:33-43.

Deroles, S. C. and Gardner, R. C., 1988a. Analysis of the T-DNA structure in a large number of transgenic petunias generated by Agrobacterium-mediated transformation. Plant Molecular Biology, 11:365-377.

Deroles, S. C. and Gardner, R. C., 1988b. Expression and inheritance of kanamycin resistance in a large number of transgenic petunias generated by Agrobacterium-mediated transformation. Plant Molecular Biology, 11:355-364.

Deus-Neumann, B. and Zenk, M. H., 1984. Instability of indole alkaloid production in Catharanthus roseus cell suspension cultures. Planta Medica, 50:427-431.

Evans, D. A., 1989. Somaclonal variation. Genetic basis and breeding applications. TIG, 5:46-50.

Feher, F., Tarczy, H., Bocsa, I. and Dudits, D., 1989. Somaclonal chromosome variation in tetraploid alfalfa. Plant Sci., 60:91-99.

Finkle, B. J., Ulrich, J. M., Rains, D. W. and Stavarek, S. J., 1985. Growth and regeneration of alfalfa callus lines after freezing in liquid nitrogen. Plant Sci., 42:133-140.

Hauptmann, R. M. and Widholm, J. M., 1982. Cryostorage of cloned amino acid analog-resistant carrot and tobacco suspension cultures, Chapter 11, pp 217-227, in: "Cell Culture and Somatic Cell Genetics", Volume 4. I. K. Vasil and F. Constabel, eds, Academic Press Inc., London.

Holden, P. R., Aitken, M., Lindsey, K. and Yeoman, M. M., 1986. Variability and stability of cell cultures of Capsicum frutescens, pp 31-243, in: "Secondary Metabolism in Plant Cell Cultures". P. Morris, A. H. Scragg, A. Stafford and M. W. Fowler, eds, Cambridge University Press, Cambridge.

Kovacs, E. I., 1985. Regulation of karyotype stability in tobacco tissue cultures of normal and tumorous genotypes. Theor. Appl. Genet., 70:548-554.

Landsmann, J. and Uhrig, H., 1985. Somaclonal variation in Solanum tuberosum detected at the molecular level. Theor. Appl. Genet., 71:500-505.

Pijnacker, L. P., Hermelink, J. H. M. and Ferwerda, M. A., 1986. Variability of DNA content and karyotype in cell cultures of interdihaploid Solanum tuberosum. Pl. Cell Reports, 5:43-46.

Ramulu, K. S., Dijkhuis, K. S. and Roest, S., 1989. Patterns of phenotypic and chromosome variation in plants derived from protoplast cultures of monohaploid, dihaploid and diploid genotypes and in somatic hybrids of potato. Pl. Sci., 60:101-110.

Reuff, I., Seitz, U., Ulbrich, B. and Reinhard, E., 1988. Cryopreservation of Coleus blumei suspension and callus cultures. J. Pl. Physiol., 133:414-418.

Seitz, U., Alfermann, A. W. and Reinhard, E., 1983. Stability of biotransformation capacity in Digitalis lanata cell cultures after cryogenic storage. Plant Cell Reports, 2:273-276.

Shillito, R. D., Carswell, G. K., Johnson, C. M., Dimaio, J. J. and Harms, C. T., 1989. Regeneration of fertile plants from protoplasts of elite inbred maize. Biotechnology, 7:581-594.

Stiekema, W. J., Heidekamp. F., Louwerse, J. D., Verhoeven, H. A. and Dijkhuis, P., 1988. Introduction of foreign genes into potato cultivars Bintje and Desiree using an Agrobacterium tumefaciens binary vector. Plant Cell Reports, 7:47-50.

Strauss, A., Fankhauser, H. and King, P. J., 1985. Isolation and cryopreservation of O-methyl threonine-resistant Rosa cell lines altered in the feed back sensitivity of L. threonine deaminase. Planta, 163:554-562.

Takeuchi, M., Matsushima, H. and Sugawara, Y., 1982. Totipotency and viability of protoplasts after long-term freeze perservation, in: "Proc. 5th Intl. Cong. Plant Tissue and Cell Culture, Tokyo, Japan". Fujiwara, A., ed., The Japanese Association for Plant Tissue Culture.

Ulrich, J. M., Finkle, B. J. and Tisserat, B. H., 1982. Effects of cryogenic treatment on plantlet production from frozen and unfrozen date palm callus. Plant Physiol., 69:624-627.

Volkova, L. A., Gorskaya, N. V., Popov, A. S., Paukov, V. N. and Urmantseva, V. V., 1986. Preservation of the main characteristics of Dioscorea mutant cell strains after storage at extemely low temperatures. Fiziologia Rasterii, 33:779-787.

Withers, L. A., 1987. Long-term preservation of plant cells, tissues and organs. Oxford Surveys of Plant Molecular and Cell Biology, 4:221-227.

Zheng, K. L., Castiglione, S., Biasini, M. G., Biroli, A., Morandi, C. and Sala, F., 1987. Nuclear DNA amplification in cultured cells of Oryza sativa L. Theor. Appl. Genet., 74:65-70.

VITRIFICATION IN PLANT TISSUE CULTURE

P.-L. Pasqualetto
Azienda Agricola Meristema SRL, Laboratorio
di micropropagazione, Via Martiri della Liberta'
n.13, 56030 Cascine di Buti (Pisa), Italy

INTRODUCTION

Vitrification is a serious problem since it can affect shoot multiplication and culture vigour and can impede the successful transfer of micropropagated plants to in vivo conditions (Hammerschlag, 1986). Affected plants exhibit tissue hyperhydricity and hypertrophy, chlorophyll a and b deficiency, lack of cell wall lignification or only a layer surrounding the cell membrane, and leaves with large intercellular spaces in the spongy mesophyll and palisade cells. Since the phenomenon apparently involves many factors (gelling agent type, hormones, organic and inorganic compounds, water potential, growth room temperature and light, ecology of container), I will consider morphological, phisiological and ecological aspects separately.

MORPHOLOGICAL AND PHYSIOLOGICAL ASPECTS

Generally after 4 to 5 days of culture, leaves on affected explants become turgescent, dark green, thick, curled and much longer and wider than normal. After 3 weeks, stems are thick, very brittle and traslucent. The descriptions of vitrified cultures given by various authors are very similar, whatever the species involved. The poor growth of vitrified shoots is accompanied by low rates of multiplication, rooting and survival on transfer to soil; wilt quickly and are very susceptible to infections. Such cultures may rapidly lose all capacity for propagation (Gaspar and Kevers, 1985). The problem is serious; most micropropagation laboratories face vitrification of their cultures and many tissue culturists have focused their efforts on pratical means of avoiding vitrification.

Cell Hyperhydricity . Few data are available on the relative water content of vitreous tissues. A comparative analysis of dry weight percentage in normal and vitreous plants of at least seven species shows a 0 to 43 % lower dry weight and a correspondingly higher wet weight in vitreous samples (Kevers et al., 1984).

Cell wall lignification . It is generally recognized that affected plants have a deficiency in lignification of the cell wall. Several histological studies have clearly demonstrated cell wall lignification

Plant Aging: Basic and Applied Approaches
Edited by R. Rodriguez _et al._
Plenum Press, New York, 1990

is lacking or limited to a thin layer opposed to the cell membrane. This finding has been correlated to: a) phenylalanine ammonia-lyase and acidic peroxidase content; b) C/N ratio; the biosynthesis of lignin and cellulose are closely associated since deficiency of both substances may result from decreased C/N ratio produced by an excess of N; c) phenolic production; soluble phenols are more abundant in vitreous than in normal Prunus plant (Phan and Letouze, 1983), while other authors have found that the addition of phloridizin and phloroglucinol to the culture media can help plants undergoing vitrification to return to normal plants (Hedegus and Phan, 1983).

Deficiency of chlorophyll . The traslucent appeareance of vitreous plants is related to a lower chlorophyll content than normal plants.

Modification of leaf structure and elemental composition . Anatomical analysis of leaf structure of vitreous "Gala" apple shoots has showed the lack of a clear differentiation between the palisade and spongy mesophyll layers, and the presence of large intercellular spaces (Pasqualetto et al., 1988). Modification of internal leaf structure in response to environmental conditions is well-known, e.g. changes in palisade mesophyll linked to higher relative humidity during development occur in many hydrophytes whose mesophyll acquires the characteristic aerenchyma (Vieitez et al., 1985). Affected leaves are generally thicker than normal due to longer and wider cells in the palisade and spongy parenchyma and to the occurrence of large intercellular spaces. A comparative study on apple has showed a difference in the number of stomata between vitreous and normal leaves, confirming that vitrification has a distinct effect on the morphology of affected plants (Pasqualetto et al., 1988). Apple cultures are normally transferred to fresh medium after 21 to 28 days of culture. If the cultures are not transferred regularly, stems elongate and leaves turn brown; thus shoots may grow poorly after transfer. This indicates that changes occur in the chemical and physical properties of the medium and in the atmosphere inside the containers, even during a normal culture cycle, and the protective conditions under which the tisuues are grown to prevent microbial contamination and retard dessication cause an accumulation of the compounds released, CO_2 and ethylene. An even more important component of the gas phase is water vapor. The atmosphere of most tissue culture containers is saturated with water vapor, so that developing shoots have an abnormal physiology and/or anatomy (Debergh, 1987). Stomatal abnormalities, reduced palisade tissue development and large intercellular spaces are characteristic anatomical features of herbaceous plants growing under conditions of high available moisture (Cutter, 1978). Vitrified leaves of apple shoots have proved to have a higher K and Ca content, indicating a cell water stress (Pasqualetto et al., 1978).

ECOLOGICAL ASPECTS

By the term "ecological aspects" I mean all the factors influencing the ecology of culture vessels and growth room conditions (light and temperature).

The concentration and brand of agar . The occurrence of vitrification is an agar-related problem. Agar should not be considered simply as a means of solidifying culture media: both the concentration and brand of agar affect the chemical and physical characteristics of a culture medium (Debergh, 1983). Moreover there are marked differences in nutrition composition among different agar brands, and striking variations in the solidity of gels among similar concentrations.

Vitrification has sometimes been overcome by substituting one gelling agent with another. For example, apple cultivars became vitreous when Gelrite or Phytagar was used to solidify the proliferation medium but not with Difco Bacto-agar. Gelrite is used commercially for _in vitro_ propagation of some ornamental plants but it induces vitrification with a lot of fruit plants. To prevent vitrification of their cultures, private laboratories test the quality of each stock of agar in a small number of jars before using it in large quantities. How do the concentration and brand of agar influence the characteristics of the medium and thus vitrification ? First of all, with increased agar concentrations the medium tends to be firmer, loosing the characteristic fluid consistency typical of lower concentrations; the availability of water is diminished and the diffusion of macromolecules is restricted, e.g. cytokinins. Water is generally recognized as a key factor in vitrification; affected plants in fact have a greater diffusion of water into the cells. The reduced availability of water and of other components may lead to the inhibition of both vitrification and shoot proliferation. Secondly, the chemical and physical properties of each agar are specific, so that one species may be influenced by such properties while others are not.

The level of BA . A cytokinin in the culture medium stimulates growth of axillary and/or adventitious buds. Debergh (1983) found that vitrification was influenced with BA levels at low agar concentrations. The vitreous condition in apple cultures can sometimes be overcome by trasferring them from a medium containing BA to one with 2iP or no cytokinin at all (Zimmerman, 1982). However, by reducing cytokinin levels in the medium or by changing type we may limit or completely solve the problem of vitrifacatioon, but at the cost of lowering or halting shoot proliferation. On the other hand, it seems that the effect of BA on vitrification can be overcome by increasing gelling agent concentration, suggesting that the effect occurs only at particular concentrations of the agar.

The ammonium ions content . One salient feature of the syndrome is that most reported cases of vitrification have occurred using MS medium, which has a particularly high content of ammonium nitrate. With _Salix babilonica_ , vitrification of shoots was reduced by varying the quantity of nitrogen in the culture medium (Daguin and Letouze, 1985). Vitrified shoots were produced in multiplication cultures of _Castanea sativa_ when MS medium was used in the subcultures, whereas normal shoots were obtained when Heller's macronutrient formula was used, with or without addition of 1 nM ammonium sulphate (Vieitez et al., 1985)
Two basic facts regarding the vitrification phenomenon are the reduction of lignin and cellulose content of tissues and the enlargement of cells. One possible correlation between these two facts and ammonium ions has been suggested in the literature. Ammonium ions are assimilated faster than other nitrogen sources such as nitrates. Both ammonium ions and lignin synthesis pathway need carbohydrates, so that a rapid uptake of the former may divert carbohydrates from the latter (Beauchesne, 1981). Deficiency of lignin and cellulose, in fact, results from a decreased C/N ratio produced by an eccess of N (Kevers et al., 1984). Both deficiencies would tend to reduce wall pressure and so favour increased absorption of water with a consequent enlargement of the tissues.

The role of K and Mg . We have tested several concentrations of K and Mg without overcoming the problem of vitrification. The lowest K level produced a higher percentage of vitrified apple shoots and affected tissue appearance so greatly that we can hypothesize the occurrence of other metabolic events not connected with the development of vitrified tissue (Pasqualetto et al., 1988).

   The role of the atmosphere inside the container . Plants and tissues cultured _in vitro_ release compounds (CO_2 and ethylene) into the atmosphere inside the container, without any exchange with the surrounding atmosphere due to the protective conditions under which the tissues are grown to prevent microbial contamination and retard dessication. Relative humidity seems to play an important role in the occurence of vitrification. Growth room conditions (light and temperature) produce what I call the "greenhouse effect". The shelf lighting heats the bottom of the containers so that the temperature inside the jars is 5 - 6 C higher than that of the growth room. This in turn evokes the formation of moisture and of water droplets on the walls of containers. As previously reported, defective deposition of wax, stomatal abnormalities, reduced palisade tissue development, large intercellular spaces and a non-continuous cuticle all seem to be caused by high relative humidity in the culture atmosphere. The fact that these problems can be overcome by increasing the vapor pressure gradient between the leaf and the atmosphere supports Debergh's hypothesis that vitrification is provoked by the high relative humidity in culture jars.

SOME REMEDIES TO OVERCOME VITRIFICATION

   A definitive solution to this problem, one able to maximize shoot proliferation and minimize vitrification, will require simultaneous attention to a number of different factors. Although growth room temperature and light sources cannot be ignored, agar, BA and ammonium ions are more often described as factors which can be adjusted to eliminate the vitreous condition in cultures. The availability of water in the culture jars seems to be a key point of the problem and agar, BA and ammonium ions are in some way related to it. Increasing agar concentration reduces the availability of water and the traslocation of macromolecules, e.g. BA. Each brand of agar influences the chemical and physical characteristics of the medium in a specific way and consequently the water status in the jar. Cytokinins can cause an increase in the size of leaf tissues by a process involving only cell enlargement, while rapid ammonium ion uptake brings about a drop in the C/N ratio, leading to a deficiency of lignin and cellulose with a consequent absorption of water into the cells. In most tissue culture containers the atmosphere is almost saturated with water vapor and as a concequence transpiration and traslocation processes are reduced to a minimum. Size and type of container, type of culture vessel closure and label, brand and concentration of agar, and climatic parameters of the culture room are all factors which can create an atmosphere less saturated with water vapor. Lowering the relative humidity of the container atmosphere by cooling the culture medium, creates conditions in the container which activate transpiration and traslocation, so that _in vitro_ shoots behave as normal plants.

IMPLICATIONS ON MASS PROPAGATION

   Loss of plants during weaning may be caused by a defective condition of the plants themselves rather than by poor acclimation. An insufficient deposition of wax on the leaf cuticle, may easily provoke a considerable loss of water with a consequent withering of the plants. Working on a commercial scale, once a suitable culture medium has been found attention should be focused on the climatic conditions of the growth room. Tissue culturist should create an environment which promotes transpiration and translocation, so that _in vitro_ shoots behave as normal plants.

CONCLUSION

Further research is needed to clarify the physiological problem of vitrification. From a practical point of view, once a suitable medium and gelling agent have been found, attention should be directed to the ecology of the container, with particular reference to the water status inside the jar.

REFERENCE

Beauchesne, G., 1981, Les milieux mineraux utilises en culture _in vitro_ et leur incidence sur l'apparition de boutures d'aspect pathologique. _C.R. Acad. Agric._ , Paris 67:1389-1397.

Cutter, E.G., 1978, Plant anatomy I, _in_ "Cells and tissues" 2n ed. Addison-Wesley, Reading, Mass. pp 168.

Daguin, F., and Letouze R., 1985, Relations entre hypolignification et etat vitreax chez _Salix babylonica_ en culture _in vitro_ . Role de la nutrition ammoniacale. _Can. J. Bot._ , 63:324-326.

Debergh, P. C., 1983, Effects of agar brand and concentration on the tissue culture medium. _Physiol. Plant._ , 59:270-276.

Debergh, P. C., 1987, Improving micropropagation. _IAPTC Newsletter_ , 51:2-10.

Gaspar, T., and Kevers, C., 1985, Cobalt prevention of vitrification process in carnation. _Plant Physiol._ , 77(suppl.):13.

Hammerschlag, F. A., 1986, Temperate fruits and nuts. _In_ :"Tissue culture as a plant production system for horticultural crops". 221-236. Zimmerman, R. H., Griesbach, R. J., Hammerschlag, F. A., and Lawson, R. H. (eds). Martinus Nijhoff Publishers, Dordrecht.

Hedegus, P., and Phan, C. T., 1983, Actions de phenols sur les malformations observes chez les porte-greffes de pommiers M-26 et O-3 cultives _in vitro_ . _Rev. Can. Biol. Exp._ , 42:33-38.

Kevers, C., Coumans, M., Coumans-Gilles, M. F., and Gaspar, T. H., 1984, Physiological and biochemical events leading to vitrification in plants cultured _in vitro_ . _Physiol. Plant._ , 61:69-74.

Pasqualetto, P.-L., Wergin, W. P., and Zimmerman, R. H., 1988, Changes in structure and elemental composition of vitrified leaves of "Gala" apple _in vitro_ . _Acta Hort._ , 227:352-357.

Pasqualetto, P.-L., Zimmerman, R. H., and Fordham, I., 1988, The influence of cation and gelling agent concentrations on vitrification of apple cultivars _in vitro_ . _PCTOC_ , 14:31-40.

Phan, C. T., and Letouze, R., 1983, A comparative study of chlorophyll, phenolic and protein contents and of hydroxycinnamate:CoA ligase activity of normal and vitreous plants (_Prunus avium_ L.) obtained _in vitro_ . _Plant Sci. Lett._ , 31:323-327.

Vieitez, A. M., Ballester, A., San-Jose, M. C., and Vieitez, E., 1985, Anatomical and chemical studies of vitrified shoots of chestnut regenerated _in vitro_ . _Physiol. Plant._ , 65:177-184.

Zimmerman, R. H., 1982, Apple tissue culture. _In_ :"Handbook of plant cell culture", Vol.2. Evans, D. A., Sharp, W. R., and Ammirato (eds). MacMillan, New York.

SECTION III

ULTRASTRUCTURAL, GENETIC AND BIOCHEMICAL CHARACTERISTICS OF
AGING AND SENESCENCE

THE MOLECULAR GENETICS OF MATURATION IN EASTERN LARCH

(LARIX LARICINA [DU ROI] K. KOCH)

Keith Hutchison[+], Michael Greenwood[*], Christopher Sherman[*],
Joanne Rebbeck[*], and Patricia Singer[+]

[+]Department of Biochemistry
[*]Department of Forest Biology
University of Maine
Orono, Maine 04469 U.S.A.

INTRODUCTION

Maturation in woody plants has recently received much attention because of the maturation-related decrease in the ability to clone selected individuals using explants from mature plants (Hackett, 1985; Greenwood, 1987). Furthermore, the phenotypic changes that accompany maturation make it difficult for the tree breeder to select superior genetic families or individuals at the seedling stage (e.g., Lambeth, 1980). The changes associated with maturation include growth rate, branching characteristics and growth habit, reproductive behavior, and the morphology and physiology of foliage. The significance of the latter will be emphasized in this brief report, with emphasis on the role of gene expression in the maturational process. Changes in foliage associated with maturation have been discussed (e.g., Zimmerman et al., 1985; Bauer and Bauer, 1980) and do not follow a completely consistent pattern among species.

We are using larch as a model system to study conifer maturation for several reasons: 1) its juvenile and mature characteristics have been described (Greenwood et al., 1989), 2) it is very responsive to all forms of vegetative propagation relative to most other conifers, and 3) its rapid growth rate and ease of propagation make it attractive as a species for plantation establishment. Furthermore, the maturational characteristics of larch are typical of those for other conifers, and its genome (DNA content is 19pg/2C nucleus) is among the smallest of the extraordinarily large conifer genomes, which range from 12.6 to 41pg of DNA/2C nucleus (Dhillon, 1987). An understanding of maturation at the molecular level would not only help elucidate the mechanism of maturation, but the feasibility of reversing it as well.

Plant Aging: Basic and Applied Approaches
Edited by R. Rodríguez *et al.*
Plenum Press, New York, 1990

141

MATERIALS AND METHODS

Our experimental system utilizes grafted scions from trees of different ages, since the developmental behavior of the scions reflects the age of the donor tree, and is described in detail in Greenwood et al., 1989. Use of grafted scions permits comparison of the morphological and physiological changes that occur with maturation without the confounding effects of size or environment. Gene expression at the molecular level has been examined in juvenile and mature foliage by extracting mRNA at different stages of foliar development, followed by preparation and cross comparison of cDNA libraries from juvenile and mature foliage (Hutchison et al., 1988a). The analysis of the expression of individual genes has been carried out by probing cDNA libraries with heterologous probes for the chlorophyll a/b binding protein (Cab) and the small subunit of ribulose-1,5 bisphosphate carboxylase (rbcS) in order to create homologous larch probes; for methodology see Hutchison et al., 1988a and 1988b. Homologous probes for both genes were used for location of transcripts of these genes in total RNA preparations using both Northern and slot blots.

The photosynthetic characteristics of juvenile and mature foliage was assessed using a Licor 6200 portable photosynthesis system, which measured net photosynthesis and stomatal conductance per unit leaf area. Chlorophyll content was measured after extraction with DMSO (for details see Greenwood et al., 1989).

RESULTS AND DISCUSSION

Chlorophyll content of the long and short shoot foliage of eastern larch increases linearly with the $\log_{10}$ of age (effect of age on chlorophyll content significant at $p < .001$, Greenwood et al., 1989), and the net photosynthetic rate is positively and significantly correlated with chlorophyll content ($r^2 = 0.39$, $p < .001$). Stomatal conductance showed no change with increasing age. Needle length of long shoot foliage declines slightly but significantly with age, but needle thickness and cross sectional area increase ($p < .001$). In addition, the specific leaf weight (dry weight per unit of projected surface area) increases with age ($p < .001$) which indicates that long shoot needles become more massive with increasing age. Therefore, maturation results in both morphological and physiological changes in the long shoot foliage. The chlorophyll content of short shoot foliage varies in exactly the same way with age as the long shoot foliage, and large quantities of short shoot foliage can be collected at similar developmental stages since all short shoots begin to expand at roughly the same time. Although we have not measured the photosynthetic capacity of short shoot needles, there is no reason to expect that short shoot foliage will behave any differently than that of long shoots. Our observations on gene expression have so far emphasized short shoots because of the availability of large amounts of uniform material for extraction of RNA.

Northern blots of total RNA using homologous probes for Cab and rbcS showed that there was little difference in expression of these genes just after the start of needle expansion, before differences in chlorophyll content were apparent. In fact, Cab expression appeared greater in juvenile material (for both long and short shoot foliage) at that time (Hutchison et al., 1988b). However, as the chlorophyll content of the mature foliage increased relative to the juvenile, the expression of both genes appears to increase relatively faster in more mature foliage. This increase is observed whether the amount of transcript is normalized to the amount of total or mRNA, and is several fold greater for the oldest material. The relatively increased levels of transcripts for both genes in the more mature foliage occur during a sharp drop in total RNA between 4 and 6 weeks after needle elongation started (Hutchison, 1989, unpublished data). These changes are not associated with changes in levels of total DNA methylation, which were exactly the same (20%) in fully expanded juvenile and mature foliage.

Clearly, maturation affects the morphology and physiology of larch needles so that the mature needles appear to be relatively more massive, contain more chlorophyll, and exhibit more net photosynthesis. These anatomical and physiological changes are associated with change in the expression of genes coding for proteins vital to the photosynthetic process, but unique gene products associated with the juvenile or mature state were not detected. At present, we do not know whether the differential expression of the Cab and rbcS is regulated at the transcriptional or post-transcriptional level. Maturation also affects the needle anatomy of a number of other woody species (Greenwood and Hutchison, 1989), but the effects of these anatomical changes on photosynthesis have only been investigated in English ivy (Hedera helix) and red spruce (Picea rubens). In Hedera, mature leaves also appear more massive (they are thicker and have a higher specific leaf weight), and net photosynthesis per unit surface area is also higher, which appears to be due to a higher stomatal frequency and conductance. In contrast to larch, total chlorophyll content is actually greater in juvenile than mature leaves (Bauer and Bauer, 1980). The mature foliage of red spruce is also thicker and more massive than juvenile foliage, but the juvenile foliage exhibits greater net photosynthesis, which appears to be due to increased stomatal conductance. Chlorophyll content is similar in both types of foliage. With the exception of a tendency for mature foliage to be more massive, the physiological differences between mature and juvenile foliage do not follow a consistent pattern (see Table 1). Clearly whatever is causing the observable differences between juvenile and mature foliage in these species does not elicit totally similar physiological responses. Therefore, one would not expect to see the same patterns of differential expression of Cab and rbcS genes among these species. Consequently, the mechanism which immediately affects the increased expression of the Cab and rbcS genes in mature larch will probably not be observed in Hedera or spruce. However, since the mature foliage does share common morphological traits, there may be a common maturational process which can have varied effects on physiological processes.

At present we have not found any obvious evidence that patterns of gene expression are anything but a consequence of whatever controls maturation. Hackett (personal communication, 1989) also reports little differential gene expression during root formation by petioles of juvenile and mature Hedera. Cross comparison of cDNA libraries from both types of petioles, with and without IAA treatment, reveals only 1 or 2 unique clones, found in mature petioles treated with IAA. Since decreased rooting seems to always be associated with maturation, one cannot conclude at this time that the increased rooting capacity of juvenile tissue is due to the expression of particular genes which is subsequently lost with maturation. Instead, there may be gene products specific to the mature state which inhibit rooting.

The role of gene expression in maturational changes involving rooting or photosynthesis, not to mention other maturational characteristics would appear to be quite subtle. Tracing a common underlying maturational cause by examining gene products associated with changes in rooting or photosynthetic

Table 1. Photosynthetic Characteristics of Juvenile and Mature Foliage of Larch, Spruce and English Ivy. (Different letters indicate means differ between J and M at $p \leq .05$.)

	Larch		Spruce		English Ivy*	
	J	M	J	M	J	M
Net photosynthesis ($\mu mol \cdot CO_2 \ m^{-2} \cdot S^{-1}$)	9.22a	11.04b	7.07a	4.94b	5.46a	8.33b
Stomatal conductance ($\mu mol \cdot m^{-2} \cdot S^{-1}$)	0.17b	0.17b	.213a	.112b	0.24a	0.38b
Specific needle (leaf) weight (g dry wt $\cdot m^{-2}$)	169a	197b	386a	525b	54a	68b
Chlorophyll content	6.5a	7.9b	4.33a	4.51a	12.5a	9.2b

*Data adapted from Bauer and Bauer (1980).

capacity would, in our opinion, be very difficult. But the maturational time courses for change by the suite of morphological and physiological characteristics in larch are similar, and the rate of change is most rapid in the first 5 years (Greenwood et al., 1989). Consequently, there may be a single (or small number) of events, several steps prior those events we report here, which affect all maturational processes.

The issue of phase change is sufficiently complex to generate disagreement as to what the phenotypes can be specifically correlated with the juvenile or mature phases. As these questions are examined more deeply, one can be left with the impression that maturation is a form of confusion with no consistent cause or affect. This hypothesis is not experimentally approachable. Much of the confusion, though, comes from comparing the maturation-associated changes across species, genera, families, etc. As an alternative approach, we can start with the assumption that the underlying mechanisms are the same in all or most plants, and it is simply the resulting phenotype that is variable. What this suggests is that the phenomenon we study should be consistently associated with phase change within the species one is studying, but that it need not necessarily be associated with maturation in all systems. A corollary to this statement is that maturation-related phenomenon need not necessarily be irreversible in all plant systems.

The phenomenon of maturation is a complex developmental problem and the genetic mechanisms of control are likely to be similarly complex at the molecular level. We believe, however, that the genetic mechanisms that ultimately control phase change are likely to be simpler, at the conceptual level, than might be suggested by the variety of phenotypic changes associated with maturation.

REFERENCES

Bauer, H., Bauer, U., 1980, Photosynthesis in leaves of the juvenile and adult phase of ivy (Hedera helix), Physiol. Plant. 49: 366-372.

Dhillon, S. S., 1987, DNA in tree species, in: "Cell and Tissue Culture in Forestry: General Principles and Biotechnology," J. M. Bonga and D.J. Durzan, eds., Martinous Nijhoff, Boston, 298-313.

Greenwood, M.S., 1987, Rejuvenation of forest trees. Plant Growth Regulation, 6:1-12.

Greenwood, M.S., Hopper, C.A., Hutchison, K. W., 1989, Maturation in larch. I. Effect of age on shoot growth, foliar characteristics, and DNA methylation, Plant Physiol. 90:406-412.

Greenwood, M.S., Hutchison, K. W., 1989, Maturation as a developmental process, in: "Clonal Forestry: Genetics, Biotechnology and Application,", M.R. Ahuja and W.J. Libby, eds., Springer Verlag, New York.

Hackett, W.P., 1985, Juvenility, maturation, and rejuvenation in woody plants. Hort. Reviews, 7:109-155.

Hutchison, K.W., Singer, P. B., Greenwood, M. S., 1988, "Molecular genetic analysis of development and maturation in larch," in: Molecular genetics of forest trees. Proc. 2nd IUFRO working group on molecular genetics S2.04.06. June 16-18, Petawawa National Forestry Research Institute, Chalk River, Ontario, Canada, P1-X-80: 26-33.

Hutchison, K.W., Singer, P.B., Greenwood, M.S., 1988, Gene expression during maturation in eastern larch, in: " Molecular Genetics of Forest Trees," J.E. Halgren, ed., Franz Kempe Symp., Umea, Sweden, June, 1988.

Lambeth, C.C., 1980, Juvenile-mature correlations in Pinaceae and implications for early selection, For. Sci. 26:571-580.

Zimmerman, R. H., Hackett, W. P., Pharis, R. P., 1985, Hormonal aspects of phase change and precocious flowering, in: "Encyclopedia of Plant Physiology," R.P. Pharis and D.M. Reid, eds., 79-115.

CELLULAR, BIOCHEMICAL AND MOLECULAR CHARACTERISTICS RELATED

TO MATURATION AND REJUVENATION IN WOODY SPECIES

W.P. Hackett, J.R. Murray, H-H. Woo, R.E. Stapfer, and R. Geneve[1]

Horticultural Science, Univ. of Minnesota, St. Paul, MN
[1]Horticulture and Landscape Architecture, Univ. of Kentucky, Lexington, KY

In the development of all woody plants from seed there is a so-called juvenile phase lasting up to 30-40 years in certain forest trees, during which flowering does not occur and cannot be induced by the normal flower-initiating treatment or conditions. In time, however, the ability to flower is achieved and maintained under natural conditions; at this stage, the tree is usually considered to have attained the adult or mature condition. The transition from the juvenile to the mature phase has been referred to as phase change by Brink (1962), ontogenetic aging by Fortanier and Jonkers (1976), or meristem aging (cyclophysis) by Seeliger (1924) and Oleson (1978). Associated with this transition are progressive changes in morphological, developmental, and physiological characteristics. Changes in such characteristics during development vary from species to species. Most change gradually during the period preceding the mature phase, resulting in transitional forms. Usually no distinct change in any one characteristic is apparent at the time the ability to flower is attained.

Phase change is of considerable theoretical importance relative to morphogenetic control, differentiation, and determination in plant development. It has practical significance for the following reasons:

1. The length of the juvenile period is inversely related to the breeding efficiency of woody perennials and to the selection of improved cultivars (Hansche and Beres 1980).

2. The ease of cuttage propagation and in vitro morphogenesis and somatic embryogenesis for all types of woody perennials is strongly affected by ontogenetic age (Heybroek and Visser 1976; Greenwood 1987).

3. The quantity and quality of productivity of a forest tree species is related to its degree of maturity (Heybroek and Visser 1976).

Although maturation has usually been defined in terms of flowering ability because it is the most consistent and late characteristic to change during development of woody species, systematic efforts to document the time course of changes in characteristics during maturation are sparse. Greenwood et al (1989) have noted that an examination of the time courses of the onset of maturation characteristics may provide some clues as to how they arise. Using a grafting experiment, Greenwood et al (1989) found that 6 vegetative maturational characteristics in larch exhibited two distinct time courses of change with increasing age when plotted as the $\log_{10}$ of age. This suggests that some of the vegetative maturational characteristics vary independently of one another. With regard to flowering, scions from juvenile trees produced more strobili per tree, 3 years after grafting, than did scions from mature trees. This suggests that the mechanism for acquisition of flowering potential is not closely related to mechanisms for changes in the vegetative maturation characteristics studied. That flowering ability may not be closely related to vegetative maturation characteristics is also suggested by observations of some eucalyptus species in which flowering occurs before change

Plant Aging: Basic and Applied Approaches
Edited by R. Rodríguez et al.
Plenum Press, New York, 1990

147

in foliage characteristics. Similarly, in <u>Hedera</u> <u>helix</u> aerial stem roots and anthocyanin accumulation can occur under certain conditions on plants that have the ability to flower.

Such observations indicate that there isn't necessarily a strict hierarchy of maturation-related characteristics, that acquisition of maturation-related characteristics is not necessarily related developmentally to acquisition of another characteristic and that there may not be a single switch from a juvenile to mature condition. This means that we should think about alternatives to a single switch mechanism. For example, there may be several switches in parallel, one for each character or for sets of characters or perhaps there are switches in series such that one character must change as a pre-requisite for a change in another character.

No matter what the mechanisms for or the timing of changes in phenotypic characteristics during development, they accumulate in what we term the mature condition. They accumulate because the characteristics are quite stable and are transmitted developmentally through cell division from one somatic cell generation to the next. However, this doesn't mean that maturation-related characteristics are permanent and non-reversible. There are many examples which indicate that they are reversible (Hackett, 1985). However, all of the characteristics may not have the same ease of reversibility and the ease of reversibility of any one characteristic might change over developmental time.

What, then, is a useful way to think about maturation in woody species from both a practical and theoretical standpoint? From what has been outlined above it seems that a reasonable way to think about maturation is that it is a very dramatic and protracted example of the process of differentiation. The integrated expression of characters during development, which results in a recognizable phenotype, is a manifestation of the process of differentiation. Anatomical and biochemical differentiation, leading to structural and functional specialization of cell, tissue and organ types, results from the differential expression of genes (Goldberg, 1987). The relatively stable differential expression of genes during particular phases of development, such as the juvenile and mature phases, is referred to as epigenetic variation (Nanney, 1958; Lewin, 1983). Epigenetic control of differentiation not only determines the difference in phenotypic characters observed in the juvenile and mature phases, but also determines the competency of the tissue to undergo secondary differentiation. For example, stem cuttings from the juvenile phase of woody plants form adventitious roots easily, while cuttings from mature phase tissue form them slowly or not at all. (Hackett, 1988).

What is most interesting and unique about maturation-related characteristics from both a practical and theoretical biological standpoint is that these characteristics are stable but reversible. This gives us the opportunity to study the control of these characteristics one at a time to determine if and/or how one characteristic is related to another and to be able to determine if there is a single switch or several switches. To be able to manipulate and maintain each characteristic independently is very attractive from a practical standpoint for forestry and horticulture.

English ivy (<u>Hedera</u> <u>helix</u> L.), a woody perennial with a juvenile phase lasting 10 or more years, exhibits a number of distinct differences in phenotypic characters between the juvenile and mature phases (Table 1). This species is a classical example of stable dimorphism. The juvenile and mature forms can be maintained as separate plants through use of cuttage propagation or they can co-exist on the same plant. Through use of asexual propagation, the same genotype can be maintained in a greenhouse as separate individuals with very different phenotypes for long periods of time (years). Treatment of the shoot of a mature plant with GA_3 will induce development of juvenile characteristics. These characteristics make it very useful for doing an experimental analysis of the kind mentioned in the previous paragraph.

For the last 10 years, we have been doing such an experimental analysis of the control of individual maturation-related characteristics using <u>H</u>. <u>helix</u>. The characteristics that we've studied are: 1) shoot apical and subapical meristem size, configuration and activity in relation to internode length and phyllotaxis; 2) root initiation potential; and 3) anthocyanin accumulation. In this paper, I will limit discussion to root initiation potential and anthocyanin accumulation.

Maturation-Related Rooting Potential

Juvenile phase <u>H</u>. <u>helix</u> plants have stem aerial roots while mature phase plants do not.

Table 1.   Phenotypic characteristics that differ between clonal juvenile and mature phase ivy (Hedera helix L.)

Characteristic	Juvenile	Mature
flowering ability	absent	present
stem orientation	plagiotropic (horizontal)	orthotropic (upright)
leaf morphology	palmately lobed	ovate and entire
phyllotaxy	distichous (alternate)	spiral
stem and petiole pigmentation	anthocyanins present	anthocyanins absent
stem adventitious roots	present	absent

We have discovered that the rooting potential of detached leaves and in vitro cultured delaminated petioles of juvenile and mature ivy is very similar to that of stem cuttings (Geneve et al, 1988; Hackett et al, 1988). Detached juvenile leaves and delaminated petioles respond to auxin by initiating roots from small, discrete areas of cells in the exterior part of the phloem of each vascular bundle and the adjacent cortex. In contrast, mature leaves and delaminated petioles respond to auxin by forming only callus as a result of cell division in these same areas as well as throughout the cortex.

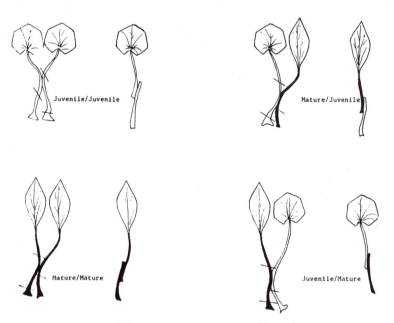

Figure 1.   Schematic representation of detached leaf approach grafting procedure used to obtain cuttings with reciprocal combinations of juvenile and mature petiole with juvenile and mature lamina.

Detached leaves and delaminated petioles have the following experimental advantages for studying rooting potential: 1) no possibility of pre-formed root initials; 2) variable rooting potential in tissues easily selected for similar physiological age and anatomical organization and having identical genetic makeup; 3) simple, uniform, fully differentiated tissue system in which source of plant growth substances can be endogenous (with lamina) or exogenous (without lamina); 4) easy in vitro manipulation of environment and precise provision of nutrients and plant growth substances as pulses or continuously fed from proximal or distal ends (delaminated petioles); 5) relatively rapid, specific and morphogenetically distinct responses to auxin; 6) ease of grafting to obtain composite cuttings; 7) tissues readily available in large amounts on a year-round basis.

Experiments have been performed using de-laminated petioles of juvenile and mature leaves and leaf cuttings composed of the four reciprocal combinations of juvenile and mature petioles with juvenile and mature lamina to study potential for root initiation. Procedures for using delaminated petioles and composite juvenile and mature leaf cuttings for studying rooting potential are described in Hackett et al, 1988. Figure 1 illustrates the use of grafting to make composite juvenile and mature leaf cuttings. The results of experiments with these systems lead to the following conclusions: (a) the morphogenetic process of root initiation is very different in easy- and difficult-to-root tissues; (b) auxin is required for root initiation in debladed juvenile petioles; (c) exogenously applied auxin and its metabolites have similar distribution patterns in juvenile and mature debladed petioles; (d) differences in ethylene metabolism do not appear to be causally related to differences in rooting potential in juvenile and mature debladed petioles; (e) root initiation is mainly a function of the rooting potential of cells localized in the petiole; (f) there is no evidence of a rooting inhibitor being transported from mature lamina; and (g) a translocatable substance(s) formed in juvenile lamina can either induce rooting initiation in mature petioles or increase rooting potential of new cells formed as a result of auxin treatment.

To investigate the molecular difference between juvenile and mature H. helix plants, the poly A(+) RNAs were isolated from juvenile and mature delaminated petioles at different time points after excision and translated in rabbit reticulocyte lysates. These translation products were analyzed by two-dimensional gel electrophoresis. At day 0 of in vitro culture, differences in only two polypeptides were detected. One polypeptide was specific in juvenile petioles and had a pI of ~9 and a size of ~25 kDa. The other polypeptide was specific in mature petioles and had a pI of ~5.3 and a size of ~28 kDa. A histological study showed that the differentiation of root initials in juvenile petioles starts at day 7 or 8. Because a primary interest was in messages related to root initiation potential, we tried to detect changes in translation products occurring before that time point. After day five of in vitro culture, we could detect only one polypeptide which was increased two or three times more in juvenile petioles than in mature petioles. Its pI was ~4 and its size was 40 kDa. These results show that the abundant mRNA species detectable via in vitro translation products are very similar in juvenile and mature petioles, both before and 5 days after auxin treatment. However, less abundant mRNAs most likely involved in synthesis of regulatory proteins wouldn't be detected by in vitro translation techniques.

To get more information on maturation-related reduced rootability, we made a cDNA library of poly A(+) RNA from auxin treated, 5 day cultured juvenile petioles and did differential screening by comparing the messages from auxin treated and non-treated juvenile and mature petioles. The initial screening of ~20,000 recombinants from the juvenile library [JA-5(S)] gave one cDNA clone which was specific in juvenile petioles. This clone (pHW101) represents a mRNA that is constitutively expressed at a higher level in juvenile than in mature petioles but is expressed at a lower level in auxin treated than non-treated juvenile petioles. It is also expressed at a higher level in juvenile than in mature leaf lamina and stem.

The disadvantage of differential screening is in its low sensitivity for the detection of low abundant messages. In our research we are also looking for low abundant messages which are also cell specific for differentiation. Therefore we have done subtraction hybridization screening to isolate low abundant messages. By using hydoxylapatite column chromatography, we could subtract about 85% of common messages. After subtraction hybridization screening of ~20,000 recombinants, we isolated a second differentially expressed clone. The second clone (pHW103) represents a mRNA that is expressed at a higher level in mature than juvenile petioles after 3 days of in vitro culture. But this clone was not expressed at all in juvenile or mature lamina, stem, or root tissues. Northern analysis shows that the mRNA size of pHW101 is ~1.4 kb and pHW103 is ~1.0 kb. Also, northern analysis shows more than one band which suggests these two clones may be in gene families. Neither of these cDNA clones seems to be closely related to the root initiation process. However, clone pHW103 which is expressed at a much higher level in mature petioles than juvenile ones, appears to be wound inducible since it isn't expressed at day 0 in any tissue tested but is expressed after excision and culture for 3 days.

This molecular analysis demonstrates that there are differences in gene expression in juvenile and mature ivy but the number of genes differentially expressed is small.

Phenylpropanoid and Flavonoid Metabolism

Because of the observed accumulation of anthocyanin in juvenile but not mature leaves and stems of H. helix, we have studied phenylpropanoid and flavonoid metabolism at the cellular and biochemical level using stem and petiole tissue and leaf discs. Biochemical and enzymological studies show that the specific activity of phenylalanine ammonia-lyase is twice as high in mature as juvenile tissue and this is reflected in 50% higher extractable phenylpropanoids in mature than juvenile tissue. So it is unlikely that early steps in the phenylpropanoid-flavonoid pathway limit accumulation of anthocyanin in mature tissues.

The above conclusion led us to concentrate on flavonoid metabolism. Juvenile phase H. helix accumulates two classes of flavonoid glycosides in the dermal tissue of stems and petioles. The flavonoids, flavonols and anthocyanin are derived from dihydroflavonols late in the flavonoid biosynthetic pathway. The anthocyanin accumulates in the hypodermal tissue consisting of 3 to 4 layers of collencyma cells. The flavonols also accumulate in the dermal tissue, however, it has not been possible to determine if they are strictly localized in the hypodermis or epidermis. Mature phase ivy synthesizes and accumulates flavonols but not anthocyanin in its anatomically similar dermal tissue. The lack of synthesis of anthocyanin in mature phase hypodermal tissue is due to the lack of activity of dihydroquercetin-4-reductase (DQR), the enzyme that catalyzes the initial step of the conversion of dihydroflavonols to anthocyanin. Leaf laminae of neither phase accumulate anthocyanin when grown at temperatures above 15°C and there is no detectable activity of DQR. These results showing that accumulation of anthocyanin in juvenile leaves and lack of accumulation in mature leaves of H. helix is due to the differential activity of DQR, is the only example we know of in which expression of a phase-related characteristic is due to the activity of a single polypeptide.

Histological studies of juvenile and mature petioles using blue light epifluoresence of fixed and stained tissues as a measure of wall bound phenylpropanoids show that fluorescence is lower in specific cells destined for root morphogenesis in juvenile petioles as compared to cells in the same location in mature petioles destined for fiber differentiation. This latter observation suggests that the two maturation characteristics anthocyanin accumulation and root potential might be metabolic related through the phenylpropanoid-flavonoid biosynthetic pathway.

Conclusions

These results provide a strong rationale for: (a) identification of a translocatable factor(s) that influences rooting potential; (b) study of enzymes specifically involved in cell wall lignin metabolism as they relate to rooting potential; (c) concluding that maturation-related differences in phenotypic characteristics are the result of differential gene expression; (d) use of the differential expression of DQR activity in juvenile and mature H. helix as a basis for investigating the molecular control of gene expression for a maturation-related characteristic.

Literature

Brink, R.A., 1962, Phase change in higher plants, and somatic cell heredity, Quart. Rev. Biol., 37-1-22.

Fortanier, E.J. and Jonkers, H., 1976, Juvenility and maturity of plants as influences by their ontogenetical and physiological aging, Acta Hort., 56:37-44.

Geneve, R.L., Hackett, W.P., and Swanson, B.T., 1988, Adventitious root initiation in de-bladed petioles from juvenile and mature phases of English ivy, J. Amer. Soc. Hort. Sci., 113:630-635.

Goldberg, R.B., 1987, Emerging patterns of plant development, Cell, 49:298-300.

Greenwood, M.S., Hooper, C.A. and K.W. Hutchison, 1989, Maturation in larch. I Effect of age on shoot growth, foliar characteristics and DNA methylation, Plant Physiol., in press.

Greenwood, M.S., 1987, Rejuvenation of forest trees, <u>Plant Growth Regulation</u>, 6:1-12.

Hackett, W.P., Geneve, R.L., and Mokhatari, M., 1988, Use of leaf petioles of juvenile and mature <u>Hedera</u> <u>helix</u> to study control of adventitious root initiation, <u>Acta Hortic.</u>, 227:141-144.

Hackett, W.P., 1988. Donor plant maturation and adventitious root formation, In: "Adventitious Root Formation in Cuttings", T.D. Davis, B.E. Haissig and N. Sankhla, eds., Dioscorides Press, Portland, Or.

Hackett, W.P., 1985, Juvenility, maturation and rejuvenation in woody plants, <u>Hortic. Rev.</u>, 7:109-155.

Hansche, P.E. and Beres, W., 1980, Genetic remodeling of fruit and nut trees to facilitate cultivar improvement, <u>Hort. Science</u>, 15:710-715.

Heybroek, H.H. and Visser, T., 1976, Juvenility in fruit growing and forestry, <u>Acta Hort.</u>, 56:71-80.

Lewin, B., 1983, "Genes", John Wiley and Sons, New York., pp. 354-355.

Nanney, D.L., 1958, Epigenetic control systems, <u>Proc. Nat. Acad. Sci.</u>, 44:712-717.

Oleson, P.O., 1978, On cyclophysis and topophysis, <u>Silvae Genetica</u>, 27:173-178.

Seeliger, R., 1924, Topophysis und zyklophysis pflanzlicher organe und ihre bedeutung fur die Pflanzenkullur, <u>Angew. Bot.</u>, 6:191-200.

UPTAKE CHARACTERISTICS OF SUGARS AND AMINO ACIDS

BY *Vitis vinifera* L. PROTOPLASTS

K.A. Roubelakis-Angelakis and P.A. Theodoropoulos

Department of Biology, University of Crete, and
Institute of Molecular Biology and Biotechnology
P.O. Box 1470, 711 10 Heraklion, GREECE

INTRODUCTION

Protoplasts offer the opportunity to take advantages of all the technologies, which are now available for plant genetic manipulation. Progress in that direction is clearly linked to the ability of protoplasts to regenerate whole plants.

In grapevine (*Vitis spp.*) parasexual breeding methods by using protoplasts could be significant because of the heterozygosity of genotypes and also the long juvenile period. Therefore, protoplast manipulation should be of great relevance for improvement of grapevine rootstocks and fruit cultivars (Roubelakis-Angelakis, 1986).

While preparing this report, aiming to review the available information on the isolation, the culture and the use of grapevine protoplasts for physiological, biochemical and molecular studies, the review by Krul (1988) was published. Therefore, a summary of our recent work is presented on the isolation of leaf protoplasts from *in vitro* grown axenic shoots of grapevine, and the characteristics of uptake of sugars and amino acids by these protoplasts (Theodoropoulos and Roubelakis-Angelakis, 1989, 1990a,b).

ISOLATION AND CULTURE OF GRAPEVINE PROTOPLASTS

Protoplasts were isolated from leaves of *in vitro* grown axenic shoots of *Vitis vinifera* L. cv. Soultanina (Fig. 1). Use of aseptically grown virus-free donor plants for isolation of protoplasts offers several advantages, especially in woody species, where plant regeneration from protoplasts is not an easy task. Some of them are the better reproducibility of the results, the juvenile stage of mother plants, the ommission of surface sterilization and antibiotics in the culture medium, and the lack of viral particles in the protoplasts which may affect their potentiality for subsequent cell wall regeneration, cell elongation and division.

The average size of grapevine protoplasts ranged from 12 to 44 μm and the yield from 25 to 30.10^6 protoplasts per g fresh tissue. Incubation of donor organs in the dark prior to the isolation of

Plant Aging: Basic and Applied Approaches
Edited by R. Rodríguez *et al.*
Plenum Press, New York, 1990

protoplasts for longer than 70 h resulted to higher yield. Also, degradation of cell walls in darkness resulted to greater although not significantly different yield compared to the cell wall degradation in the light. The ratio of cell wall degrading enzymes, Cellulase/ Macerozyme to tissue were 100/15 U per ml of incubation medium and approx. 30 ml per g fresh tissue. Optimum reaction time was 16-18 h (Theodoropoulos and Roubelakis- Angelakis, 1990a).

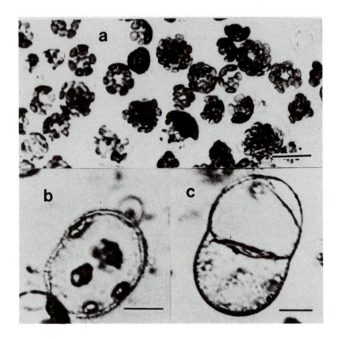

Fig. 1. Protoplasts isolated from leaves of aseptically grown plants cv. Soultanina. a. Protoplasts immediately after isolation; b. Cell wall formation and elongation after 7 days of culture; and c. First division. Scale = 20 μm.

Grapevine protoplasts have been isolated from several donor organs and cultivars (for review see Krul, 1988). However, plant regeneration has not been achieved yet. Conditions of protoplast isolation and culture may be among the factors, which determine the potentiality of protoplasts for regeneration. We tested several parameters such as plating density and addition of coconut milk, coconut meat extract and glutamine to the culture medium. The higher plating densities seemed to favour viability of protoplasts. Addition to the culture medium of 10% (v/v) coconut meat extract caused an increase in survival rate whereas glutamine had no effect (Theodoropoulos and Roubelakis-Angelakis, 1990a).

USE OF PROTOPLASTS FOR UPTAKE STUDIES

Despites extensive studies over a quarter of a century, there is still no consensus about the number and the nature of uptake mechanisms in plants (Nissen, 1989 and references therein). Recently, it was proposed that solutes are transported across the plasmalemma by multistate entities having carrier-like properties at low external concentrations and channel - like properties at high concentrations (Nissen, 1989).

Cultured protoplasts are exposed to a wide range of molecules. During culture of protoplasts, cell wall synthesis, cell elongation, cell division and plant regeneration largely depend on the potentiality of protoplasts for uptake and metabolism of the supplied nutrients. Carbohydrates and nitrogenous compounds are two of the main groups of constituents, which are included in the culture media.

Use of protoplasts for the study of physiological and biochemical processes, including transport characteristics of biological molecules offers several advantages compared to cell, tissue or organ systems: bulk diffusion and tissue penetration barriers are absent; the cell membrane is readily accessible for challenging with various analogues or membrane-modifying reagents or binding probes to facilitate carrier identification; and the cell wall associated hydrolases are absent (Guy et al., 1980; Guy et al., 1981; Lin et al., 1984). Use of protoplasts for uptake studies has also several disadvantages: the use of the cell wall degrading enzymes may alter the cell membrane properties; cell wall degradation may be incomplete in some cells or non uniform; donor tissue has to be carefully selected for anatomical, physiological and biochemical uniformity; and reproducibility of the results must be carefully tested and statistically evaluated.

UPTAKE OF SUGARS AND AMINO ACIDS BY GRAPEVINE PROTOPLASTS

Uptake of D-(U-^{14}C) glucose was performed as already described (Theodoropoulos and Roubelakis-Angelakis, 1990a). The incubation medium consisted of MES 0.02 M, mannitol 0.7 M and $CaCl_2 \cdot 2H_2O$ 0.01 M, pH 5.7. The substrate concentrations ranged from 2 μM to 15 mM.

Uptake of ^{14}C- and ^{3}H-L-arginine was also studied as described before (Theodoropoulos and Roubelakis-Angelakis, 1989). The concentration of arginine ranged from 2 μM to 15 mM.

In addition, unlabeled compounds, sugars and amino acids, were tested as substrate inhibitors or competitors. All unlabeled compounds were added at 100-fold the concentration of the labeled ones, 30 sec prior to adding the labeled substrate.

The use of leaf grapevine protoplasts to study the characteristics of uptake of sugars and amino acids was primarily done to test the functioning and the physiological integrity of plasma membrane. Uptake of labeled glucose and arginine from a 2 μM external concentration was linear for at least 60 min. The uptake rates were 130 and 100 pmoles per 10^6 viable protoplasts per h for glucose and arginine, respectively (Theodoropoulos and Roubelakis-Angelakis, 1989, 1990a). If we take into consideration that the average size of mesophyll protoplasts was $1.2 \cdot 10^{-14}$ m^3 and if we assume that the uptaken substrates were stored in single pools and were not metabolized, the internal concentrations of labeled glucose and arginine would be 10.8 and 8.3 μM, respectively. Concentration dependent uptake of labeled glucose by grapevine protoplasts was linear for concentrations higher than 1.5 mM; at lower concentrations a saturating pattern was observed (Theodoropoulos and Roubelakis-Angelakis, 1990b). At arginine concentrations from 0 to 15 mM, uptake dependence had a saturable component up to approximately 5 mM and a low affinity, non-saturable component for higher concentrations of external labeled arginine (Theodoropoulos and Roubelakis-Angelakis, 1989).

Use of the uncoupler carbonyl cyanide m-chlorophenylhydrazone (CCCP) at 0.1 mM concentration inhibited significantly both, the glucose and arginine uptake systems. The respective inhibition was 88.3% and 94.4%. The inhibitory effect may indicate that proton motive force is the main driving force for both tested compounds for entering plasmalemma. In addition, glucose uptake was not affected by NaF even at high concentrations (40 mM) whereas chlorhexidine at concentrations from 0.05 to 0.5 mM resulted to damage of cell membrane (Theodoropoulos and Roubelakis-Angelakis, 1990b). Both, NaF and chlorhexidine are inhibitors of the phosphotransferase system (PTS).
 During accumulation of sugars by the PTS system, phosphate is transferred from phosphoenol pyruvate to sugar and the latter enters the cell membrane phosphorylated, whereas when the carrier is energized by proton motive force or ATP, sugars enter the cell in an unchanged form.

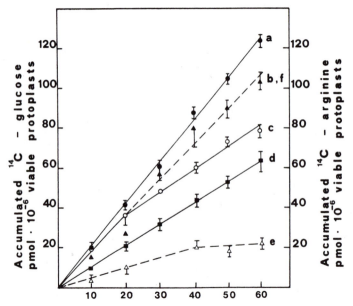

Fig. 2. Time course of accumulation of radiolabeled D-glucose (—) and L-arginine (- - -) by grapevine mesophyll protoplasts. Freshly isolated protoplasts were assayed in the light (a, b) or in the dark (c). Protoplasts were kept for 24 h in either dark or in the light (d). Protoplasts were kept for 24 h in the dark (e) and in the light (f). The external concentration of labeled substrates was 2 μM in all cases. Other conditions of the assays are described in the text. Means ± s.d. were plotted.

The uncoupler 2,4-dinitrophenol (DNP), and the plasma membrane ATPase inhibitors, N,N-dicyclohexyl carbodiimide (DCCD) and orthovanadate strongly inhibited arginine uptake by grapevine protoplasts, whereas the thiol-reacting compound N-ethylmaleimide (NEM) and arsenate, inhibitor of the ATP synthesis via glycolysis, inhibited uptake to a lesser extend than uncouplers and ATPase inhibitor (Theodoropoulos and Roubelakis-Angelakis, 1989).

The survival of cultured protoplasts largely depends on the conditions of culture. Usually, cultured protoplasts are kept in darkness. Grapevine leaf protoplasts were incubated in either the light or in the dark for 24 h, and their uptake capacity was assayed in the light. Uptake rate of ^{14}C-glucose by both, light and dark kept protoplasts decreased by approximately 50% (Fig. 2 d). Incubation of protoplasts in light for 24 h had no effect on the uptake rate of ^{14}C-L-arginine (Fig. 2f) whereas incubation in the dark resulted to an approximately 80% reduction of uptake rate (Fig. 2e). The differential effect of light incubation of grapevine protoplasts for 24 h on the uptake rate of glucose and arginine is not easily understood without further study. Actually, since leaf protoplasts photosynthesize, it was expected that incubation in light would not alter the rate of glucose uptake whereas incubation in dark would result to increasing glucose uptake rates, due to depletion of endogenous glucose. In pea protoplasts it was proposed that more than one mode of energy coupling for sugar transport may operate and a light-dependent factor acts synergistically to convert the transport system to its fully activated high affinity form (Guy et al., 1981). On the other hand, the catabolism of arginine is enhanced by increasing the carbohydrate content in leaf tissues. Thus, it is possible that incubation in light maintains high carbohydrate content in protoplasts which results to enhancement of arginine catabolism and therefore the observed uptake rates might reflect sink processes rather than entry mechanisms.

Use of the unlabeled sugars, fructose, mannose, galactose, xylose, arabinose, and lactose at concentrations 100-fold higher than that of ^{14}C-glucose showed that other hexoses than glucose inhibited glucose accumulation by approximately 60% whereas pentoses and disaccharides had no effect on glucose uptake. The results suggested that the carrier is not glucose specific, but probably hexose specific and the stereospecificity is closely related to C-1 of glucose molecule. The uptake rates and the specificity for the carrier of structural analogues of glucose, 3-O-methyl glucose, and 2-deoxyglucose were significantly lower compared to that of glucose (Theodoropoulos and Roubelakis-Angelakis, 1990b). A common glucose and fructose uptake system has been reported in other plants (Gogarten and Bentrup, 1984) whereas the uptake of glucose into beet root protoplasts was very specific (Getz et al., 1987).

Furthermore, the specificity of L-arginine uptake system was investigated by testing the effect of several basic, acidic and neutral amino acids at 100-fold concentrations. Arginine uptake was not affected by the acidic amino acids, glutamic and aspartic, whereas it was partially inhibited by basic and neutral amino acids, lysine, histidine, glycine, leucine, and methionine (Theodoropoulos and Roubelakis-Angelakis, 1989). In tobacco cells (Harrinhton and Henke, 1981) and in pea protoplasts (Dureja et al., 1986) two amino acid permeases were also reported.

The structural analogue of arginine, α-aminoisobutyric acid was not actively accumulated into grapevine protoplasts and methylamine, a structural analogue of ammonium did not exhibit inhibitory effect on arginine uptake.

Efflux of glucose from grapevine protoplasts following preloading with radiolabeled glucose showed a biphasic pattern which could be ascribed to compartmentation of the substrate into protoplasts.

SUMMARY

Leaves from virus-free axenic shoot cultures were proven an efficient source for isolation of grapevine (*Vitis vinifera* L. cv. Soultanina) protoplasts. Grapevine protoplasts accumulated ^{14}C-D-glucose and ^{14}C-L-arginine from a 2 μM external concentration at a rate of 130 and 100 pmol per 10^6 protoplasts per h, respectively. Kinetics analysis revealed biphasic uptake curves for both glucose and arginine. Incubation of protoplasts in the dark for 24 h caused a 50% and 80% reduction in glucose and arginine uptake rates respectively, whereas incubation in the light affected only glucose but not arginine uptake. The glucose carrier was related to the stereospecificity of the molecule whereas that of arginine seemed to be charge -dependent. Use of inhibitors indicated that energization of the uptake system for both substrates was mainly obtained via the proton motive force.

REFERENCES

Dureja, I., Mukherjee, S.G., and Prasad, R., 1986, Mechanism of L-arginine transport by pea protoplasts. J. Exp. Bot., 37: 549-555.

Getz, H.P., Knauer, D., and Willenbrink, J., 1987, Transport of sugars across the plasma membrane of beet root protoplasts. Planta, 171:185-196.

Gogarten, J.R., and Bentrup, F.W., 1984, Properties of a hexose carrier at the plasmalemma of green suspension cells from *Chenopodium rubrum* L. In: Membrane Transport in Plants, Cram, W.J., Janacek, K., Rydova, R., and Sigler, K., eds, Czechoslovak Academy of Sciences, Praha, pp. 183-188.

Guy, M., Reinhold, L., and Laties, C.G., 1978, Membrane transport of sugars and amino acids in isolated protoplasts. Plant Physiol., 61:593-596.

Guy, M., Reinhold, L., and Rahat, M., 1980, Energization of the sugar transport in the plasmalemma of isolated mesophyll protoplasts. Plant Physiol., 65:550-553.

Guy, M., Reinhold, L., Rahat, M., and Seiden, A., 1981, Protonation and light synergistically convert plasmalemma sugar carrier system in mesophyll protoplasts to its fully activated form. Plant Physiol., 67:1146-1150.

Harrington, H.M., and Henke, R.R., 1981, Amino acid tranport into cultured tobacco cells. I. Lysine transport. Plant Physiol., 67:373-378.

Krul, W.R., 1988, Recent advances in protoplast culture of horticultural crops: small fruits. Sci. Hortic., 37: 231-246.

Lin, W., Schmitt, M.R., Hitz, W.D., and Giaquinta, R.T., 1984, Sugar transport into protoplasts from developing soybean cotyledons. I. Protoplast isolation and general characteristics of sugar transport. Plant Physiol., 75: 936-940.

Nissen, P., 1989, Multiphasic uptake of potassium by corn roots. Plant Physiol., 89: 231-237.

Roubelakis-Angelakis, K.A., 1986, Protoplast Technology: A possibility for grapevine improvement. Proc. Symp. Biotechnology and Agriculture in Medit. Basin. Athens, June 1986, pp. 1-8.

Roubelakis-Angelakis, KA., and Zivanovitch, S.B., 1989, Morphogenetic responses of grapevine (*Vitis spp*) genotypes to plant growth regulators and culture media. J. Am. Soc. Hort. Sci., accepted.

Theodoropoulos, P.A., and Roubelakis - Angelakis, K.A., 1989, Mechanism of arginine transport in *Vitis vinifera* L. protoplasts. J. Exp. Bot., 40 (220): 1223-1230.

Theodoropoulos, P.A., and Roubelakis-Angelakis, K.A., 1990a. Progress in isolation and culture of leaf protoplasts from *in vitro* axenic shoots of *Vitis vinifera* L. shoot cultures. Plant Cell, Tissue Organ Cult., in press.

Theodoropoulos P.A., and Roubelakis - Angelakis, K.A., 1990b, Glucose transport in *Vitis vinifera* L. protoplasts. Plant Physiol., submitted.

IMPROVEMENT OF MICROPROPAGATION METHODS LINKED TO BIOCHEMICAL

PROPERTIES DURING "IN VITRO" CULTURES

Anne-Marie Hirsch and Dominique Fortune

Laboratoire d'Histophysiologie Végétale-URA-CNRS-1180
Université Pierre et Marie Curie
12, rue Cuvier, 75005 Paris, France

In vitro culture is nowadays an important biotechnology method making possible plant micropropagation and creation of new varieties of economic interest. Our laboratory was one of the first in which these techniques were developed, under Gautheret's impetus. At the beginning, the main interest of in vitro culture methods was essentially a fundamental one, as it offers new possibilities for tissues to be studied in absence of correlations with the plant and to investigate the morphogenic properties of the cell. This last point remains of value, because it offers the possibility to regenerate a whole plant from isolated and in vitro rejuvenated cells able again to produce meristematic tissue and shooting apices. Recent research in this area is naturally devoted to biochemical markers of morphogenic potentialities of cultured cells and tissues, but a new field of investigation is gaining ground, namely the study of variability of the hereditary structures during development and of the control of genetic expression. As a matter of fact, if stable enough alterations of the hereditary information, both nuclear and cytoplasmic, occur in certain cellular strains, the plants originated from these cells will keep the trace. Thus, the "somaclonal variation", which can appear when several micropropagation methods are performed, could be due to the existence of cellular strains with different genotypes. In vitro culture then unveils the preexistent differences in their hereditary organisation and expression.
We shall first summarize the methods of micropropagation set in use in our laboratory.

I MICROPROPAGATION METHODS

In vitro plant multiplication can be performed by microcutting. This method consists in a transplantation of nodes in order to obtain rooting. It allows a rapid reproduction of large numbers of individuals showing particular qualities. This method has been applied in our laboratory to Kiwifruit (1), grape (2) and Jojoba. Velayandom and al. (1) stressed the importance of the choice of cytokinins and of the mineral solution in the bud bursting medium.
The most efficient cytokinin is zeatin which, in association with a relatively poor mineral solution, Heller's (3), gives the best result (fig.1).

Plant Aging: Basic and Applied Approaches
Edited by R. Rodríguez *et al.*
Plenum Press, New York, 1990

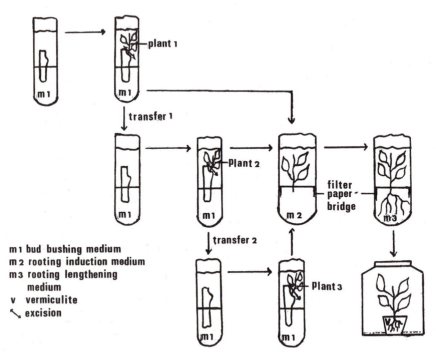

m1 bud bushing medium
m2 rooting induction medium
m3 rooting lengthening
    medium
v vermiculite
↖ excision

B-Regeneration of kiwifruit by tissue culture of stamen filament :
Tripathi and Saussay (1980)

Picture 1-Stamen lengthening

Picture 2-Rooting clones formation

Picture 3-Bud differentiation

Picture 4-Whole plant regeneration

Picture 5-Plant regeneration from
buds neoformation on leaf tissues

Picture 6-Plant regeneration from
somatic embryos originated in cultured cells of flower receptacle

Picture 7-Budding callus lines
established from endosperm of
seeds collected in fruitful males

Picture 8-Ovary development of
staminate flower in vitro
initial diameter:1mm
final diameter:15mm

Picture 9-Ovary development of
pistillate flower in vitro
initial diameter:1.5mm
final diameter:30mm

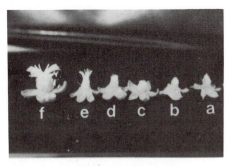

Picture 10-Flower morphology
studied in the orchard (P Blanchet)
a:male, f:female, b,c,d,e:males
with trend to hermaphrodism

In contrast, juvenile aspects appear in leaves, e.g.long shape and indented edges, when grown on Murashige and Skoog solution (4). An original micropropagation method has been evolved in our laboratory by Tripathi and Saussay (5) who observed that staminal filaments from male and female Kiwi flowerbuds are strongly rhizogenic in vitroculture. Isolated roots derived from male and female clones are producing globular callus which synthetizes chlorophyll and gives raise to buds when cultured during 45 days. In a next step, the callus give birth to leaves, and roots appeared amongst neoformed buds. Whole plants are generated when these buds are cultured in presence of auxin (pictures 1, 2, 3, 4).

Creation of new varieties cannot succeed without passing through the stage of callus, where cellular rejuvenation is accompanied by modifications of the genome, called "somaclonal variation". The somaclonal variation comes to expression when the plant undergoes regeneration from callus ; the result of the regeneration may occur under form of buds ; in the case of Kiwifruit, whole plants are very easily regenerated in our laboratory from buds grown on leaf tissues (pictures 5). A similar regeneration has also been achieved with somatic embryos derived from floral receptacles (picture 6). Unique, who is now working on Jojoba, is devicing a regeneration method by somatic embryogenesis.

In several Kiwi orchards of southwestern France, there are fruitful male plants, from which tissular strains have been established. We have harvested seeds from these fruits and for more than 3 years, we have established tissular strains from cultured endosperm and embryos 7). These strains, originated from a particularly juvenile material, are provided with remarkable organogenic and embryogenic properties, remaining constant all along successive subcultures for longer than 3 years. Numerous individuals have been easily regenerated from these strains and are now under study at the I.N.R.A. as part of a varietal amelioration program of Kiwi.

II CONTRIBUTION OF IN VITRO CULTURE TECHNIQUES TO THE CONTROL OF

SEX EXPRESSION IN DIOECIOUS PLANTS

Durand and al. (6), working on Mercurialis annua, suggested that the experimental control of the sex expression by phytohormones in diœcious plants leads to the repression of the genes which induced the normal organogenetic program and to the activation of the genes which induced the opposite program. Tissue culture seemed to be an useful tool to gain better understanding of diœcism. Our material was Kiwifruit, and tissue culture technique made also possible the study of the monœcious trends of Kiwifruit. Our attempts to produce viable pollen by mean of pistillate flower anther culture were unsuccessful : a swelling of the anthers was observed associated with a change of coloration, pollen tetrad formation occured but the microspores didn't divide. On the other hand, we succeeded in vitro-development of the staminate flower's ovaries, otherwise abortive. Gynæcial development stops in vivo after the stigma initiation in the staminate flowers. In vitro, the ovary's successful development was obtained when the ovaries were collected after stigma initiation. The morphogenic responses of the incubated ovaries varied with the degree of flower differentiation and also with the growth factor combinations added to the nutrient medium.

The most important development, a fifteenfold increase (picture 8)

was obtained for ovaries harvested on the 24th day before blossoming.

The pistillate flower ovaries were also cultivated in vitro, and developed parthenocarpic fruits. The maximal development, a twentifold increase (picture 9), was observed when the ovaries were harvested and incubated at earlier stages than these from staminate flowers.

These results, on the whole, show that it seems quite possible to obtain a hermaphrodite cultivar from male plant, a similar result not being totally excluded from female plants.

P.Blanchet, conducting a comprehensive survey of natural populations of "five years old seedlings" in the orchards of southwestern France noticed that there was in fact a full range of variation between sheer pistillate or staminate flowers (picture 10) during the three first blossoms of young kiwifruits (7).

Blanchet's observations allow us to understand how fruitful male plants in certain orchards could appear, and open prospects for obtention of hermaphrodite cultivars from these plants.

As for us, in collaboration with P.Blanchet, a peroxidasic test was evolved, allowing detection of potential fruitful males before first blossoming ; this test resulted from our precedent fundamental research in the field of peroxidase related to dioecism of kiwifruit (8).

III BIOCHEMICAL MARKERS AS EARLY INDICATORS OF TISSUES MORPHOGENIC

CAPACITIES

Rooting-Testing the morphogenic capacities of the tissue is very important all along the different stages of micropropagation. Especially, one needs to know if stem explants are sufficiently rejuvenated for rooting, e.g. in micropropagation of certain gymnosperms. In our laboratory, determination of peroxidasic activity was used as a practical test of rooting capacities of the stem explants of Kiwifruit and mostly of Sequoia sempervirens (9). Moncousin (10) was among the first who used the peroxidasic test for rooting of Cynara cardonculus. It consists in a peroxidase determination of stem explants after each subculture, which allows to count how many subcultures on a cytokinin-rich medium are needed before passage on a rhizogenic, auxin-rich medium.

Shooting ability-Brown and al. (11-12) found that there is a specific osmotic requirement for shoot formation : shoot forming tissue maintained a greater osmotic potential than proliferating tissue. Carbohydrates are specifically involved in the osmotic requirement. The osmotic adjustment takes place very early in the culture, as a consequence of mitochondrial activity for ATP production.

Somatic embryogenesis-Somatic embryogenesis draws more attention these last years as a micropropagation method, leading to numerous studies in the area of biochemical markers, mainly of enzymatic nature. Already in 1973, Lee and Dougall (13) showed that glutamate-deshydrogenase was a reliable marker of the different stages of somatic embryogenesis in wild carrot cells. Following year, Wochok and Burleson (14) made obvious the presence of anodic isoperoxidases, characteristic of wild carrot embryogenic tissues.

Everett and al. (15) showed in maize tissue that embryogenic cultures could be distinguished from callogenic cultures through esterase and glutamate-deshydrogenase (GDH). With regard to GDH, the enzymatic pattern of embryogenic callus is characteristic of zygotic embryos, while budding callus, whatever their origin, have the same pattern than stem tissues of the plantlet. The cellular strain, recognized as embryogenic according to histological observations, expresses a GDH pattern which is characteristic of zygotic embryos developed in vitro. Sanchez de Jimenez and al. (16) showed on the same material, e.g. maize callus tissue, that measurement of GDH activity was sufficient to characterize embryogenic tissues, making useless the determination of the isoenzymatic pattern.

As somatic embryogenesis induction needs most often the addition of a strong auxin, 2-4-D, to nutrient medium, it becomes easy to understand that peroxidases, and especially anodic isoperoxidases, play a regulating role for the catabolism of auxin, and are controlling the endogenous auxin/cytokinin ratio in the tissues. In matter of GDH, Loyola-Vargas and al. (17) showed on maize callus tissue that there would exist GDH isoenzymes specific for the different tissues. The isozyme pattern could be modified according to tissue nutritional needs and to differentiation stages of the tissues. Stuart and Strickland (18, a and b) showed on Medicago sativa L. that impairments of nitrogen metabolism were outstanding factors of embryogenesis. It is possible, in a variety of species, to enhance somatic embryogenesis in callus by putting them under osmotic shock, e.g. through addition of glucose to the nutrient medium. Handa and al. (19) have found that this osmotic shock was accompanied by accumulation of free proline in the cells. Two similar embryo specific proteins have also been found in embryogenic cultures of carrot (20) and pea (21) and are used as molecular markers.

IV GENETIC VARIATIONS

Genetic variations (somaclonal and gametoclonal variation) are found in plants regenerated in vitro (22). Such variations are source of new varieties and are due to (I) the genomic changes which accompany the differentiation in the plant (23) and to (II) the preexisting alterations of genome which are expressed in culture. In a recent work (24) carried on Tobacco cells, it could be evidenced that in vitro culture unveils preexisting differences in the initial cells used as material for protoclones obtention. For instance, more than 50% of the plants descending from leaf protoplasts undergo mutations concerning a variety of characters, especially reproductive (sterility in males). In several protoclone strains, among them sterile males, alterations could be evidenced in the same region of mitochondrial DNA.

REFERENCES

1- L. VELAYANDOM, A.M. HIRSCH AND D. FORTUNE -1985-
Propagation du Kiwi, Actinidia chinensis L. Planchon, par microbouturage in vitro de nœuds-C.R. Acad. Sci. Paris, série III, n° 12, 597-600.

2- M. MONNIER, O. FAURE and A. SIGOGNEAU -1989- Somatic embryogenesis of Vitis -I.A.P.T.C.- The impact of biotechnology in Agriculture, July 10-12th Amiens, France.

3- R. HELLER -1953- Recherches sur la nutrition minérale des tissus végétaux cultivés in vitro-Ann. Sc. Nat., 14, 1-223.

4- T. MURASHIGE and F. SKOOG -1962- A revised medium for rapid growth and bioassays with tobacco tissue cultures-Physiol. Plant., 15, 473- 497.

5- B.K. TRIPATHI and R. SAUSSAY -1980- Sur la multiplication végétative de l'Actinidia chinensis L. Planchon (chinese gooseberry) par culture de racines issues de filet staminaux-C.R. Acad. Sci. Paris, 291, 1067- 1069.

6- R. DURAND and D.B. DURAND -1984- Sexual differentiation in higher plants-Physiol. Plant., 60, 267-274.

7- A.M. HIRSCH, D. FORTUNE and P. BLANCHET -1989- Study of diœcism in kiwifruit, Actinidia deliciosa Chevalier-Acta Hort., under press.

8- A.M. HIRSCH and D. FORTUNE -1984- Peroxidase activity and isoperoxidase composition in cultured stem tissue callus and cell suspension of Actinidia chinensis-Z. Pflanzenphysiol., 113, 129-139.

9- B. VERSCHOORE-MARTOUZET -1985- Etude de la variation topophysique au cours du clonage de Sequoia sempervirens (Endlicher) -Thèse-Paris.

10- C. MONCOUSIN -1982- Peroxidase as tracer for rooting improvement and expressing regeneration during in vitro multiplication of Cynara cardunculus-In Fujiwara A.-Ed. Plant tissue culture-Jap. Assoc. Plant Tissue Culture, Tokyo.

11- D.C.W. BROWN, D.W.M. LEUNG and T.A. THORPE -1979- Osmotic requirement for shoot formation in tobacco callus-Physiol. Plant., 46, 36-41.

12- D.C.W. BROWN and T.A. THORPE -1980- Changes in water potential and its components during shoot formation in tobacco callus Physiol. Plant. 49, 83-87.

13- D.W. LEE and D.K. DOUGALL -1973- Electrophoretic variation in glutamate deshydrogenase and other isozymes in wild carrot cells cultured in presence and in absence of 2-4-D-In Vitro, 8, 347-352.

14- Z.S. WOCHOK and B. BURLESON -1974- Isoperoxidase activity and induction in tissues of wild carrot : a comparison of proembryos and embryos-Physiol. Plant., 31, 73-75.

15- N.P. EVERETT, M.J. WACH and D.J. ASHWORTH -1985- Biochemical markers of embryogenesis in tissue culture of the maize inbred B 73-Plant Science, 41, 133-140.

16- E. SANCHEZ de JIMENEZ, M. VARGAS, R. AGUILAR and E. JIMENEZ -1988-
Age dependent responsiveness to cell differentiation stimulus in maize callus culture-Plant. Physiol. Biochem., 26(6), 723-732.

17- V.M. LOYOLA-VARGAS and E. SANCHEZ de JIMENEZ -1984- Differential role of glutamate deshydrogenase in nitrogen metabolism of maize tissue-Plant Physiol., 76, 536-540.

18- D.A. STUART and S.A. STRICKLAND -1984-
somatic embryogenesis from cell cultures of Medicago sativa L.
     a- The role of amino-acid addition to regeneration medium.
Plant. Sci. Lett., 34, 165-174.
     b- The interaction of amino-acids with ammonium- Ibidem, 175-181.

19- S. HANDA, R.A. BRESSAN, A.K. HANDA, N.C. CARPITA and P. HASEGAWA
-1983- Solutes contributing to osmotic adjustment in cultured plant
cells adapted to water stress-Plant Physiol.,73, 834-843.

20- Z.R. SUNG and R. OKIMOTO -1983- Coordinate gene expression
during somatic embryogenesis in carrots-Proc. Nat. Acad.Sci, 80,
2661-2665.

21- S. STIRN and H.J. JACOBSEN -1987- Marker proteins for embryonic
differentiation patterns in pea callus-Plant Cell Rep., 6, 50-54

22- D.A. EVANS and W.R. SHARP -1986- Applications of somaclonal
variation. Biotechnol.,4, 528-532.

23- W. HALPERIN -1986- Attainment and retention of morphogenic
capacity in vitro-In "Cell culture and somatic cell genetics of
plants", vol.3, I.K. Vasil Ed., p.3-47, Acad. Press New-York.

24- X.Q.LI, P. CHETRIT, C. MATHIEU, F. VEDEL, R. de PAEPE, R. REMY
and   F. AMBARD -1988- Regeneration of cytoplasmic male sterile
protoclone of Nicotiana sylvestris with mitochondrial variations.
Curr. Genet., 13, 261-266.

AGE DEPENDENCE OF DIFFERENT COMPONENTS OF VARIANCE

Manfred Huehn[1] und Jochen Kleinschmit[2]

[1] Institute of Crop Science and Plant Breeding
University of Kiel
D-2300 Kiel, F.R. Germany

[2] Lower Saxony Forest Research Institute
Dept. of Forest Tree Breeding
D-3513 Staufenberg-Escherode, F.R. Germany

INTRODUCTION

Forest tree breeding in most species faces comparatively long testing times. Therefore efforts have been made since long to get reliable early predictions of tree performance. These predictions are biased by different biological influences which restrict the reliability. Some of the most important ones are:

1. Maternal influences due to seed weight, speed of germination etc.

2. Genotype x site interactions

3. Different developmental rhythms due to individual genotype combined with genetic differences in maturation stage.

In 1923, Cieslar already described the significant influence of seed size on growth of oak progenies up to an age of 10 years. This influence is especially strong in species with big seed, it already disappears during the first year in species with small seed like birch. This fact has to be taken into account for early predictions. Seed size influences can be eliminated by analysis of covariance.

Genotype x site interactions bias early predictions from nursery evaluations, especially if the nursery site and climatic conditions deviate considerably from the later on planting site. However, they have consequences, too, if results from few testing sites shall be extrapolated to larger areas. If the nursery site and the testing site are comparable in soil and climate, this results in much more precise predictions as Schober and Fröhlich (1967) could demonstrate for a larch provenance experiment established by Oelkers. In this case, provenance performance at age 32 could be predicted reasonably well from the height at age 1.

Contrary to this, the oldest Douglas fir provenance experiment from 1912 planted on different sites in the United States showed considerable rank

Plant Aging: Basic and Applied Approaches
Edited by R. Rodríguez *et al.*
Plenum Press, New York, 1990

changes even after age 17 (Silen, 1965). On clonal level, rank changes on different sites are even more expressed than on provenance or progeny level. This is due to the lack of genetic variation and therefore reduced adaptive potential. It is however interesting that single clones show high stability over a wide range of sites (Figure 1). This seems to be correlated with a high degree of heterozygozity. Adaptive potential is an important genetic trait (St. Clair, 1986).

Differences in developmental rhythm result in rank changes with increasing age. These rank changes can be described by correlation coefficients. Changes in social position of trees have been described for Norway spruce and Scots pine by Busse (1930), Liebold (1965), Leibundgut (1971), Abetz (1972) and others. Studies in advanced stands more than 30 years old show that rank changes in pure stands are usually less expressed than generally expected and that most changes occur in the middle height class with a trend to decrease. The dominant trees remain dominant, the small ones remain small.

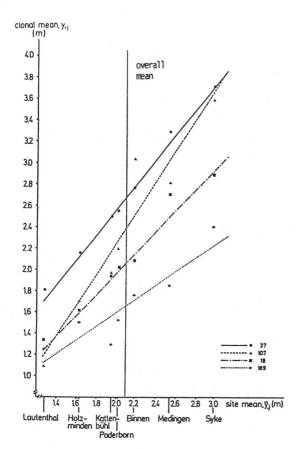

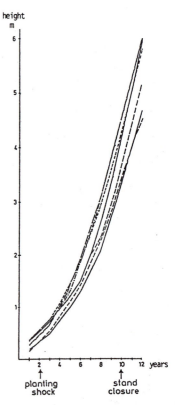

Figure 1. Plot of clonal means for four Norway spruce clones against site means of 7 contrasting sites with corresponding regression lines

Figure 2. Height growth curves of 7 Norway spruce clones (clonal means): Rank changes due to planting shock and stand closure

During stand development there are 2 critical phases where most of the rank changes occur (Fig. 2): Directly after plantation establishment substantial rank changes occur due to planting shock. With the closure of the stand and increasing intraspecific competition a second maximum of rank changes can often be seen due to selection pressure. The direction of changes is from above to below (Delvaux, 1981; own studies). This process ends up in a stabilized ranking since environmental conditions (light, root extension) support the resulting differentiation.

From a competition point of view, these questions have been early studied by Stern (1968), Sakai and Hatekayama (1963) and Huehn (1969, 1970, 1971).

From the point of view of early selection, many authors addressed this problem (e.g. Schmidt, 1962; Nanson, 1968, 1974; Burdon and Sweet, 1975) which is of central importance for forest tree breeding. Numerous early selection experiments have been carried out, generally resulting in genetic gain for advanced age from an intensive early selection. However, the gain was quite different from one species to the other. As well seed size as selection and testing conditions had implications. But not only seed size and site heterogeneity from year to year, also differences in resistance to pathogenes and differences in individual response (differential gene activities?) influenced prediction preciseness. In clonal propagation, an additional source of variation is the difference in the maturation stage of different cycles of propagation (St. Clair et al., 1985). In Norway spruce, branch length seems to decrease with increasing maturation state and initial growth potential may be affected by the propagation cycle.

Summarizing these introductory remarks, we can conclude that rank changes in growth of populations and of individual trees within populations going along with maturation are a general phenomenon. These rank changes have environmental and genetic reasons as well. These are in close interrelationship. These rank changes have implications on heritability estimates and efficiency of early selection.

AGE DEPENDENCE OF DIFFERENT COMPONENTS OF VARIANCE

The observation of changes in statistical parameters like variances and covariances is the easiest and most obvious approach to these phenomena, although they give no biological explanation of the reasons for these changes. Some of them have been discussed in the introduction.

So we can regard these statistical parameters only as a first and very superficial approach to the problem. Efficiency of early selection is very much dependent on the size of juvenile-mature correlations and the stability of heritability estimates. This efficiency has far reaching consequences for any tree breeding program. The problems have been repeatedly approached by different authors. Some of the more recent papers by Namkoong et al. (1972, 1976), Franklin (1979), Robinson et al. (1984), Loo et al (1984), Foster (1986), Gill (1987), Riemenschneider (1988), Magnussen (1988) and Lambeth et al (1983) show that the situation may be different from one species to the other. Mostly worse predictions were received for wood quality from very juvenile wood characteristics and for height growth if the juvenile environment was very different from the environment of the more mature stage, e.g. greenhouse as contrasted to forest site conditions.

In the following chapters we will look to variances of a set of Norway spruce clones on contrasting sites, to correlations in clonal mixtures, functional relationships between different ages and the combination of existing functions.

Age Dependence of Variance Components

In a Norway spruce clonal test, established with 5 clones on 5 extremely contrasting sites in 1967 and remeasured for plant height 10 times until 1981 we estimated the relative importance of the components of variance for "locations", "clones", "locations x clones ", "blocks", and "experimental error" dependent on the time. The results show that in this study "blocks" and "interactions" are of minor importance, accounting to less than 5 % of the total variation. "Clones" account for roughly 10 %. These three sources of influence remain more or less constant over time. Considerable changes occur in the components for "locations" and "experimental error". The first one increases quickly until it reaches a plateau at about 70 %, the last one decreases correspondingly and ends up after few years at 15 %. All values are quite stable at the end of the time of measurement (Figure 3).

Since all plants of the field test have grown initially in the same nursery, the differences in growth at the time of plantation establishment can be explained by clonal influences and experimental error mainly. From this time onwards, the influences of the location (climate, soil conditions) on the further on growth of the clones can act. These are initially only disturbed by the planting shock. The clones now start to collect the site information, which is variable in space and over the years and reflects this information in health and growth (physiological homeostasis). The annual influences can act quite differently on the different clones, as to be seen e.g. from the variance components for annual ring width, estimated by Lewark (1981) for these experiments, ranging from 7 % to 53 % for clones, depending on the year. During further growth development, the annual variation however levels out to the location mean. Since the volume of the trees increases steadily, the relative contribution of the annual environmental variation to the growth is finally so small that this has nearly no implications on the proportion of variability (Huehn et al., 1987).

Of special interest are the implications for forest tree breeding

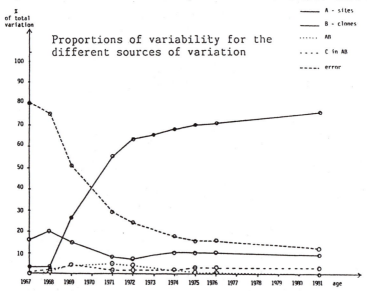

Figure 3. Proportions of variability for the different sources of variation (estimated from a Norway spruce clonal test on 5 contrasting sites)

methodology which can be derived from the dependence of "within - site clonal h²" over time for the trait tree height.

For this ratio

$$\sigma^2_{\text{clone}} / \sigma^2_{\text{clone}} + \sigma^2_{\text{error}}$$

the numerical values in the different years of measurement are:

0.16 (1967); 0.21 (1968); 0.23 (1969); 0.21 (1971); 0.23 (1972); 0.30 (1973); 0.35 (1974); 0.38 (1975); 0.38 (1976); 0.42 (1981).

These numerical values show a tendency towards an increase with time (Huehn et al., 1988). There are some interesting analogies to results in the recent paper of Foster (1986) with loblolly pine and to the previous paper of Franklin (1979) with Douglas fir, ponderosa pine, loblolly pine and slash pine. See also the study of Riemenschneider (1988) with Pinus banksiana, where the age related fluctuations in heritability show no clear tendency with time.

For the trait "total height increase = growth" too, the variance components for "clones","locations", "clones x locations", "blocks" and "experimental error"have been estimated and converted into proportions of variability. The main results are: The sources of variation "clones", "clones x locations" and "blocks" are all of minor importance $\leq$ 10%. The predominant sources "locations" and "error" show no clear tendency dependent on time. Both continue with moderate fluctuations (Huehn et al., 1988). Compared to the development of the components derived from the overall growth, the components derived from the annual growth fluctuate considerably and the fluctuation does not decrease with age. Even the ranking between error and locations may change.

Whether or not a stabilization of these proportions of variability may be possibly reached at higher ages cannot be answered by these experimental investigation. Furthermore, the sample of clones as well as the sample of sites are, of course, both extremely restricted. They do not represent the available variability of clones and sites. Therefore, all attempts to generalize the obtained empirical results must be handled very cautiously.

Correlations Dependent on Number of Clones in Clonal Mixtures (Huehn and Kleinschmit, 1986 and Huehn, 1987)

An increasing interest in clonal forestry and essential improvements of methods of vegetative propagation lead to intensive breeding work of multiclonal varieties.

The aim here is to give some theoretical approaches and numerical results concerning the necessary number of clones in clonal mixtures with regard to juvenile-mature correlations.

Relations between the juvenile-mature correlations of clonal mixtures and their number of clones can be simply described quantitatively and conclusions can be derived.

If we denote the juvenile-mature correlation based upon the values of the single clones (= clone means at juvenile and mature ages) by r_E, then we may ask whether or not a low r_E can be increased by using mixtures of clones. Introducing r_M = juvenile-mature correlation based upon the means of mixtures of different clones we usually will have $r_M > r_E$.

Furthermore, r_M will increase with an increasing number of clones in the clonal mixtures.

If we have $r_M > r_E$, an increase of the juvenile-mature correlation and, therefore, an improvement of the efficiency of early testing can be obtained by using mixtures instead of single clones. But what are the necessary numbers of clones so that r_M will be sufficiently high to enable an early selection of such clonal mixtures with a sufficient precision and efficiency?

The main purpose of this contribution is to give an explicit expression for the function $r_M = r_M$ (n) with n = number of clones in the clonal mixture. For any given required numerical value of r_M this equation $r_M = r_M$ (n) can be solved for n.

Additionally, experimental results on clonal tests of Norway spruce shall be reported which are in complete agreement with these theoretical results.

To see how the correlations change, groups of clones of different size have been formed (1-100 clones). By this procedure the genetic variance slowly increases within the groups and the expected gain between groups decreases. The correlations between ages 3 and 12 of 0.30 using single clones increases very rapidly if the group size increases: 0.80 with groups of 10 clones, 0.90 with groups of 20 clones, 0.95 with groups of 50 clones, 0.99 with 65 clones and 1.00 for groups of 100 clones (Kleinschmit, 1983).

Simultaneously, the differences in height between the group means and therefore the gain decrease. However, this gain can be realized at an early stage if the correlations are sufficiently high. In this experimental study, the juvenile-mature correlation reaches an approximate constant value for groups of 20-30 clones. Increasing this number results in no further significant changes of the correlation. Huehn (1987) developed a function for r_M = juvenile – mature correlation based upon the means of mixtures of different clones:

$$r_M = \frac{n^k\ r_E}{1+(n^k-1)r_E}$$

with k = const. k can be estimated from empirical data or derived by theoretical assupmtions. For a given required numerical value of r_M the necessary clone numbers n can be calculated with (1) by solving for n (for given r_E and k). Some results are presented in Table 1.

Table 1. Necessary numbers n of clones in clonal mixtures for different numerical values of r_E, r_M and k.

k r_E	r_M = 0.70					r_M = 0.80					r_M = 0.90				
	0.6	1.0	1.4	1.8	2.2	0.6	1.0	1.4	1.8	2.2	0.6	1.0	1.4	1.8	2.2
0.10	160	21	9	6	4	393	36	13	8	6	1517	81	24	12	8
0.20	42	10	5	4	3	102	16	8	5	4	393	36	13	8	6
0.30	17	6	4	3	3	42	10	5	4	3	160	21	9	6	4
0.40	9	4	3	3	2	20	6	4	3	3	77	14	7	5	4
0.50	5	3	2	2	2	11	4	3	3	2	39	9	5	4	3
0.60	3	2	2	2	2	6	3	3	2	2	20	6	4	3	3

Table 2. Empirical r_M-values (Kleinschmit, 1983) and theoretical estimates for r_M for different clone numbers n.

n	empirical r_M-values	estimates of r_M (by (1) with k=1)
1	0.31	0.31
10	0.81	0.82
20	0.91	0.90
50	0.95	0.96
65	0.99	0.97
100	1.00	0.98

There is some experimental evidence (Huehn, 1987) that the simplification k=1 may be a sufficient approximation.

The application of (1) (for k=1) to experimental results (Kleinschmit, 1983) gives a nearly perfect fit (Table 2). The experimental values are given in Fig. 4.

To characterize the approximate numerical magnitude of necessary clone numbers in clonal mixtures which may be relevant for practical applications, we assume k=1 and a required numerical level of r_M = 0.90. From Table 1 one obtains: n = 36 for r_E = 0.20, n = 21 for r_E = 0.30, n = 14 for r_E = 0.40, n = 9 for r_E = 50 and n = 6 for r_E = 0.60.

Extensions and generalizations of these findings have been discussed from a theoretical point of view by Huehn (1987).

Age-age correlations of Norway spruce clones

Based on an experimental study (see: Huehn et al., 1987) with five Norway spruce clones grown under five extreme variable site and climatic conditions with height measurements at nine different ages, age-age correlations were calculated and discussed.

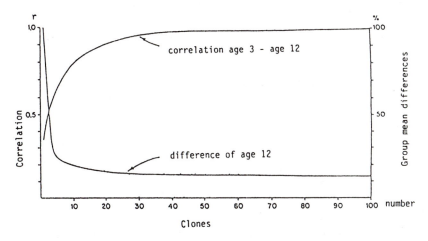

Figure 4. Changes of correlation and differences between group means for height with group size for Norway spruce clones

The objectives are:

1. to derive some results on the interrelationships of phenotypic variances and phenotypic covariances in different ages (time trends in variances and covariances)

2. to develop a new formula for age-age correlations dependent on age.

The experimental results of the standard deviations and the age-age covariances as well as the age-age correlation coefficients are presented in Fig. 5. Results can be summarized as follows:

1. The graphs indicate a strong linearity of the standard deviations and, additionally, of the covariances dependent on the age (Fig. 5).

2. The correlation coefficients show a rapid decrease in the first years, while this decrease diminishes and finally tends to more or less asymptotic behaviour towards a specific level (0.23)(Fig. 5).

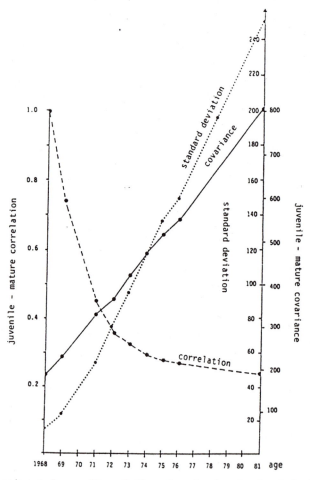

Figure 5. Experimental results of the phenotypic standard deviations, the age-age covariances and the age-age correlation coefficients

This linear dependence of the standard deviations and the age-age covariances on the age can be demonstrated by calculating simple linear regression. These numerical results and a detailed discussion of the resulting conclusions have been presented in Huehn and Kleinschmit (1988). For all theoretical derivations and all numerical calculations we, therefore, refer to this publication.

The main result of these investigations is the following explicit formula for age-age correlations dependent on age:

$$r = A \cdot \frac{1 + B \cdot age}{1 + C \cdot age} \qquad (2)$$

with specific constants A, B and C and age = difference between late (=mature) age minus early (=juvenile) age (in years).

This formula leads to an extraordinary good agreement between observed and predicted age-age correlations. This fit is better than with the formula given by Lambeth (1980) $r = a + b \ln x$ where $x = t/T$ with t = juvenile age and T = mature age.

These numerical results are summarized in Table 3 (see: Huehn and Kleinschmit, 1988)

These results, of course, are only valid for the specific experimental situation of the material which has been analysed in this study. Generalizations need a broader base of experimental material and comparison.

THEORETICAL EXTENSION: Combination of (1) and (2)

For all following theoretical investigations we assume equal proportions of the clones in the clonal mixture.

Additionally, all theoretical studies are based on a certain, given time interval from time T_1 (= defined as 'juvenile age') until time T_2 (= defined as 'mature age'). Extended experimental and theoretical studies on the question "What is the optimum selection age?" have been published

Table 3. Experimental values of the age-age correlations and the theoretically expected values (predicted by (2) and predicted by Lambeth's formula).

age	age-age correlations		
	exp. value	prediction by (2)	prediction by Lambeth's formula
1	0.74	0.74	0.66
3	0.45	0.40	0.48
4	0.36	0.35	0.42
5	0.33	0.32	0.37
6	0.30	0.30	0.33
7	0.28	0.28	0.30
8	0.27	0.27	0.27
13	0.24	0.24	0.15

in the literature: See, for example Kang (1985). But, in the derivation of (1) both ages are assumed to be given and they are related with each other where this relation is described quantitatively by r_E and r_M.

The correlation coefficient from (1) r_E can be explicitly expressed by using (2).

Combination of both formulae leads to an expression which can be solved for 'age'. One obtains the following explicit formula:

$$\text{'age'} = \frac{An^k\ (1-r_M)\ -\ (1-A)\ r_M}{(C-AB)\cdot r_M\ -\ ABn^k\ (1-r_M)} \qquad (3)$$

This estimate of 'age', therefore, depends on 6 parameters: n, k, r_M, A, C, B.

A necessary and sufficient condition for a positive numerator of (3) is A > r_E which reduces to C > B. The denominator will be positive if the condition Cr_E > AB holds.

In most situations of practical applications the first condition A > r_E will be valid.

Estimates of selection ages can be only calculated by this 'age'-formula (3) if the denominator is positive, that means for Cr_E > AB.

Numerical example:

We apply the same numerical example as in the discussion of (2).

The numerical estimates of the unknown constants A, C and B are:

$$\hat{A} = 2.96 \quad \hat{C} = 4.06 \quad \text{and} \quad \hat{B} = 0.26$$

There is some experimental evidence that the simplification k = 1 may be a sufficient approximation in the field of practical applications (Huehn and Kleinschmit, 1986; Huehn, 1987). In this numerical example we, therefore, use k = 1.00.

Furthermore, we have to require a certain numerical level for the correlation coefficient r_M. This value for the juvenile-mature correlation of clonal mixtures which have been composed by n different clones must be sufficiently high to ensure an efficient early selection. For the numerical calculations of this numerical example we use: r_M = 0.60, 0.70, 0.80 and 0.90.

With increasing 'age' (=difference (years) between late and early ages) the correlation coefficient r_E between the juvenile and mature ages, of course, decreases and a certain required numerical level of r_M can be only realized by increasing the number n of clones in the clonal mixture.

Some numerical results are summarized in Table 4.

The consequence of these findings in practical terms is to use clonal mixtures to improve prediction. Numbers of 30 - 50 clones seem to be sufficient under normal conditions. From a propagation point of view we need as early information as possible to be able to propagate the material vegetatively. As long as we have no efficient methods of rejuvenation for all tree species, we have to live with a certain level of error. This can be balanced by genetic variation of the material under propagation.

Table 4. Some numerical parameter values satisfying (1) and (2) for 'age' (in years) (= difference between late and early ages), r_E (= juvenile – mature correlation based upon the values of the single clones), r_M (= juvenile – mature correlation based upon the means of mixtures of different clones) and n (= number of clones in the clonal mixture).

'age'	r_E	required r_M	necessary n
2	0.49		8
3	0.40		12
4	0.35		16
5	0.32		18
6	0.30		20
7	0.28		22
8	0.27	0.90	24
10	0.26		26
13	0.24		28
17	0.23		30
23	0.22		32
35	0.21		34
66	0.20		36
2	0.49		4
4	0.35		6
5	0.32		8
7	0.28	0.80	10
12	0.25		12
21	0.22		14
66	0.20		16
4	0.35		4
8	0.27	0.70	6
19	0.23		8
8	0.27		4
66	0.20	0.60	6

SUMMARY

Forest tree breeding largely depends on early selection. Early selection is influenced by rank changes.

Biological reasons for rank changes are summarized. These are before all differences in genetical reaction to maturation and to environmental variation. Beside these maternal influences (seed size e. g.) can induce rank changes.

Populations have genetic variation as a buffer system which diminishes rank changes. Rank changes are therefore more expressed in clones.

Statistical approaches to describe these phenomena are discussed using clonal material. Age dependence of components of variance is analyzed. Site influences are prevailing by far. These stabilize roughly ten years after plantation establishment.

The influence of clonal mixtures on correlation coefficients is discussed. Correlation coefficients increase quickly with an increasing number of clones and reach r = 0.90 already with 20 clones in a mixture.

Based on the empirical study, an explicit formula for age-age correlations is given which shows an excellent fit to the empirical data. Covariances show a strong linearity dependant on age. Correlation coefficients decrease rapidly during the first years and show then asymptotic behaviour towards a specific level. Finally a theoretical extension is given which allows the calculation of selection ages and necessary clonal numbers.

LITERATURE

Abetz, P., 1972, Zur waldbaulichen Behandlung der Kiefer in der nord-badischen Rheinebene, Allgemeine Forstzeitschrift, 27: 591 - 594.

Burdon, R. D. and Sweet, G. B., 1975, The problem of genetic inter-pretation of early growth measurements, Forest Research Institute Rotorua, New Zealand, 26 p.

Busse, J., 1930, Vom Umsetzen unserer Waldbäume, Tharandter Forstliches Jahrbuch, 81: 118 - 130.

Cieslar, A., 1923, Untersuchungen über die wirtschaftliche Bedeutung der Herkunft des Saatgutes der Stieleiche, Centralblatt für das ge-samte Forstwesen: 97 - 149.

Delvaux, J., 1981, Différenciation Sociale. Schweiz. Zeitschr. Forstwes. (132) p. 733-749.

Foster, G. S., 1986, Trends in genetic parameters with stand development and their influence on early selection for volume growth in loblolly pine, Forest Science, 32: 944 - 959.

Franklin, E. C., 1979, Model relating levels of genetic variance to stand development of four North American conifers, Silvae Genetica, 28: 207 - 212.

Gill, J. G. S., 1986, Juvenile-mature correlations and trends in genetic variances in Sitka spruce in Britain, Silvae Genetica, 35 (5-6): 189-194.

Huehn, M., 1969, 1970, 1971, Untersuchungen zur Konkurrenz zwischen ver-schiedenen Genotypen in Pflanzbeständen, I-V, Silvae Genetica, 18: 186 - 192; 19: 22 - 31, 77 - 89, 151 - 164; 20: 218 - 220.

Huehn, M. and Kleinschmit, J., 1986, A model for juvenile mature correla-tion of clonal mixtures dependent on the number of clones, Proceed. IUFRO Conference on Breeding Theory, Progeny Testing and Seed Orchards, Williamsburg, Virg., Oct. 1986, 71 - 84.

Huehn, M., Kleinschmit, J. and Svolba, J., 1987, Some experimental results concerning age dependency of different components of variance in testing Norway spruce (Picea abies [L.] Karst.) clones, Silvae Genetica, 36: 68 - 71.

Huehn, M., 1987, Clonal mixtures, juvenile-mature correlations and necessary number of clones, Silvae Genetica, 36: 83 - 92.

Huehn, M. and Kleinschmit, J. 1988: Juvenile mature correlations of Norway spruce (Picea abies (L.) Karst) clones, Lambeth's formula and some new results. IUFRO Tagung Tjörnarp, Sweden, 15p.

Huehn, M., Kleinschmit, J. and Svolba, J., 1988, Readers reaction and authors reply to " M. Huehn, J. KLeinschmit and J. Svolba: Some experimental results concerning age dependency of different components of variance in testing Norway spruce (Picea abies [L.] Karst.) clones", in Silvae Genetica 36 (2): 68 - 71, 1987. Silvae Genetica, 37: 247 - 249.

Kang, H., 1985, Juvenile selection in tree breeding; Some mathematical models, Silvae Genetica, 34: 75 - 84.

Kleinschmit, J., 1983, Concepts and experiences in clonal plantations of conifers. Proceedings 19th Meeting Canadian Tree Improvement Association on Clonal Forestry, 26 - 56.

Lambeth, C. C., 1980, Juvenile-mature correlations in Pinaceae and implications for early selection, Forest Science, 26: 571 - 580.

Lambeth, C. C., van Buijtenen, J. P., Duke, S. D. and McCullough, R. B., 1983, Early selection is effective in 20-year-old genetic tests of loblolly pine, Silvae Genetica, 32: 210 - 215.

Leibundgut, H., 1971, Ergebnisse von Durchforstungsversuchen 1930 - 1965 im Sihlwald, Mitteilungen der Schweiz. Anstalt für das Forstliche Versuchswesen, 47: 305 - 316.

Lewark, S., 1981, Untersuchungen von Holzmerkmalen junger Fichten (Picea abies [L.] Karst.), Dissertation Univ. Göttingen, 192 p.

Liebold, E., 1965, 1. Bemerkungen zu den Ursachen des Umsetzens der Einzelbäume in Waldbeständen. 2. Die Erkennbarkeit der Wuchspotenz des Einzelbaumes im gleichaltrigen Kiefernreinbestand, Archiv für Forstwesen, 14: 611 - 617 and 1123 - 1131.

Loo, J. A., Tauer, C. G. and Van Buijtenen, J. P., 1984, Juvenile-mature relationships and heritability estimates of several traits in loblolly pine (Pinus taeda), Canadian Journal of Forest Research 14 (6): 822 - 825.

Magnussen, S., 1988, Minimum age-to-age correlations in early selections, Forest Science 34 (4): 928 - 938.

Namkoong, G., Usanis, R. A. and Silen, R. R., 1972, Age related variation in genetic control of height growth in Douglas fir, TAG, 42: 151 - 159.

Namkoong, G. and Conkle, M. T., 1976, Time trends in genetic control of height growth in ponderosa pine, Forest Science, 22: 2 - 12.

Nanson, A., 1968, La valeur des tests précoces dans la sélection des arbres forestiers en particulier au point de vue de la croissance, Dissertation, Faculté des Sciences Agronomiques de l'Etat, Gembloux.

Nanson, A., 1974, Frühtests von Fichtenherkünften, Travaux Serie E, No. 6, Station de Recherches des Eaux et Forêt, Groenendaal-Hoeilaart.

Riemenschneider, D. E., 1988, Heritability, age-age correlations, and inferences regarding juvenile selection in Jack Pine, Forest Science 34 (4): 1076 - 1082.

Robinson, J. F., van Buijtenen, J. P. and Long, E. M., 1984, Traits measured on seedlings can be used to select for later volume of loblolly Pine, Reprint from the Southern Journal of Applied Forestry 8 (1): 59 - 63.

Sakai, K. I. and Hatakeyama, S., 1963, Estimation of genetic parameters in forest trees without raising progenies, Silvae Genetica, 12: 152 - 157.

Schmidt, H., 1962, Der Frühtest als Hilfsmittel für die genetische Beurteilung von Waldbäumen, Forstwiss. Centralblatt, 81: 128 - 148.

Schober, R. and Fröhlich, H. J., 1967, Der Gahrenberger Lärchenprovenienzversuch, Schriftenreihe der Forstlichen Fakultät der Universität Göttingen. J. D. Sauerländer's Verlag, 37/38.

Silen, R., 1965, A 50 year racial study of Douglas fir in Western Oregon and Washington, Forest Service Proceed., Western Forest Gen. Assoc.: 6 - 7.

St. Clair, J. B., Kleinschmit, J. and Svolba, J., 1985, Juvenility and serial vegetative propagation of Norway spruce clones (Picea abies Karst.). Silvae Genetica, 34: 42 - 48.

St. Clair, J. B. and Kleinschmit, J., 1986, Genotype environment interaction and stability in ten-year height growth of Norway spruce clones (Picea abies Karst.), Silvae Genetica, 35: 177 - 186.

Stern, K., 1968, Vollständige Varianzen und Kovarianzen in Pflanzenbeständen. IV. Phänotypische Korrelationen zwischen Wuchsleistungen in verschiedenen Altersstufen, TAG, 38: 66 - 73.

INTERACTIONS BETWEEN GENOTYPE AND DEVELOPMENTAL FACTORS MODIFYING

PEROXIDASE EXPRESSION

Anne-Marie Hirsch

Laboratoire d'Histophysiologie Végétale-URA-CNRS-1180
Université Pierre et Marie CURIE
12, rue Cuvier, 75005 Paris, France

The function of the peroxidase isozymes in higher plants is not yet well understood. As an introduction to a comprehensive analysis of the interactions between genotype and developmental factors modifying peroxidase expression, it seems necessary to introduce some current notions about the genetics of peroxidases.

GENETICS OF PEROXIDASE ISOZYMES

The main contribution on this problem is owed to the group of Wijsman, Hendricks and Co (1,2), from the Institute of Genetics (University of Amsterdam), working on Petunia (Solanaceæ) and using the gel electrophoretic method. According to Gottschalk (3), the basal chromosome number in Solanaceæ is six; Petunia has an additional chromosome (n=7) ; this low basal chromosome number, associated with a short generation time, have been reasons to retain Petunia as a model for these studies. Genetically, the several isozymes can be shared over a discrete number of structural genes and genetics are an easy way to discover what enzymes are modifications of primary structural compounds, despite the fact that several of these isozymes show only minor differences. The formation of "mozymes", isozymes having a peptide part in common, is normally an age-dependent phenomenon. The allozyme balance is also dependent on the age of the plant. The presence of variable bands is highly dependent on the age of tissue and plant. All these data explain that even in Petunia, the zymograms are often complex : they are post-translational modifications of the gene expression. Finally, Wijsman and Hendrijks claimed that no more than 10 structural genes are active in Petunia ; zymograms and gene loci of peroxidase isozymes are represented in Fig.1. The group of Amsterdam geneticists suggested to introduce a rational nomenclature of isoperoxidases : the structural gene with a capital letter, every allele with an own number and after the point, the mozyme number. In fact, in the literature, the number of different abbreviations of peroxidase genes is disconcerting ; one reason for this diversity is that several important crop organisms are of hybrid descent and it should be preferable to speak of "homoeoenzymes" of one constituent gene.
In Petunia and in all studied Solanaceæ representants, it has been

shown that isoperoxidases are monomeric proteins, with exceptions in
Graminaceæ : Shahi (4) described the Px2 gene in rice and
demonstrated that the hybrid has 3 bands, both parental
contributions and a heterodimeric band right between them.

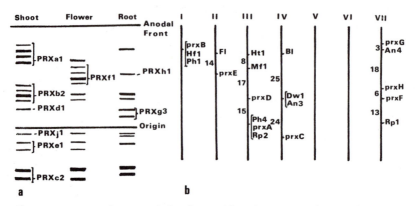

Zymograms (a) and gene loci (b) of peroxidase isoenzymes in petunia

Fig. 1. Peroxidase bands in <u>Petunia</u> (in shoots, flowers and roots)
with map of the genes. (HENDRIKS T, R VINKENOOG and HJW WIJSMAN—1985—
Theor. Appl. Genet. 70(6) : 595-598)

Ainsworth and Co (5) working on cereals (wheat, rye and barley)
demonstrated that no alleles are fairly common among peroxidases,
due merely to the fact that peroxidases are tested with artificial
substrates creating artefacts not related with the in vivo
situation. This stresses the importance of the immunological assay
which alone will decide whether the relevant peroxidase molecule is
present. Finally, following conclusions can be drawn about genetical
regulation of peroxidase synthesis : it is impossible to ascertain
the number of sites changed in differentially regulated genes ;
there is only suggestive evidence of separate sites. This short
review of peroxidase isoenzyme genetics studies will just keep clear
how difficult the analysis of isoperoxidase zymograms is.

VARIATIONS OF BASIC AND ACIDIC PEROXIDASES IN DIFFERENTATION,

PHYSIOLOGICAL AND ADAPTATION PROCESSES

In spite of all the difficulties above mentionned, the
peroxidases appear to be the key molecules for a rapid adaptation of
whole plants to changes in their environment. The variations of
acidic and basic peroxidases will be considered in different ways :
- cell wall genesis and plasticity.
- somatic embryogenesis
- cambium formation,rooting, shooting and flowering processes.
- sex differentiation.
- stress environment adaptation.

Cell wall genesis and plasticity

Lamport (6) reviewed the possible role of peroxidases in cell
wall genesis and plasticity and proposed a scheme of possible
control of cell wall rheology by the levels of extensin monomers,
these levels being controlled by peroxidases in dependance of a pH
value effect.

Catesson and co (7) precised the nature, localization and specificity of peroxidases involved in lignification processes : these peroxidases take part in the final step of lignin biosynthesis. The authors studied the localization and the pattern of isoperoxidases having a high affinity towards syringaldazine in sycamore phloem and xylem (fig.2). Relationships between peroxidase activities and cell wall plasticity were studied by Goldberg and Co (8). Their conclusions were following : significant modifications appeared in cell wall peroxidasic fractions along the growth gradient of the mung bean hypocotyl. The changes are the most important in epidermis tissue : specific isoperoxidases were developed in parallel to the reduction of cell wall plasticity. These events corresponded to a significant increase of covalently bound anionic isoperoxidases having high affinity for various chromogens. These cell wall isoperoxidases were inhibited by acidic pH and might have been involved in cross linkings of phenolic compounds inside the cell wall.

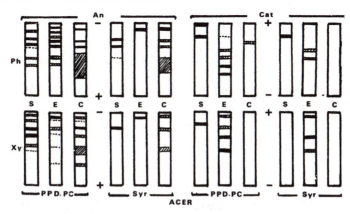

Fig. 2. Isoenzyme patterns of the three peroxidase fractions (S:soluble; E:ionic; C:covalent) extracted from sycamore phloem (Ph) and xylem (Xy). An:anionic and Cat:cationic peroxidases. Staining of polyacrylamide gels either with PPD.PC:p-phenylene-diamine-pyrocatechol or Syr:syringaldazine (CATESSON AM, A IMBERTY, R GOLDBERG and Y CZANINSKI:Molecular and physio logical aspects of plant peroxidases-1986-Université de Genève. ISBN 288164-001-x : 189-199)

Somatic embryogenesis

Somatic embryogenesis occurs when embryos are generated from somatic cells which previously have undergone dedifferentiation. This technique ensures simultaneously a certain degree of genetic variation together with the possibility of rapid obtention of large numbers of plantlets. Somatic embryogenesis has presently a place of choice among the biotechnological processes for vegetal cell regeneration. For practical reasons, an early characterization of the embryogenic capacities of the callus would be appreciated ; isozymes could be very useful for this purpose. Wochok and Burleson (9) first showed the presence of isoperoxidases in embryogenic tissues of wild carrot. Krsnick and Co (10) pointed out in pumpkin tissues that peroxidases could be used as early indicators of somatic embryoid differentiation. Recently, in 1988, Sanchez de Jimenez and Co (11), studying the loss of morphogenic capacities of aged maize callus tissue, induced embryoid structures by increasing the osmotic pressure in the nutrient medium devoid of hormones, and only in a certain type of callus, with isodiametric cells, while an other type of callus with elongated cells failed to produce embryoid structures.

These authors pointed out differences in soluble peroxidasic activity and iso-peroxidasic pattern between embryogenic and non-embryogenic callus. Specific changes were noticed in the cathodic isoperoxidases.

Cambium formation, rooting, shooting and flowering processes

In our laboratory (12, 13), the study of cambium formation, xylem formation and rooting meristematic cells initiation, correlated to peroxidase variations was carried out using Jerusalem artichoke tubers as in vitro experimental material. Changes in the soluble peroxidase levels were noticed during the histological sequence leading to the root formation. Increases of soluble peroxidase activity appeared suddenly just before the cambium rooting meristematic cells initiation. Changes in isoperoxidasic pattern occured at the same moment ; one band (a) of great electrophoretic mobility appeared just before the cambium initiation (on the 8th day after incubation (fig.3). Root formation in cuttings of explants from different sources was studied by Gaspar (14). Rooting is the result of successive inductive and initiative phases characterized first by increased, subsequently by decreased activity of basic peroxidases, whereas the activity of acidic peroxidases regularly increased during both periods (fig.4). Thorpe and Co (15), studying the de novo shooting formation in vitro, demonstrated that this organogenesis phenomenon is controlled by the endogenous phytohormone level and principally by the ratio auxin/cytokinin. IAA is highly involved and as a consequence, the peroxidases. Changes in the number of cathodic and anodic isoperoxidases have been used as an indirect measure of IAA levels in organ forming tissues. These changes indicated that during shoot primordium formation, there is a requirement for a low and steady level of auxin, a continuous reduction in endogenous auxin taking place during floral bud formation.

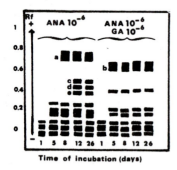

Fig. 3. Peroxidases pattern during cambium, xylem and rooting meristematic cells initiation in Jerusalem artichoke tubers cultured in vitro.(HIRSCH AM-1975- C.R.Acad.Sc.Paris 280 : 829-832)

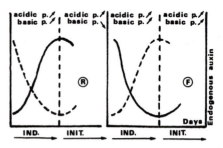

Fig. 4. Schematic representation of variations in peroxidase (p.) and auxin levels during the course of the inductive (IND.) and initiative (INIT.) phases of rooting (R) and flowering (F) by non-induced explants. (GASPAR Th, C PENEL, T THORPE and H GREPPIN : Peroxidases 1982-Université de Genève)

Sex differentiation

The analysis of the isoperoxidases reflects the differential gene expression and diœcious plants seem to be good models to study the biochemical markers of sex differentiation. Two plants of economical importance were studied in our laboratory : Kiwi and Jojoba. The determination of the soluble total peroxidase activity and the analysis of specific isoperoxidases markers were carried out in male and female flower-buds of Kiwi. The soluble total peroxidase activity was suddenly reduced just before flowering, in male and female flower-buds, although peroxidase activity remained higher in the male flower-buds than in the female ones (16). Such results confirm the report of Durand and Co (17). These authors, working on a diœcious plant, Mercurialis annua, claimed that the sexual expression is controlled by auxin/cytokinin balance and peroxidases act as regulators of this balance.

In order to know if the sexual genes regulate exclusively the sexual organogenesis or interfere with the general metabolism, stem callus lines were established with the aim to study the peroxidases.

An interesting question which has not yet been examined was following: are undifferentiated calli devoid of any sexual morphogenesis able to express differences in their isoperoxidase pattern ? If any differences are detected, are they stable in successive subcultures and in cell suspensions obtained from these subcultures ? The study of peroxidase activity of cultured stem tissue, callus subcultures and cell suspensions of cultured stem were made. The release of isoperoxidases in the nutrient liquid medium was also investigated. The peroxidase activity was low in stem segments of both male and female plants ; the number of isoperoxidases was also low. Different kinds of cultured stem callus were obtained : "growing", "rooting" and "budding" stem calli. The soluble peroxidase activity (S.P.A.) and the isoperoxidase composition were estimated in the three different kinds of calli, S.P.A. values were always higher in callus tissue produced from female than from male plants (fig.5). In the three types of callus, two isoperoxidases (a and b), characteristic of callus tissue produced by female plants were isolated (fig.6). Callus cell lines were established by means of successive subcultures in different media. Several shooting callus lines were maintained during 4 years and their isoperoxidase composition determinated : the two characteristic bands (a and b) from female callus tissue were still present on zymograms. Cell suspensions from these shooting callus lines were made : S.P.A. from the suspensions produced by the female callus lines was always higher than that produced by the male callus lines. The isoperoxidase pattern of cell suspensions and of residual nutrient medium were determinated (fig.7). The isoperoxidases were more numerous in cell suspensions than in callus lines and the permanence of both a and b isoperoxidases of female callus tissue and subcultures was found again in the corresponding cell suspensions. The S.P.A. of the residual medium was weak ; however, almost all the characteristic isoperoxidases of the cell suspensions were also found in the residual medium. In conclusion, tissue culture of diœcious plants seemed to us an excellent tool for the investigation of the hormone regulation of sexual differentiation. Sexual genes are not only efficient at the moment of the differentiation of the sexual organs (flower-buds) but also in undifferentiated callus tissue and cell suspensions devoid of any sexual morphogenesis. The level of the soluble peroxidase activity and the isoperoxidase pattern might express the action of the sexual genes on the hormone metabolism (18,19). Such results were confirmed on Gingko biloba (Gymnosperm) by chinese authors (20).

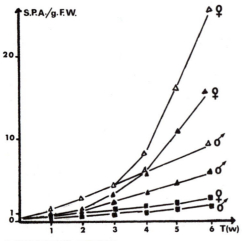

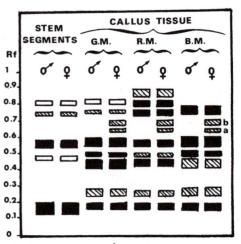

▲ "GROWING" MEDIUM
  (Kin 1mg/l and 2-4D 0.01mg/l)
■ "ROOTING" MEDIUM
  (IAA 10mg/l and 2-4D 0.01mg/l)
△ "BUDDING" MEDIUM
  (IAA 0.1mg/l and Kin 1mg/l)
♂ callus tissue obtained from ♂ plant
♀ callus tissue obtained from ♀ plant

Fig. 5. Soluble peroxidase (S.P.A.)
activity of stem cultured tissue.
(HIRSCH AM and D FORTUNE-1984-
Z. Pflanzenphysiol. 113 : 129-139)

■ black bands   ) represent
▨ hatched bands) decreasing
□ white bands   ) degrees of
                     enzyme activity

G.M. "GROWING" MEDIUM
R.M. "ROOTING" MEDIUM
B.M. " BUDDING" MEDIUM
a,b isoperoxidase fractions
     characteristic of female
     callus tissue

Fig. 6. Peroxidase pattern of stem
segments and of stem cultured
tissue (callus tissue). (ibidem)

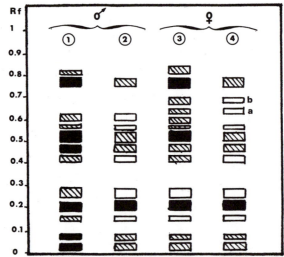

① cell suspensions obtained from ♂ shooting callus
  lines maintained on "BUDDING" MEDIUM
② nutritious residual medium of ♂ cell suspensions
③ cell suspensions obtained from ♀ shooting callus
  lines maintained on "BUDDING" MEDIUM
④ nutritious residual medium of ♀ cell suspensions

Fig. 7. Peroxidase pattern of cells in suspension
and peroxidase release in the medium. (ibidem)

These authors showed that peroxidase isozyme patterns are sex associated, more bands having been evidenced in the young leaves of female than of male plants. Moreover, the investigation of the isoperoxidases in the leaves of trees of unknown sex, i.e. before first blossom, evidenced two types of patterns : the first one agrees with the females, the other with the males, without any intermediary type.

Stress environment adaptation

Peroxidases appear to be key molecules in a rapid adaptation of the whole plants to changes of their environment. Adaptation to absence or excess of NaCl is an excellent example of compliance under stress. Peroxidase activities were studied in Suaeda maritima, cultured in absence or in presence of NaCl at optimal or infraoptimal concentrations (21). From the experimental data, it can be emphasized that a strong increase of both basic and acidic peroxidases explains altogether growth inhibition and hyperlignification. It has been suggested by the authors that in an halophytic plant, absence of NaCl can be considered as a chemical stress by deficiency and that halophilia would be a need of salt rather than an increased tolerance threshold to an hypersaline medium. Adaptation to salinity is linked to an osmotic tuning of the membranes, which in turn depends on the regulation of ion transport and accumulation via plasmalemma and tonoplasts. Ca^{++} ions (22) have a regulation role on binding of basic peroxidases to membranes; fatty acids may contribute both to peroxidasic activation and to conformational changes resulting in additional strong binding sites.

REFERENCES

1- VAN DEN BERG B.M. and H.J.W. WIJSMAN -1981- Genetics of the peroxidase isozymes in Petunia. Part1 : Organ specificity and general geneticaspects of the peroxidase isozymes. Theor. Appl. Genet. 60:71-76
2- HENDRICKS T., B.M. VAN DEN BERG and A.W. SCHRAMM -1985- Cellular location of peroxidase isozymes in leaf tissue of Petunia and their affinity for concanavalin A sepharose. Planta 164:89-95.
3- GOTTSCHALK W. -1953- Die Chromosomen Struktur der Solanaceen unter Berücksichtigung phylogenetischer Fragestellungen. Chromosoma 6:539-626
4- SHAHI B.B., Y.E. SHU and H.I. OKA -1969- Analyses of genes controlling peroxidase isozymes in Oryza sativa and O.perennis. Jpn. J. Genet. 44:321-338.
5- AINSWORTH C., H.M. JOHNSON, E.A. JACKSON, T.E. MILLER and M.D. GALE
-1984- The chromosomal locations of leaf peroxidase genes in hexaploid wheat, rye, and barley. Theor. Appl. Genet. 69:205-210.
6- LAMPORT D.T.A. -1986- Roles for peroxidases in cellwall genesis. Molecular and physiological aspects of plant peroxidases
199-208. Université de Genève. ISBN 288164-001-X.
7- CATESSON A.M., A. IMBERTY, R. GOLDBERG and Y. CZANINSKI -1986- Nature, localization and specificity of peroxidases involved in lignification processes. 189-198. ibidem.
8- GOLDBERG R., A. IMBERTY, M. LIBERMAN and R. PRAT -1986- Relationships between peroxidasic activities and cell wall plasticity. 209-220. ibid.

9- WOCHOK Z., B. BURLESON -1974- Isoperoxidase activity and induction incultured tissues of wild carrot, a comparison of proembryos and embryos Physiol. Plant. 31:73-75.

10- KRSNIK-RASOL M., S. JELASKA and D. SERMAN -1982- Isoperoxidases : early indicators of somatic embryoid differentiation in pumpkin tissue. Acta Botan. Croat.41:33-39.

11- SANCHEZ de JIMENEZ E., M. VARGAS, R. AGUILAR and E. JIMENEZ - 1988- Age dependant responsiveness to cell differentiation stimulus in maize callus cultures. Plant Physiol. Biochem.26 (6):723-732.

12- HIRSCH A.M. -1975- Evolution des activités péroxydasiques et phosphatasiques au cours des phénomènes de rhizogénèse des fragments de rhizomes de Topinambour cultivés in vitro-C.R. Acad. Sci. Paris. D280:829-832.

13- HIRSCH A.M. -1976- Isoperoxydases et phénomènes de cambiogénèse et de rhizogénèse de fragments de rhizomes de deux variétés de Topinambour en culture in vitro. Actes du 101ème Congrès National des Sociétés Savantes, Lille 1976. Sciences 1:507-518.

14- GASPAR Th. -1981- Rooting and flowering, two antagonistic phenomena from a hormonal point of view. In "Aspects and prospects of plant growth regulators". Monograph 6.Jeffcoat B.(Ed)-British Plant Growth.
Regulator Group. Wantage.39-49.

15- THORPE T.A., M. TRAN THANH VAN and Th. GASPAR -1978- Isoperoxidases in epidermal layers of Tobacco and changes during organ formation in vitro Physiol. Plant.44:388-394.

16- HIRSCH A.M., D. BLIGNY and B.K. TRIPATHI -1977- Biochemical properties of tissue cultures from different organs of Actinidia chinensis-Acta Horticulturae 78:75-85.

17- DURAND B., R. DURAND, G. KAHLEM, A. CHAMPAULT and M. DELAIGUE - 1972- Gènes, hormones, protéines et différenciation du sexe chez Mercurialis annua. Mém. Soc. Bot. (Colloq. Morphol.)191-199.

18- HIRSCH A.M. and D. FORTUNE -1984- Peroxidase activity and isoperoxidases composition in cultured stem tissue,callus and cell suspensions of Actinidia chinensis-Z.Pflanzenphysiol.113:129-139.

19- HIRSCH A.M and D. FORTUNE -1986- Peroxidases as markers of organogenesis and diœcism in tissue of Actinidia chinensis L. Planchon cultured in vitro-Molecular and Physiological Aspects of Plant peroxidases-p.361-366.ISBN 288164-001-X- Université de Genève Ed.

20- ZHONG H.W., Z.H. YANG, G.I. ZHU and Z.X. CAO -1982- Peroxidase isozyme pattern, as biochemical test to distinguish the sex of individual plant in Ginkgo biloba L.-Scientia Silvae Sinicae, 18(1):1-4.

21- HAGEGE D., C. KEVERS, J. BOUCAUD and Th. GASPAR -1988- Activités peroxydasiques, production d'éthylène, lignification et limitation de croissance chez Suaeda maritima cultivé en l'absence de NaCl- Plant Physiol. Biochem.26(5):609-614.

22- GASPAR TH., C. PENEL, F.J. CASTILLO and H. GREPPIN -1985- A two step control of basic and acidic peroxidases and its significance for growth and development-Physiol. Plant.64:418-423.

ORGANOGENESIS: STRUCTURAL, PHYSIOLOGICAL AND BIOCHEMICAL ASPECTS

Trevor A. Thorpe

Plant Physiology Research Group, Department of Biological Sciences, University of Calgary, Calgary, Alberta, Canada T2N 1N4

INTRODUCTION

The process of de novo organogenesis in cultured tissues is a complex one, in which extrinsic and intrinsic factors play a role (Thorpe, 1980, 1988). Organized development can be regulated through manipulation of the culture medium and the culture environment, and by judicious selection of the inoculum. Manipulation of these factors allows cells that are quiescent or committed only to cell division to undergo a transition, which at the molecular level involves selective gene activation. This activity is reflected by biochemical, biophysical, and physiological changes, which lead to structural organization within the cultured tissues. Much is known about the manipulation of the factors regulating de novo organogenesis (Thorpe, 1980; Evans et al., 1981), but virtually nothing is known about regulation at the molecular level (Brown and Thorpe, 1986; Thorpe, 1988). Even less information is available on the process in woody plants. Understanding de novo organogenesis is not only of interest to developmental morphologists and plant physiologists, but also to those involved in the exploitation of tissue culture technology for plant improvement because regeneration of plantlets is central to this activity. Regeneration of plantlets in vitro is most commonly achieved via organogenesis in a procedure requiring several distinct stages (see Thorpe and Harry, this volume). In general, shoot-bud induction and shoot primordium formation are the first stages in plant regeneration. This article presents information on these stages as obtained primarily from studies carried out with excised cotyledons of Pinus radiata D. Don (radiata or Monterey pine) in my lab (Thorpe, 1988).

TISSUE CULTURE SYSTEM

The radiata pine explant system consists of cotyledons excised from aseptic, dark-germinated seed approximately one day after radicle emergence (Aitken et al., 1981). At this stage the radicle is between 0.5 and 2.0 cm long and the cotyledons between 3 and 5 mm. Younger or older cotyledons lead to a reduced production of adventitious shoots (Aitken et al., 1981; Aitken-Christie et al., 1985). Each seed consists of seven to ten cotyledons, and about 90% of seeds behave identically in culture.

Plant Aging: Basic and Applied Approaches
Edited by R. Rodríguez *et al.*
Plenum Press, New York, 1990

The excised cotyledons are generally cultured in sterile plastic petri dishes on a modified Schenk and Hildebrandt medium (Reilly and Washer, 1977) containing 3% sucrose and 25 µM N^6-benzyladenine (BA) in the light (16 hr, ~ 80µmol.m^{-2}.s^{-1} from Sylvania Gro-lux fluorescent tubes) at 27±1°C for 21 days (Aitken et al., 1981). At the end of this period meristematic tissue is formed along the entire length of the cotyledon on the side in contact with the medium. For shoot development, these cotyledons are transferred to BA-free medium with 2% sucrose under the same culture conditions. Cotyledons cultured in the absence of BA serve as non-shoot-forming controls.

STRUCTURAL ASPECTS OF BUD FORMATION

In the absence of BA the cotyledons rapidly elongate, reaching a length of about 24 mm by day 21 (Thorpe, 1988). At excision anticlinal cell divisions occur throughout the cotyledons. These cease after day 2 in culture in the absence of BA (Yeung et al., 1981). Protein bodies disappear by day 2, there is a reduction in stored starch and lipid, the cells become very vacuolated and plastids become fully developed by day 3 (Douglas et al., 1982; Villalobos et al., 1985). The stomatal complexes begin to differentiate by day 1 and are fully developed by day 5, at which time large intercellular air spaces begin to appear in the mesophyll. Thus in many respects these cotyledons develop similar to non-excised cotyledons, i.e. they become essentially photosynthetic organs.

In contrast to the above, the BA-treated cotyledons do not elongate and cell division becomes restricted to the epidermal and subepidermal cell layers on the side in contact with the medium by day 3 (Villalobos et al., 1985). Organized structures, termed promeristemoids consisting of six to eight cells, which arise from a single subepidermal cell as a result of both anticlinal and periclinal divisions, can be seen by day 5. Cells within each promeristemoid are tightly packed together, with little or no intercellular spaces, but prominent plasmodesmata are present between the relatively thin cell walls within each promeristemoid. The breakdown of storage lipid, and protein also occur as in BA-free cotyledons, but chloroplast development is slower. Newly synthesized starch can be observed early in culture, but this gradually disappears as the promeristemoids develop. By day 10 these cotyledons are nodular, due to the development of the promeristemoids into meristemoids, which give rise to shoot primordia by day 21.

The position at which promeristemoids develop is probably influenced in part by physiological gradients of nutrients (including cytokinin) from the medium into the tissue. The promeristemoids do not arise from a typical meristematic cell. The cells contain vacuoles and have abundant chloroplasts. Densely plasmatic cells, as found in herbaceous species during organized development, are conspicuously absent in radiata pine cotyledons. The cells become densely plasmatic later in culture. However, the developmental sequence leading to multiple shoot formation is quite similar in various conifer explants, such as mature embryos, cotyledons, and epicotyls (Thorpe and Patel, 1986). The sequence includes the formation of (a) meristemoids (also referred to as meristematic bud centers or meristematic tissue), (b) bud primordia, and (c) adventitious shoots with well-organized apical domes and needle primordia.

PHYSIOLOGICAL ASPECTS OF BUD FORMATION

Exogenous cytokinin is required for shoot formation, the optimum level being 25 µM with an exposure time of 21 days (Aitken et al., 1981; Biondi and Thorpe, 1982). BA must be present during the first three days in culture, but exposure to light can be delayed until day 10 (Villalobos et al., 1984a).

After 21 days in darkness no shoots are formed if BA-treated cotyledons are subsequently transferred to light. The addition of other phytohormones or growth regulators, including auxins, gibberellic acid, abscisic acid, cAMP, aromatic amino acids, and growth retardants (CCC and AMO-1618), to the medium during the 21-day culture period tend to reduce cytokinin-induced meristematic tissue and promote callus production (Biondi and Thorpe, 1982). These findings do not mean that endogenous phytohormones play no role in bud formation. Indeed we have some data which indicate that the contents of endogenous and the metabolism of exogenous indoleacetic acid and abscisic acid change during different stages of bud formation (Macey, Reid and Thorpe, unpubl.).

The role of ethylene and its interaction with CO_2 in bud induction has been examined (Kumar et al., 1987). The excised cotyledons were cultured in Erlenmeyer flasks, which were sealed with serum caps on various days in culture. The caps were also removed at different stages of the morphogenetic process. Second, the gases within the flasks were absorbed by traps, singly or in combination, during the various stages of differentiation. Third, the cultures were incubated under continuous flow of constant gas mixtures. It was found that shoot-forming cotyledons produce considerable amounts of C_2H_4 and CO_2, and the frequency of organogenesis can be correlated with the concentrations of these two gases. The highest number of buds per explant are obtained when the flasks accumulate about 5 to 8 μl 1^{-1} of C_2H_4 and about 10% CO_2 in the headspace during the first 15 days in culture. When both gases are eliminated from the flasks, differentiation is completely inhibited. However, excessive accumulation beyond day 15 causes partial dedifferentiation of the shoot buds, and trapping the gases or allowing their diffusion after 15 days have no adverse effects on bud formation.

The effects of C_2H_4 and CO_2 are possibly synergistic and necessary in order for the cytokinin (BA) in the medium to bring about the switch in morphogenesis from the normal maturation of the cotyledons to shoot bud differentiation. The first ten days in culture (when C_2H_4 and CO_2 exert their influence on differentiation) coincide with a period of intense, localized cell division leading to the formation of meristematic domes. Thus, the stimulatory action of C_2H_4 on morphogenesis could be mediated through modification of the cell division process (Constabel et al., 1977). From the known interactions of CO_2 and C_2H_4, three possible roles of CO_2 can be proposed. First, in the early stages it may be acting primarily to enhance the biosynthesis of C_2H_4; second, at later stages (beyond day 10 or 15) it may be important in antagonizing the action of C_2H_4; and third, CO_2 may have a totally independent role in metabolism (Kumar et al., 1987). This last role has been examined (see later).

BIOCHEMICAL ASPECTS OF BUD FORMATION

We have approached this topic through the use of autoradiographic, histochemical, and biochemical (including precursor incorporation) techniques. In most cases, bud-forming (+BA) cotyledons are compared and contrasted to non bud-forming (-BA) tissues, in an attempt to determine what aspects of metabolism can be directly correlated with and presumably are causative to shoot bud formation.

Autoradiograms of precursor incorporation of [³H]uridine into RNA, [³H]leucine into protein, and [³H]thymidine into DNA in the epidermal and subepidermal cells from the cotyledonary face that was in contact with the medium revealed that during the first two days in culture the patterns of incorporation are similar in both BA-treated and BA-free cotyledons (Villalobos et al., 1984b). By days 3 to 5 labelling becomes concentrated in the epidermal and subepidermal cell layers of the BA-treated cotyledons, while very little incorporation occurs in BA-free tissues. [³H]thymidine labelling of plastids occurs, but only in BA-treated tissues are labelled nuclei found after day 2. The BA-free cotyledons show a dramatic increase in the rate of RNA and, to a lesser extent, protein synthesis during the first 24 hr in culture. During that period the rate of synthesis in the BA-treated cotyledons decrease. After 24 hr the rate also decreases in the BA-free cotyledons, so that by day 3 both BA-free and BA-treated cotyledons exhibit the same rate of RNA and protein synthesis. In contrast, the DNA synthetic rates decrease, being greater in the BA-treated tissues. In both cultures a secondary rise in synthetic rate is observed at day 3. Thus the change in the rates of macromolecular synthesis precede the histological localization of labelling patterns and the subsequent elongation/maturation of the BA-free cotyledons and the differentiation of the BA-treated tissues.

Histochemical analysis of the cotyledons confirmed the presence of the various polymeric reserves and their earlier and more rapid decline in the BA-free tissues (Patel and Thorpe, 1984). Increased staining for DNA, RNA and both nuclear and cytoplasmic proteins is observed in the shoot-forming cell layers of the BA-treated cotyledons in contrast to the BA-free ones. Increased staining for several enzymes, including acid phosphatase, ATPase and succinate dehydrogenase, is also observed in the shoot-forming cell layers. Lipase activity, confined to the shoot-forming layers early in culture, is detected in cells away from the meristematic region later in culture, indicating the pattern of the mobilization of the lipids. These findings suggest that the newly synthesized proteins may be enzymatic, related in part to energy metabolism.

Fatty acid and sterol analyses indicate that both quantitative and qualitative changes in different classes of lipids occur during bud formation (Douglas et al., 1982). The most dramatic change is in the rapid and nearly linear degradation of triglycerides, which is far greater than that needed for the observed increase in polar lipids. This suggests that the excised cotyledons, like germinating seeds, rely heavily upon the stored lipid reserves for energy production, particularly during the first three days in culture when respiration rates are highest (Biondi and Thorpe, 1982). Most of this metabolism early in culture has recently been shown to be wound-related (Joy and Thorpe, unpublished). The observed changes noted above and others (e.g. increases in membrane lipid, carotenoids and chlorophyll) appear to play only an indirect role in organogenesis, as they are similar to those that developing seedlings undergo.

A major study on primary metabolism in relation to organogenesis is being undertaken. Initial studies involved feeding the radiata pine cotyledons with [¹⁴C]glucose, [¹⁴C]acetate or [¹⁴C]bicarbonate at different stages of the shoot-forming process (Obata-Sasamoto et al., 1984). In experiments with the labelled glucose and acetate, $^{14}CO_2$ is released and incorporation of label into ethanol-insoluble and mainly into ethanol-soluble fraction occurs. The latter can be further fractionated into lipids, amino acids, organic acids and sugars. In general, there is a tendency towards a high rate of incorporation of label in BA-free cotyledons during the period of rapid elongation (day 3), and during meristematic tissue fractionation (days 10, 21) in BA-treated cotyledons.

In contrast, a similar study with [¹⁴C]bicarbonate indicated that the incorporation of label into the ethanol-soluble fraction is always higher in the BA-free cotyledons than in the BA-treated ones (Obata-Sasamoto et al., 1984). This was confirmed in a later study (Kumar et al., 1988), which also shows higher activity of ribulose bisphosphate carboxylase beyond day 5 in BA-free cotyledons in comparison with BA-treated ones. In contrast, phosphoenolpyruvate carboxylase activity is higher in the latter cotyledons than in the former ones during the first 10 days in culture. In agreement the label in the malate and aspartate fractions as a percentage of total ^{14}C incorporation is 3x higher in the bud-forming cotyledons than in the BA-free ones. These findings clearly indicate the relative importance of non-autotrophic CO_2 fixation to photosynthetic fixation during organogenesis.

In further studies [^{14}C]glucose was supplied to the cotyledons at different days in culture for 3 hr, followed by a 3 hr chase with [^{12}C]glucose (Bender et al., 1987a, b). The incorporation of ^{14}C into individual soluble metabolites, as well as into protein, was followed. The major labelled metabolites are malate, citrate, glutamate, glutamine, and alanine. The synthesis of glutamine strongly increases in the BA-treated cotyledons, suggesting a positive influence of BA on nitrogen incorporation prior to differentiation. At days 10 and 21, metabolic patterns are qualitatively similar in shoot-forming and non-shoot-forming cotyledons. However, respiratory metabolism and amino acid synthesis is strongly enhanced in the shoot-forming cultures. An increased incorporation of label into protein due to the cytokinin treatment is detected early in culture (days 0 and 3). Labelled amino acids are incorporated into protein to different degrees, but this is not influenced by the hormonal treatment. Later in culture, protein synthesis is strongly enhanced in the shoot-forming cultures. The radioactivity present in the lipid fraction at days 0 and 3 indicates the synthesis of metabolically stable lipids, as the radioactivity of this fraction increases during the chase period. In contrast, later in culture, there is considerable turnover in the lipid fraction, indicating that lipid synthesis for both structural and nonstructural components is taking place. Similar studies using [^{14}C]acetate are in progress (Joy, Bender and Thorpe, unpubl.). These indicate that similar selective labelling of amino acids and organic acids is taking place.

Spermidine is the major polyamine in radiata pine cotyledons, but [^{14}C]putrescine is mainly metabolized into gamma-aminobutyric acid, aspartate and glutamate, with concomitant release of $^{14}CO_2$, after pulse feeding for 2 hr on different days in culture (Kumar and Thorpe, 1989). However, when the cotyledons are cultured on medium containing [^{14}C]putrescine for extended periods (up to 10 days), the higher polyamines, spermidine and spermine, are labelled by day 2 (Biondi et al., 1988). The activity of arginine decarboxylase, a principal enzyme in putrescine synthesis, is higher in BA-treated tissues than in BA-free tissues early in culture, indicating the possible importance of putrescine biosynthesis in de novo organogenesis.

CONCLUSIONS

Information arising from the study of de novo organogenesis in the gymnosperm Pinus radiata, is in general agreement with studies carried out with mainly herbaceous species, such as tobacco (Brown and Thorpe, 1986; Thorpe, 1980). The process begins with changes in a single cell, in most cases, which becomes activated to give rise to a meristemoid from which shoot primordia arise. Phytohormones, both exogenous and endogenous, play a key role, and use of modified Ti-plasmids have supported this view (Brown and Thorpe, 1986). The data suggest that different gene products independently suppress shoot and root formation and that the T-DNA gene products act in an analogous way to auxin- and cytokinin-like growth regulators. Furthermore, it appears that plants have separate genetic programs for shoot and root development and that both of these programs must be internally coordinated.

Carbohydrate has been shown to have a dual role in both energy (ATP) production and osmotic adjustment in tobacco (Brown and Thorpe, 1988). Various biochemical measures have supported the idea that primordium formation is a high energy-requiring process, in which accumulated starch and free sugars from the medium are utilized. The process is metabolically very demanding, as there is also a higher requirement for reducing power (NADPH) compared to growing tissue. Studies carried out on nitrogen assimilation and amino acid (particularly aromatic amino acid) metabolism also indicate, not unexpectedly, that N-metabolism is also important in the differentiation process. The results obtained using excised cotyledons of radiata pine support those obtained with tobacco callus, with obvious differences, e.g., the utilization of stored lipid in energy production in cotyledons, as opposed to the accumulation of starch and its utilization in tobacco. Nevertheless, the requirement for the production of energy is clear in both tissues.

Much more work has to be done with both woody and herbaceous species, so that generalizations about the process of organogenesis can be made. Much of the research on de novo organogenesis still follows empirical approaches. This is understandable, as today we have so little information on the basic aspects of differentiation. With expanded interest in the application of tissue culture and recombinant DNA technologies to woody plants, the need for this understanding increases. Furthermore, the exploitation of these technologies for tree modification and improvement demands the ability to regenerate plantlets from the modified cells (Thorpe, 1988). Thus, the understanding of the structural, physiological, biochemical and molecular biological bases of de novo organogenesis will make this work more rational. Unfortunately, there are still too few researchers carrying out studies at these levels, and therefore only rather slow progress is being made.

ACKNOWLEDGEMENTS

The author acknowledges with gratitude the contributions of colleagues, graduate students, postdoctoral fellows, visiting scientists, and research assistants to the research reported here. The author also gratefully acknowledges research funding from the Natural Sciences and Engineering Research Council of Canada (Operating and Strategic Grants). Finally, he acknowledges with thanks the gift of the radiata pine seed from the Forest Research Institute, Rotorua, New Zealand.

REFERENCES

Aitken, J., Horgan, K. J., and Thorpe, T. A., 1981, Influence of explant selection on the shoot-forming capacity of juvenile tissue of Pinus radiata, Can. J. For. Res., 11:112-117.

Aitken-Christie, J., Singh, A. P., Horgan, K. J., and Thorpe, T. A., 1985, Explant developmental state and shoot formation in Pinus radiata cotyledons, Bot. Gaz., 146:196-203.

Bender, L., Joy IV, R. W., and Thorpe, T. A., 1987a, Studies on [^{14}C]-glucose metabolism during shoot induction in cultured cotyledon explants of Pinus radiata, Physiol. Plant., 69:428-434.

Bender, L., Joy IV, R. W., and Thorpe, T. A., 1987b, [^{14}C]-Glucose metabolism during shoot bud development in cultured cotyledon explants of Pinus radiata, Plant Cell Physiol., 28:1335-1338.

Biondi, S., and Thorpe, T. A., 1982, Growth regulator effects, metabolite changes and respiration during shoot initiation in cultured cotyledon explants of Pinus radiata, Bot. Gaz., 143:20-25.

Biondi, S., Torrigiani, P., Sansovini, A., and Bagni, N., 1988, Inhibition of polyamine biosynthesis by dicyclohexylamine in cultured cotyledons of Pinus radiata, Physiol. Plant., 72:471-476.

Brown, D. C. W., and Thorpe, T. A., 1986, Plant regeneration by organo-genesis, in: "Cell Culture and Somatic Cell Genetics of Plants," Vol. 3, I. K. Vasil, ed., Academic Press, Inc., New York, pp. 49-65.

Constabel, F., Kurz, W. G. W., Chatson, K. B., and Kirkpatrick, J. W., 1977, Partial synchrony in soybean cell suspension cultures induced by ethylene, J. Cell Res., 105:263-268.

Douglas, T. J., Villalobos, V. M., Thompson, M. R., and Thorpe, T. A., 1982, Lipid and pigment changes during shoot initiation in cultured explants of Pinus radiata, Physiol. Plant., 55:470-477.

Evans, D. A., Sharp, W. R., and Flick, C. E., 1981, Growth and behavior of cell cultures: Embryogenesis and organogenesis, in: "Plant Tissue Culture: Methods and Applications in Agriculture," T. A. Thorpe, ed., Academic Press, New York, pp. 45-113.

Kumar, P. P., Reid, D. M., and Thorpe, T. A., 1987, The role of ethylene and carbon dioxide in differentiation of shoot buds in excised cotyledons of Pinus radiata in vitro, Physiol. Plant., 69:244-252.

Kumar, P. P., Bender, L., and Thorpe, T. A., 1988, Activities of ribulose bisphosphate carboxylase and phosphoenolpyruvate carboxylase and ^{14}C-bicarbonate fixation during in vitro culture of Pinus radiata cotyledons, Plant Physiol., 87:675-679.

Kumar, P. P., and Thorpe, T. A., 1989, Putrescine metabolism in excised cotyledons of Pinus radiata cultured in vitro, Physiol. Plant., (in press).

Obata-Sasamoto, H., Villalobos, V. M., and Thorpe, T. A., 1984, ^{14}C-Metabolism in cultured cotyledon explants of radiata pine, Physiol. Plant., 69:490-496.

Patel, K. R., and Thorpe, T. A., 1984, Histochemical examination of shoot initiation in cultured cotyledon explants of radiata pine, Bot. Gaz., 145:312-322.

Reilly, K. J., and Washer, J., 1977, Vegetative propagation of radiata pine by tissue culture; plantlet formation from embryonic tissue, New Zealand J. For. Sci., 7:199-206.

Thorpe, T. A., 1980, Organogenesis in vitro: Structural, physiological and biochemical aspects, Int. Rev. Cytol. (Suppl.), 11A:71-111.

Thorpe, T. A., 1988, Physiology of bud induction in conifers in vitro, in: "Genetic Manipulation of Woody Plants," J. W. Hanover and D. E. Keathley, eds., Plenum Publishing Corporation, pp. 167-184.

Thorpe, T. A., and Patel, K. R., 1986, Comparative morpho-histological studies on the rates of shoot initiation in various conifer explants, New Zealand J. For. Sci., 16:257-268.

Villalobos, V. M., Leung, D. W. M., and Thorpe, T. A., 1984a, Light cytokinin interactions in shoot formation in cultured cotyledon explants of radiata pine, Physiol. Plant., 61:497-504.

Villalobos, V. M., Oliver, M. J., Yeung, E. C., and Thorpe, T. A., 1984b, Cytokinin-induced switch in development in excised cotyledons of radiata pine cultured in vitro, Physiol. Plant., 61:483-489.

Villalobos, V. M., Yeung, E. C., and Thorpe, T. A., 1985, Origin of adventitious shoots in excised radiata pine cotyledons cultured in vitro, Can. J. Bot., 63:2172-2176.

Yeung, E. C., Aitken, J., Biondi, S., and Thorpe, T. A., 1981, Shoot histogenesis in cotyledon explants of radiata pine, Bot. Gaz., 142: 494-501.

CHANGES IN CARBOHYDRATE METABOLISM DURING TRANSIENT

SENESCENCE OF MUSTARD (SINAPIS ALBA L.) COTYLEDONS

H.I. Kasemir

Biologisches Institut II
Schänzlestr. 1
7800 Freiburg, FRG

INTRODUCTION

   Higher plants very often display a transient senescence
which occurs under unfavorable environmental conditions (for
example in stress situations such as light stress or nitrogen
deficiency). This adaptive senescence can be reversed as soon
as the stress conditions cease. In fact, adaptive senescence
is a characteristic of the flexibility of the plant with
respect to the choice between preserving an organ or giving
it up which in turn may depend on the actual source-sink
situation (Kelly and Davies, 1988). It can be assumed that
this process of "differential sacrifice" takes place even in
one and the same organ at the level of organelles by means of
different controls of either protein degradation or synthesis. In this case, the metabolic state within the cell may be
a decisive tool for regulation (Trippi et al., 1989). By
determining the activity of ribulose-1.5-bisphosphate car-
boxylase/oxygenase (Rubisco, E.C. 4.1.1.39) and, in some
experiments, the mRNA of its small subunit (SSU) we have
tried to find out with mustard seedlings whether chloroplast
degradation during adaptive senescence can be correlated with
parameters of carbohydrate metabolism such as the levels of
glucose, sucrose, and starch. As an experimental approach we
used the reversible transformation of chloroplast into amyloplasts which can be induced by continuous irradiation (cWL)
and accelerated by nitrogen deprivation.

MATERIAL AND METHODS

   Plant material, growth conditions and determination of
Rubisco activity, isolation of total RNA and in vitro translation of SSU-mRNA were performed according to Kasemir et al.
(1988). Glucose, sucrose and starch were determined in accordance with Bergmeyer (1984). The data shown in the figures

Plant Aging: Basic and Applied Approaches
Edited by R. Rodríguez et al.
Plenum Press, New York, 1990

are means of at least 6 independent experiments. Standard
error values lay between 5 and 10%. If not otherwise indicated the seedlings were either treated with 1/10 strength
Hoagland solution or with distilled water leading, in prin-
ciple, to the same results.

RESULTS

Variation in Light Conditions. Figure 1 demonstrates the
fact that under cWL the chloroplasts of developing mustard
seedlings were transformed into amyloplasts within a certain
period of time which strongly depended on the light energy
flux (10 or 30 Wm^{-2}). In contrast, when seedlings were grown
under short day conditions within the experimental period up
to 21 d after sowing no transformation was observed.
The development of amyloplasts under cWL was accompanied
not only by the accumulation of starch (Fig. 2) but also by
high levels of glucose (gluconeogenesis) and between 9 and 18
d after sowing by increasing sucrose accumulation (Fig. 3).
Comparable results were obtained when the seedlings were kept
under long day (16 h light/8 h dark) conditions (data not
shown).

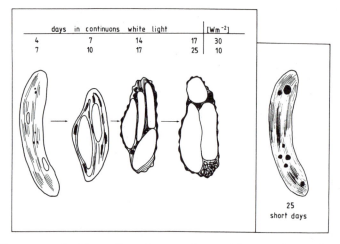

Fig. 1 Representative drawings of sections through plastids
of mustard cotyledons. Seedlings treated with 1/10 strength
Hoagland solution were irradiated with cWL (10 or 30 Wm^{-2}) or
were kept under short day conditions (8 h light/16 h dark).
The numbers indicate days after sowing. either in cWL of 10
Wm^{-2} (lower line) or 30 Wm^{-2} (upper line).

Under cWL Rubisco activity increased up to 6 days after
sowing (Fig. 4) and then a nearly linear decline in enzyme
activity was observed which was less pronounced when seedlings were kept in cWL of lower light energy flux. In contrast to the effects of long-term light conditions only a
slight decline in Rubisco activity appeared under an 8 h
light/16 h dark photoperiod between 6 and 21 days after
sowing. Irrespective of whether the seedlings were kept under
the short day regime from the time of sowing onwards (Kasemir

et al., 1988), or were transferred to these conditions 6 d after sowing (Fig. 4) a similar time course was found with respect to chlorophyll (data not shown). Thus, the lability of Rubisco and chlorophyll appears to depend on the absence of a regular light/dark photoperiod. Similar results were obtained with rice leaves (Kar, 1985).

As shown in Fig. 5 quantitative estimations of SSU-mRNA revealed an optimum curve with a peak 6 d after sowing, irrespective of whether the seedlings were kept under cWL or were exposed to short days. However, the decrease under short day conditions was not as rapid as under the other conditions. As

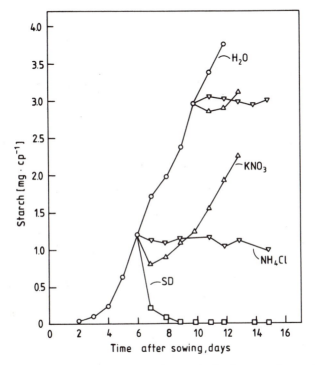

Fig. 2 Accumulation of starch in cotyledons of mustard seedlings which were kept in cWL (o, 30 Wm^{-2}) or transferred to short day conditions ($\square$, 8 h light of 30 Wm^{-2}/16 h dark). At 6 or 10 d after sowing the seedlings were treated with 15 mM solutions of KNO_3 ($\triangle$) or NH_4Cl (∇). cp: pair of cotyledons.

an example, the SSU-mRNA level in seedlings kept in cWL decreased from 100 to approximately 5% between 6 and 9 d after sowing, whereas in plants grown under short-day conditions the decrease was only 50%.

Variation of the N-Status. When senescing mustard seedlings were supplied with 15 mM KNO_3 or NH_4Cl 6 or 12 d after sowing an immediate reaccumulation of Rubisco can be found (Fig. 4). No such effect is seen after the addition of KCl.

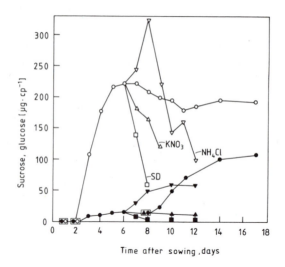

Fig. 3. Accumulation of glucose and sucrose in cotyledons of mustard seedlings which were kept in continuous white light (o, 30 Wm^{-2}) or were transferred to short days (□ ■ , 8 h light/16 h dark) 6 d after sowing. At the same time another batch of seedlings was treated with 15 mM solutions of KNO_3 (△▲) or NH_4Cl (▽▼). cp: pair of cotyledons. Open symbols: glucose; black symbols: sucrose.

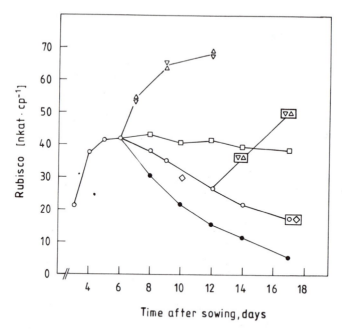

Fig. 4. Rubisco activity as a function of time in mustard cotyledons which were kept in continuous white light (o, 10 Wm^{-2} or •, 30 Wm^{-2}) or were transferred to short days (□, 8h light/16 h dark) 6 d after sowing. At 6 or 12 d after sowing seedlings were treated with 15 mM solutions of KNO_3 (△), NH_4Cl (▽) or KCl (◇). cp: pair of cotyledons.

At present it is still an open question whether a new increase in SSU-mRNA occurs after nitrogen application or whether the tiny amount of remaining mRNA (Fig. 5) is sufficient to sustain Rubisco resynthesis. Independently, the level of chlorophyll increased in a similar way as that of Rubisco (Kasemir et al., 1988).

Regarding the levels of carbohydrates the following results were obtained: after the application of NH_4Cl 6 d after sowing the level of starch (Fig. 2) did not further increase but remained constant. On the other hand, after the application of KNO_3 the level of starch decreased during the first day and then continued to increase (Fig. 3). A remarkable decrease in glucose was found with the exception that during 2 d after NH_4^+ application the level temporarily increased. In contrast, sucrose increased between 6 and 10 d after sowing and remained constant thereafter.

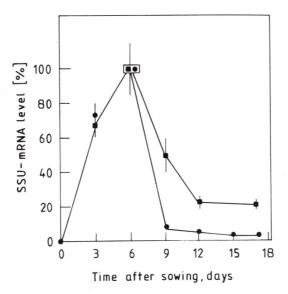

Fig. 5. Relative amounts of *in vitro* translatable SSU-mRNA. Seedlings were either kept in continuous white light 30 Wm^{-2} (●) or under short day conditions (■).

DISCUSSION

As indicated by the constant level of Rubisco (Fig. 4) and the low levels of carbohydrates (Figs. 2, 3) mustard seedlings under short day conditions were in a well-balanced metabolic situation although their N-status (1/10 strength Hoagland solution or even distilled water) was low.

In this stage the chloroplasts remained completely intact (as far as the ultrastructural results are concerned) and might be considered as a source for the rest of the cell. When seedlings were exposed to light stress, however, the metabolic situation changed completely. Depending on the light energy flux the cotyledons started senescence (Fig. 4) as soon as the chloroplasts accumulate huge amounts of starch in the stromal matrix (Fig. 2) and developed into

amyloplasts (Fig. 1) which are known to posses an
unidirectional gluconeogenic pathway (Ngernprasirtsiri et
al., 1989). Accordingly, the amyloplasts of mustard
cotyledons accumulated glucose ten times higher than sucrose
(Fig. 3). The low level of sucrose may be correlated with a
glycolytic breakdown in the cytoplasm whereby the resulting
C-compounds were utilised for the synthesis of starch
(Ngernprasirtsiri et al., 1989). In this connection the
transformed plastids should be considered as sinks within the
cells of mustard cotyledons.

Recently Luna and Trippi (1986) have shown with oat
leaves that accumulated sugars accelerated senescence by
increasing photooxidative processes associated with the
increase of membrane permeability. Since senescence in
mustard cotyledons was hastened by increasing light energy
flux (Fig. 4) a similar mechanism may proceed under our
experimental conditions. As soon as the light stress was
terminated 6 d after sowing by transferring the seedlings to
short day conditions the levels of carbohydrates decreased
rapidly (Figs. 2, 3) and concommittantly amyloplasts were re-
transformed into chloroplasts (data not shown) resulting in
an immediate stop of Rubisco decrease (Fig. 4). An additional
increase of Rubisco can only be registered if the seedlings
were treated with N-compounds (NH_4Cl, KNO_3, Fig. 4). In this
case the correlation between the levels of carbohydrates
(Figs. 2, 3) and Rubisco remains obscure at present. Very
probably the reversion of N-deficiency concerned primarily
the metabolism of amino acids and proteins (Coleman et al.,
1988). It is still an open question to what extend the level
of SSU-mRNA reacts to changes in N-status. Thus, in future
the involvement of regulation at the level of mRNAs as
indicated in Fig. 5 has to be investigated in more detail.

Acknowledgement - The investigation was supported by grants
from the Project Europäisches Forschungszentrum für Maßnahmen
zur Luftreinerhaltung (PEF), Kernforschungszentrum Karlsruhe,
FRG. The technical assistance of C. Nolte was gratefully
acknowledged.

REFERENCES

Bergmeyer, H.U., 1984, Methods of Enzymatic Analysis, Vol.
        VI, Metabolites 1: Carbohydrates, Verlag Chemie, Wein-
        heim, pp. 2-10 and 96-103.
Coleman, L.W., Rosen, B.H. and Schwarzbach, S.D., 1988, Pre-
        ferential loss of chloroplast proteins in nitrogen
        deficient *Euglena*, Plant Cell Physiol., 29:1007.
Kar, M., 1985, The effect of photoperiod on chlorophyll loss
        and lipid peroxidation in excised senescing rice leaves,
        J. Plant Physiol., 123:389.
Kasemir, H., Rosemann, D. and Oelmüller, R., 1988, Changes in
        ribulose-1,5-bisphosphate carboxylase and its translata-
        ble small subunit mRNA levels during senescence of mus-
        tard (*Sinapis alba*) cotyledons, Physiol. Plant., 73:257.
Kelly, M.O. and Davies, P.J., 1988, The control of whole
        plant senescence, in: CRC Critical Review of Plant
        Science, Vol. 7, Issue 2, pp. 139-173.

Luna, C.M. and Trippi, V.S., 1986, Membrane Permeability - Regulation by exogenous sugars during senescence of oat leaf in light and darkness, <u>Plant Cell Physiol.</u>,27:1051.

Ngernprasirtsiri, J., Harinasut, P., Macherel, D., Strzalka, K., Takabe, T., Akazawa, T. and Kojima, K., 1988, Isolation and characterization of the amyloplast envelope-membrane from cultured white-wild cells of sycamore (*Acer pseudoplatanus* L.), <u>Plant Physiol.</u>, 87:371.

Ngernprasirtsiri, J., Takabe, T. and Akazawa, T., 1989, Immunochemical analysis shows that an ATP/ADP-translocator is assosiated with the inner-envelope membranes of amyloplasts from *Acer pseudoplatanus* L., <u>Plant Physiol.</u>, 89:1024.

Trippi, V.S. and De Luca d'Oro, G.M., 1985, The senescence process in oat leaves and its regulation by oxygen concentration and light irradiance, <u>Plant Cell Physiol.</u>, 26:1303.

Trippi, V.S., Gidrol, X. and Pradet, A., 1989, Effects of oxidative stress caused by oxygen and hydrogen peroxide on energy metabolism and senescence in oat leaves, <u>Plant Cell Physiol.</u>, 30:157.

STUDY OF GENE EXPRESSION DURING IN VITRO CULTURE
OF TOBACCO THIN CELL LAYERS BY TWO-DIMENSIONAL
ELECTROPHORESIS OF PROTEINS

Kiêm Tran Thanh Van(*),
Michel Zivy (**),
Alain Cousson (*),
et Hervé Thiellement (**)

(*) Institut de Physiologie Végétale. CNRS,
91190 Gif sur Yvette. France
(**) Laboratoire de Génétique des Systèmes Végétaux
CNRS-INRA-UPS. La Ferme du Moulon,
91190 Gif sur Yvette. France

INTRODUCTION

Compared to animal systems, plant systems have tremendous potential for organ regeneration, embryogenesis from somatic cells, pollen grains, and ovules, for early sexual reproduction in in vivo conditions and in vitro conditions. Advanced progress already made in genetic manipulation technology would have led to important plant improvement if its feasibility was not limited to a small number of species. This limit is due to the difficulties in controlling in vitro plant regeneration via protoplasts, cells, tissues and/or organs culture of recalcitrant species such as monocotyledons, leguminous, and woody species which also have great economic importance. It is generally observed in recalcitrant species that regeneration of organs can be induced only during the juvenile phase. For example, mature tissues of woody plants lose their ability to differentiate organs in vitro . It is important to understand the phenomenon of maturation in recalcitrant species as it is important to understand the mechanisms of organ regeneration in recalcitrant and nonrecalcitrant species. Progress is slow, because of the lack of regeneration /maturation mutants and because of the complexity of the eucaryotic genome. In species such as Arabidopsis for which the genome size is relatively reduced and for which several mutants are available, it still lacks developmental and regeneration mutants. There exist mutants with altered flower shape; however, these mutations are not suitable for studying the control of organogenesis.

Abbreviations :

2D-PAGE : two-dimensional polyacrylamide gel electrophoresis
rubisco : ribulose bispohosphate carboxylase/oxygenase
LSR : large subunit of rubisco
TCL : thin cell layer
d_0, d_n : day 0, day n

Plant Aging: Basic and Applied Approaches
Edited by R. Rodríguez *et al.*
Plenum Press, New York, 1990

Developmental "variants" (and not mutants) have been obtained in hypohaploïd of Nicotiana tabacum (Tran Thanh Van et al. 1978) and Nicotiana plumbaginifolia (Tran Thanh Van et al. 1985, Cannon et al. 1989).New variants of the juvenile type, both green and albinos, which do not develop into mature plants have recently been isolated in our laboratory. More drastic selection procedures and mutagenesis would allow us to select juvenile/mature mutants in order to analyze the process of maturation/senescence. As for species easy to regenerate in vitro such as tobacco (Tran Thanh Van et al. 1981), Petunia (Mulin et al. 1988), or carrot (Sung et al. 1983), not all patterns of morphogenesis can currently be induced in vitro . Biochemical/molecular changes during morphogenetic processes are not yet systematically described nor fully understood.

In this paper, we report changes in protein synthesis of Nicotiana tabacum thin cell layers during in vitro culture in conditions inducing root or flower differentiation and in conditions leading to the absence of morphogenesis. The latter pathway is conceived in order to account for changes induced by in vitro culture conditions themselves (Cousson et al. 1989).

Thin cell layers excised from the surface of the floral branches of tobacco (Nicotiana tabacum) can be induced in vitro to form either flowers, roots, vegetative buds, or callus, by varying glucose, indolebutyric acid and kinetin concentrations (Tran Thanh Van, 1973, 1981). The experiments presented here were conducted to study changes in gene expression during in vitro culture, when morphogenesis was oriented towards the formation of flowers or roots : total proteins were studied by two-dimensional polyacrylamide gel electrophoresis (2D-PAGE) (O'Farrell,1975). As changes in proteins were expected before visible organ development, several stages preceding organogenesis were studied. Some modifications occurring during in vitro culture can be due to certain phenomena that are not directly related to morphogenesis (adaptation to the culture medium, to the excision, etc…); in order to distinguish these changes from those involved directly in organ formation, TCLs conditioned to form roots or flowers were compared to TCLs cultured in a culture medium which does not induce morphological change.

MATERIAL AND METHODS

Plant material and method of in vitro culture : The TCLs were excised from basal parts of floral branches of Nicotiana tabacum L. cv Samsun when the inflorescence contained approximately 20 green fruits. The floral branches were surface-sterilized by (7% $Ca(OCl)_2$ for 10 min), and then rinsed three times in sterile water. Twenty TCLs were floated on liquid medium (40 ml) in Petri dishes (diameter 10cm). The basal medium was composed of Murashige and Skoog's macro- and micro-elements, myo-inositol (100 mg/l), thiamine-HCl (0.4 mg/l) and glucose (30 g/l) (Tran Thanh Van, 1980). For root formation, kinetin at 2×10^{-7}M and indolebutyric acid at 7×10^{-6}M were added, and the medium pH was adjusted to 5.8. For flower formation, kinetin and indolebutyric acid were both at 3×10^{-7}M and the medium pH adjusted to 3.8. The absence of morphogenesis was obtained by increasing the TCL density by a factor of 2.5 (i.e. 50 TCL were floated on 40 ml of medium).

Protein extraction : 50 to 100 mg of TCL were taken for each extract (enough for several gels) and kept in liquid nitrogen until use. TCLs were taken after 2, 4, 8, and 10 days of in vitro culture in media for the formation of roots (R), of flowers (F), or for the absence of morphogenesis (AM). TCLs were also taken at day 0 (d_0), i.e. before in vitro culture. Proteins were extracted and denatured according to Zivy (Damerval et al., 1986) : TCLs were dry-crushed in a liquid nitrogen cooled mortar, and the powder was resuspended in acetone with 10% trichloroacetic acid and 0.07% 2-mercaptoethanol. After protein precipitation and centrifugation, the pellet was rinsed in acetone with 0.07% 2-mercapto-ethanol. It was then dried under vacuum and the proteins were solubilized in a solution containing 9.5 M urea, 5mM K_2CO_3, 1.25% SDS, 0.5% dithiothreitol, 2% LKB Ampholines pH 3.5 to 9.5, 6% Triton X-100. Forty μl of this solution were used per mg of dried pellet. After centrifugation the pellet was discarded.

2-D PAGE : performed according to Colas des Francs et al. (1985), except the ampholyte mixture for isofocusing was 80% Pharmalyte pH 5-8, 10% Pharmalyte pH 3-10, 10% Servalyte pH 3-10. The 2D-gel dimensions were 160x165x1mm. They were bound to Gelbond PAG. Acrylamide concentration was 11%.

Protein revelation : proteins were silver stained according to Oakley et al. (1980) in the apparatus described by Granier et al. (1986).

Gel comparisons : each type of morphogenetic pathway was represented by 3 gels from different extracts. Differences were only recorded when they were found in all 3 gels.

RESULTS

1) Global effect of in vitro culture : common evolution of the patterns in 3 different morphogenetic pathways, decrease of the large subunit of Rubisco.

For most proteins, the evolution from d_0 to d_{10} was similar in AM, F, and R TCLs. The patterns at d_0 and at d_{10} were extremely different. At d_0, a few polypeptides, including the large subunit of ribulose bisphosphate carboxylase/oxygenase (Rubisco), were very abundant (fig. 1A). Most of these polypeptides decreased dramatically up to d_{10} (fig. 1B). This decrease began early; patterns at d_4 were intermediate between those from d_0 and those at d_{10}. In the same period, numerous spots appeared or increased in intensity. A group of 5 polypeptides (spots 4 to 8) of similar molecular weights but with different pi's (arrows in fig. 1B) had a particular behavior : they were not observed at d_0, but had appeared, and were relatively abundant by d_2, then stayed approximately at the same level up to d_{10}.

2) Differences between the patterns of TCLs in different morphogenetic pathways

First differences were observed at d_8, and they were clearer at d_{10}. Examples of variations are shown in figure 2. At this stage, 22 spots were more intense and 46 less intense in R compared with AM. In F, these numbers were 7 and 32 respectively. Some of these spots varied similarly in R and in F (tab. 1). The large subunit of Rubisco (LSR) was among the polypeptides that specifically decreased in R (fig. 2C and 2D).

Except for a few spots such as LSR and spots 9 and 10 (fig. 2G, 2H, and 2I), the variations in protein amounts were of relatively low amplitude, and no presence/absence variation was observed, i.e. no spot specific to one or to the other differentiation pathway was observed.

Spots 9 and 10 are those that showed the most important quantitative variation. Small amounts of these polypeptides were already present at d_4 in the 3 morphogenetic pathways, and from d_8 their amounts in R and F were much higher than in AM and relatively higher in R than in F. The position of spots 9 and 10 in the first dimension was not stable; this might be due to an unstable interaction

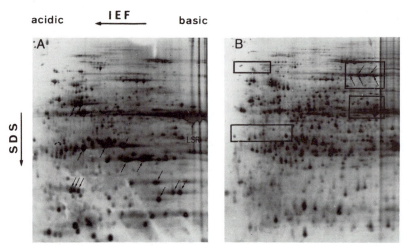

acidic   **I E F**   basic

SDS

Figure 1.   Evolution of the protein pattern during 10 days of in vitro culture.
A: TCL at d_0 ; B : TCL in the non-morphogenetical pathway at d_{10}. Arrows in A show polypeptides decreasing markedly from d_0 to d_{10}. The arrows in B (in a square) shows the group of spots 4,5,6,7 and 8, that were absent at d_0 and appeared from d_2. The squares show parts of the gel that are en-larged in fig. 2.

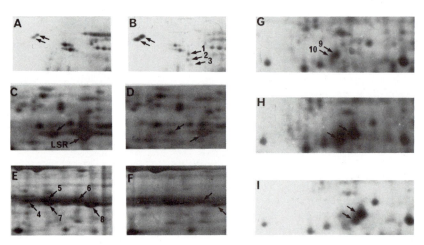

Figure 2.   Examples of variations between TCLs in the different morphoge-netic pathways at d_{10}.
A,C,E : F TCL ; B,D,F : R TCL ; G,H,I : respectively AM,R, and F TCLs. Arrowed spots show spots varying according to the morphogenetic pathway. Numbered spots are cited in the text. LSR : large subunit of Rubisco. The position of spots 9 and 10 was not stable; they were sometimes found in a more basic position.

of these polypeptide with the ampholytes, or to posttranslational modifications.

DISCUSSION

1) Changes common to the 3 morphogenetic pathways

A very important modification of the protein pattern was observed after 10 days of in vitro culture. It is likely that the large decrease in intensity of the spots corresponding to polypeptides very abundant at d_0 corresponds to a stop or decline of their synthesis at the beginning of in vitro culture. As the large subunit of Rubisco was among these polypeptides, it is possible that at least some of them were also chloroplastic polypeptides. The intensity increase of a large number of spots observed during the same period could simply be relative; these protein probably correspond to "house keeping" proteins, whose synthesis stayed relatively stable but whose relative amount in the cells increased by virtue of the large decrease of the others. Thus, the majority of the changes observed in 10 days of in vitro culture probably correspond to the decrease of some functions initially exercised by the cells before being excised from the plant, such as photosynthetic activity. However, the behavior of a series of polypeptides (spots 4 to 8) did not fit with this interpretation, since they accumulated in a relatively large quantity in the 3 morphogenetic pathways before the above mentioned decrease; these polypeptides, absent in original plant tissue, seem to be actively synthesized shortly after the beginning of in vitro culture. They may be related to the wound of excision, and/or to a reaction to in vitro culture conditions (growth substances, ionic components, pH). Interestingly, at least 2 of these polypeptides became less abundant in the R TCLs than in the AM and F TCLs at d_{10} (see fig. 2E and 2F).

2) Differences between proteins patterns of TCLs in 3 different morphogenetic pathways

A large part of the differences observed at d_{10} allowed a discrimination between AM on the one hand and F and R on the other. These changes possibly correspond to the intense mitotic activity observed in F and R TCLs.

In addition, changes specific to F (11 spots) as well as to R (40 spots) were found. They allowed us to discriminate without ambiguity the two types of TCL at d_{10}, i.e. before macroscopic morphological changes : they can thus be considered as early markers of differentiation.

From d_0 to d_8 or d_{10}, the evolution of these polypeptides was similar in the 3 morphogenetic pathways. Some of them decreased in quantity : for them, the difference between the morphogenetic pathways could be due to the restart of the synthesis of the protein in one pathway, while it continues to be slowly degraded in the others. Another possibility is that the protein is more actively degraded in one condition than in the others. This might be the case of the large subunit of Rubisco, which continued to decrease in R TCLs, whereas it stopped decreasing or decreased less in AM and F TCLs.

Others differences were observed in spots that increased in intensity or appeared from d_0 to d_{10}. As discussed above, they do not necessarily correspond to polypeptides actually increasing in absolute quantity. However it is clear for some of them, that they were actively synthesized during in vitro culture : for example, spots 1, 2, and 3 appeared by d_8, and continued to increase up to d_{10} in R TCLs, although no important variation of large spots occurred at this stage. This is also very clear for spots 9 and 10, which accumulated in very large amounts in F and specially in R TCLs.

Kay et al. (1987) observed differences for the activity of some peroxydase isozymes (EC 1.11.1.7) between TCLs induced to produce flowers, vegetative buds, or callus according to Tran Thanh Van (1980). The activity of peroxydases is known to be generally

Table 1. Number of spots varying according to the morphogenetic pathway at d_{10}.

Compared to AM	In F and R	In R only	In F only
More intense	4	18	3
Less intense	24	22	8

correlated to growth regulators and differentiation. We have no evidence that these enzymes are revealed by our method.

CONCLUSION

In the present study five polypeptides (spot 4 to 8) which appeared by d_2 in three morphogenetic pathways : flowers, roots and absence of morphogenesis could be considered as stress proteins. No polypeptide was found specific to one morphogenetic pathway : only quantitative differences were found. They concerned numerous polypeptides. This can be due to the fact that silver stain reveals accumulated polypeptides, whether they are newly synthesized or not. It is likely that some changes at the level of protein synthesis were masked by the presence of proteins previously synthesized.

Experiments currently in progress using radioactive labelling will allow study of newly synthesized polypeptides. These experiments will give new results on the synthesis of polypeptides related to organogenetic induction, and will perhaps allow us to detect differences between morphogenetic pathways before the 8th day of in vitro culture. Differences in mRNA were detected in TCL induced to form flowers or vegetative buds at d_7 (Meeks-Wagner et al. 1989) when kinetin or zeatin were used respectively (Tran Thanh Van et al. 1981). In order to understand the mechanisms of control of morphogenesis, the function of specific proteins is yet to be determined. Preliminary results showed that different cell wall enzymes (ß1-3 glucanase, pectylmethylesterase...) are induced in different morphogenetic programs (unpublished results in collaboration with G. Noat, S. Mutafstchiev, and L. Richard).Further analysis is being carried out in order to define the functions of these enzymes. It has been postulated that one of these functions may be the release of oligosaccharides from the cell wall (Tran Thanh Van et al. 1985, Mutafstchiev et al. 1988). Some of these enzymes are the same ones which were considered up to now to be related to stress or defense mechanisms. Since wounding and cell division occur in both types of morphogenetic pattern, comparison of these patterns allows selection of genes implied in each type of morphogenesis. Furthermore, comparison with other morphogenetic programs such as absence of morphogenesis, callus, or root differentiation will allow determination of genes expression specific and/or common to different morphogenetic pathways and to different stages of a given morphogenetic pathway.

Absence of morphogenesis is conceived in order to study evolution of cell/tissue in in vitro culture conditions without interferences with organogenesis. In conclusion, TCL can be considered as a model system for systematic study of molecular markers of in vitro morphogenetic differentiation. By their small size and their great potential for organogenesis, TCL are also appropriate system for systematic study of gene insertion via agrobacterium, microjectiles bombardment or DNA uptake.

Acknowledgements: The authors thank D. Quigg for his help in preparing and editing this manuscript.

REFERENCES

Cannon G., Tran Thanh Van K., Heinhorst S., Trinh H., and Weissbach
    A.,1989, Plastid DNA in tobacco hypohaploïd plants. Plant
    Physiology, 90:390.
Colas des Francs C., and Thiellement H., 1985, Chromosomal
    localization of structural genes and regulatores in wheat by
    2D electrophoresis of ditelosomic lines. Theor. Appl. Genet.,
    71:31.
Cousson A., Toubart P.,Tran Thanh Van K., 1989, Control of
    morphogenesis in tobaco thin cell layer by culture medium pH.
    Can.Journ.of Bot.(in press).
Damerval C., de Vienne D.,Zivy M., and Thiellement H., 1986,
    Technical improvements in two-dimensional electrophoresis
    increase the level of genetic variation detected in wheat
    seedling proteins. Electrophoresis, 7:52.
Granier F. and de Vienne D., 1986, Silver staining of proteins :
    standardized procedure for two-dimensional gels bound to
    polyester sheets. Anal. Biochem.,155,45.
Kay L. E. and Basile D. V., 1987, Specific peroxydase isoenzymes
    are correlated with organogenesis. Plant Physiol., 84:99.
Meeks-Wagner D.R., Dennis E.S., Tran Thanh Van K., and Peacock W.J.,
    1989, Tobacco genes expressed in in vitro floral initiation
    and their expression during normal plant development. The Plant
    Cell., 1: 25.
Mulin M. and Tran Thanh Van K., 1989, Obtention in vitro of flowers
    from thin epidermal cell layers of partial somatic hybrid
    between Petunia hybrida . Plant Science., 62:113.
Mutafstchiev S., Cousson A.,Tran Thanh Van K.,1988, Effect of chemical
    factors in the regulation of cell growth, in "Monograph 16 British
    group on plant growth regulators, M. Jackson.
Oakley B. B., Kirsh D. R., and Morris N. R.,1980, A simplified silver
    stain for detecting proteins in polyacrylamide gels. Anal.
    Biochem.,105:361.
O'Farrell P. H.,1975, High resolution two-dimensional
    electrophoresis of proteins.J. Biol. Chem., 250:4007.
Sung Z.R. and Okimoto R., 1983, Coordinate gene expression during
    somatic embryogenesis in carrots. Proc. Natl. Acad. Sci. USA.
    80: 2661
Tran Thanh Van K.,1973, In vitro control of de novo flower, bud,
    root, and callus differentiation from excised epidermal tissue.
    Nature, 246:44.
Tran Thanh Van K.,1980, Thin cell layers : control of morphogenesis
    by inherent factors and exogenously applied factors. in Int.
    Rev. Cytol.,11A, Vasil I., ed., Academic Press,175.
Tran Thanh Van K.,1981, Control of morphogenesis. Ann. Rev. Plant
    Physiol., 32:291.
Tran Thanh Van K. and Marcotte J.L.,1981, Influence of the nature of
    cytokinins on the control of organogenesis in tobacco thin cell
    layer. Proceedings of XIIl Int. Botanical Congress, Sydney,
    Australia.
Tran Thanh Van K., Toubart P., Cousson A., Darvill A., Gollin D.,
    Chelf P., and Albersheim P., 1985, Manipulation of the
    morphogenetic pathways of tobacco explants by oligosaccharides.
    Nature, Vol. 314,6012:615.

AN EXPERIMENTAL MODEL FOR THE ANALYSIS OF PLANT/CELL DIFFERENTIATION :

THIN CELL LAYER

CONCEPT, STRATEGY, METHODS, RECORDS AND POTENTIAL

Kiem Tran Thanh Van
Luc Richard
Cyrille Amin Gendy

Laboratoire de Physiologie Végétale
CNRS
91190 Gif sur Yvette, France

THIN CELL LAYER CONCEPT

Plant differentiation including morphogenesis, growth, development and plant senescense result from complex interaction between different organs, tissues and cells. The analysis at the molecular level of the mechanisms of differentiation is difficult due to the lack of knowledge on i) the nature of the morphogenetic signal, ii) the perception and transduction mechanisms and iii) the localization of target cells and of responsive cells.

Thin cell layer (TCL) system composed of one or a few cell layers are simplified experimental system in which organ interaction is suppressed and a minimum of tissue / cell interaction is maintained. They are different from isolated cells and protoplasts for which the cell wall previously structured during the ontogenesis of the plant is disrupted or destroyed. It has been shown that cell wall is an important recipient of biologically active factors (Albersheim et al., 1983, Tran Thanh Van et al., 1985).

STRATEGY

In order to answer to the question of how plant cell sense and process signals for differentiation, our strategy is to select for signal(s) which induce specific responses in defnded target cell(s).
In a monolayer system such as epidermis TCL, the target cells are also the responsive cells. In TCL system, the responsive cells are well defined and are not scattered all through a heterogenous organism, organ or callus. Chemical signals such as kinetin, zeatin, dihydrozeatin, and oligosaccharides induce specific differentiation pattern which we refer to as specific program of differentiation.

METHODS

1. Classical TCL method

TCL method have been previously described (Tran Thanh Van, 1973, Tran Thanh Van et al., 1974, Tran Thanh Van 1980[a], Tran Thanh Van, 1983 and Tran Thanh Van et al. 1985 [a, b]).

Thin cell layer (TCL) are excised longitudinally or transversally from different organ types. Their small size (from some microns up to 1 mm x 5 mm) and the reduced cell number allow TCL to be uniformely exposed to signal molecules or to foreign DNA for gene insertion.

Longitudinal TCL (lTCL) are used when definited cell type (epidermal, subepidermal, cortical, cambial or medullary cells) are to be analysed. lTCL can be excised from stem, leaf vein, floral stalk, petiole, pedicel, bulb scale etc... As for transverse TCL (tTCL), other organs can also be used : leaf blade, root, rhizome, floral organs (sepal, petal, anther, filament, pistil), young spiklets, meristem, stem nodes etc... Micro-thin cell layer (μTCL) of some microns thick were obtained using a vibrotome (Hackett et al, 1989, in these proceedings).

Several species have been successfully tested in our laboratory (Nautilocalyx, Begonia, Torenia, tobacco, Cichorium, Brassica, Malva, Psophocarpus, Vicia, Soja, Lens, Catharanthus, Saintpaulia, Petunia, Tomato, Antirrhinum, Beta, Phalaenopsis, Triticum, Iris, Bambusa...). TCL have been used to obtain callus or bud regeneration (Douglas Fir, Sequoia, Sequoiadendron, chinese willow, Eucalyptus, Salix...)

The culture medium substrates used were solidified (agar or gelrite) or liquid. The culture containers were test tubes (of 15 cm and 2.5 cm of diameter) or petri dishes (of 10 cm, 5 cm or 2 cm of diameter) according to the culture medium volume and to the atmosphere volume to be offered to the explant. Light intensity as well as ethylen concentration have been shown to be critical to the expression of the programme of differentiation (Tran Thanh Van 1980[b], Tran Thanh Van 1981, Mutafstchiev et al. 1987).

2. TCL modified method

In order to study the kinetic of different biosynthetic pathways using precursor / inhibitor incorporation and in order to reduce to the minimum the intrinsic variations of the environmental factors as well as those inherent to the biological material, a modified method has been developed in our laboratory (Richard 1987).Multiwell dishes of 24 wells of 1,6 cm diameter ("Costar 3424" dishes of 12,7 cm length x 8,5 cm width x 1,8 cm height) were used as container. Each well containing definited volume of culture medium (500 μl to 1 ml) for definited atmosphere / culture medium volumes ratio, received one TCL explant. An elastic plastic film isolated each individual well in order to prevent gaz exchange between different wells and contamination risk. The dish was sealed with 2 layer(s) of "urgo" film.

This modified method offers a number of advantages. There is a random distribution of TCL explants in each dish. Due to the reduced dimensions of each multiwell dish, a greater number of TCL explants are exposed to more homogenous environmental conditions (light, temperature, humidity, CO_2, O_2, ethylen...) leading to a less variable background. In this way, the influence on the program of differentiation of factors or chemical(s) added to the culture medium

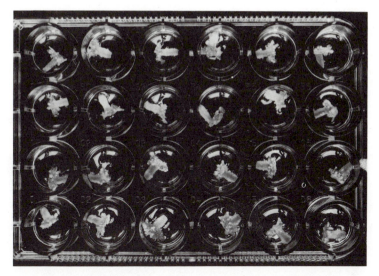

Figure 1.   24 thin cell layer (TCL) explants cultured in a multiwell
dish.

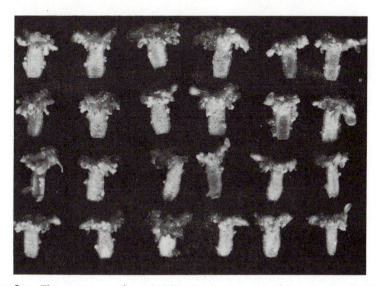

Figure 2.   The same serie of TCL explants outside of the multiwell
dish.

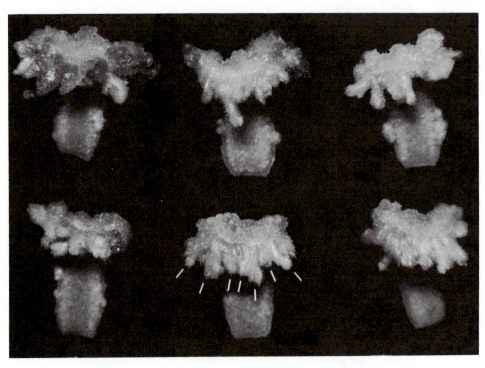

Figure 3.   Close up of 6 of the same series of explants showing flowers indicate by white marks.

can be detected in a more sensitive way. Another major advantage consists of the possibility to change the medium (by syringe), withoutremoving the TCL explant, therefore the change in gaz atmosphere was negligeable. Furthemore, experiments can be designed with a lesser number of TCL explants (8 to 12 per experiment instead 2 x 25 or 2 x 20 for experiment using test tubes or petridish respectively) and a greater number of repetitions (4 to 6 instead of 3) for statistical test analysis.

This method have been made available to other joint research groups (in particular to R. Meeks-Wagner since 1986 then in the Canberra group, to A. Havelange and P. Denis, 1986 of the Liège group and to D. Mohnen and P. Toubart in 1987 of the Athens group during their visit to our group...), who have confirmed its efficiency. However, compared to solidified agar medium in test tubes, the average number of organs formed per TCL explant was lower. This number varied according to the environmental conditions applied to the donor plants and to TCl explants. In our experimental conditions, the average number of flowers per TCL is 20 and 10 on solidified and on liquid medium respectively. This number decreased strongly when the donor plants and TCL were grown under stress conditions (ozone pollution, water and temperature stress, reverse light regime, vibrations in the growth chambers).

2.1 Conditions for controlled morphogenetic differentiation on tobacco TCL cultured in multiwell dish

Morphogenetic Programs

Parameters	Flower	Vegetative Bud	Root	Absence of Morphogenesis
AIB (µM)	1	3	14	0.5
Kin (µM)	1	0.5	0.4	0.5
pH	5.0	5.0	5.0	6.0
Culture Medium Volume(ml)	1	1.5	1.5	1.5

- Basal Murashige Skoog medium containing 30 g/l of glucose
- Light intensity = 100 µE ± 0,8 / m^2/s
- Photoperiod : 24 h
- Temperature 25 C° ±2
- 3 repetitions, 12 TCL explants per experiment

The homogeneity of morphogenetic responses of tobacco TCL cultured in multiwell dish is illustrated in Fig. 1, 2 and 3 for flower program.

2.2. Effect on morphogenetic differentiation of IBA/kinetin concentration and ratio at different culture medium pH.

Four combinations of 2 concentrations of IBA and Kinetin, three ratios (0.5,1 and 2) and 3 culture medium pH values (4.0, 5.0 and 6.0) were selected from previous study (Cousson et al. 1988). Fig. 4 shows that the number of flowers formed per TCL was higher when [IBA] = [Kinetin] = 1 µM. This holds true for the 3 pH tested ; pH 4.0 being the optimum (10 ± 1 flowers per TCL). For [IBA] = [Kinetin] = 0.5 µM, the number of flower is lower at pH 4.0 (4 ± 1 flowers per TCL) and 5.0, at pH 6.0, no flowers nor any other organ are obtained. We refer to it as an "Absence of morphogenesis program". pH had a determining effect depending upon the concentration and ratio of growth substances (when the [IBA] = 0.5 µM and [Kinetin] = 1 µM).Thus both pH and growth substances play an important role in the control of differentiation.

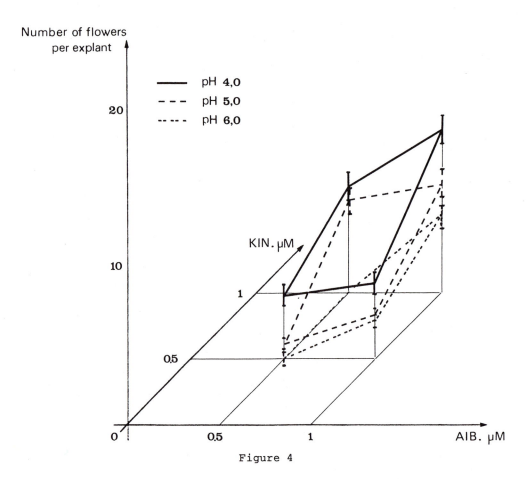

Figure 4

2.3 Effect of oligosaccharides (cell-wall released fragments)

The impact on differentiation of exogenous supply of oligosaccharides can only be tested once the effect of pH as well as IBA/kinetin concentration and ratio determined. As previously shown

(Mutafschiev et al. 1987), the dose effect of oligosaccharides on tobacco TCL differentiation depend on the culture medium pH. It was shown that supply of oligosaccharides (released by endopolygalacturonase) decreased the number of root formed per explant on root induction medium and induced flower formation on the same medium depending upon the variation of pH during the culture period (buffered or unbuffered pH).

Our working hypothesis is as follows : growth susbtances (auxins, cytokinins), external and internal pH and ionic concentration can, among other roles, activate specific cell wall enzymes which release oligosaccharides.

RECORDS

The TCL method has led to the conception and realisation of pure morphogentic programs, with two characteristics : intensity and short scale timing of the responses (Tran Thanh Van 1980[b], 1981).It has allowed the determination of reversibility and irreversibility stages. These records are
prerequisite conditions for the study of mechanisms of differentiation via i) the determination of molecular marker(s), ii) the identification and selection of gene(s) specific to the process of morphogenetic induction, iii) the search for gene function and regulation. The sensitivity of TCL method has led to i) the selection of specific signal(s) : kinetin / zeatin / dihydrozeatin or ions (A. Gendy, unpublished results) and ii) the purification of biologically active chemicals such as oligosaccharides.

1. Records of TCL method in the findings of molecular markers of differentiation.

Several research groups as well as our group have identified molecular makers during the time course of tobacco TCL differentiation. Peroxydases (Thorpe et al.1978, Kay et al.1987), polyamins (Tiburcio et al. 1988, Tiburcio et al.1989) as well as a protein bound to spermidine during flower differentiation (Apelbaum et al.1988). cDNA clones (Meeks-Wagner et al. 1989) and proteins (Tran Thanh Van et al. 1989, in these proceedings, Richard et al., unpublished results) have been isolated.

2. Advantages of thin cell layer method for the study of transformation via Agrobacterium or direct gene transfer.

The transformation of tobacco and Petunia by Agrobacterium is an efficient process. However, systematic analysis of transformation process has not been undertaken. In particular, several basic aspects are yet to be studied : the impact of antibiotic supply itself on the morphogenetic expression as well as the influence of wounded cells on early steps of Agrobacteria attachment to the cells, the changes in the cell wall and cell membranes structures and properties, and the possible relationship between tDNA insertion and the state of cellular differentiation (cell cycle).
In our laboratory, systematic study of the influence of different antibiotics (carbenicillin, kanamycin, cefotaxim, vancomycin, rifampicin,...) used in the transformation process has been developed (A. Gendy in part in collaboration with D. Valvekens, Gent group). Preliminary results have shown that changes in the thin cell layer

tobacco morphogenetic differentiation occured depending upon the concentration of antibiotic. Supply of 150 mg / l of cefotaxim to vegetative bud inducing medium decreased the number of bud per TCL and induced flower formation, this on 100 % of explants. At 250 and 500 mg/l, bud were inhibited and only flowers were obtained respectively. Therefore, study of the influence and role of antibiotic on cell differentiation in each plant material is an important step in transformation study.

Compared to tobacco leaf disc where target cells and responsive cells are limited to the edge around the leaf disc and therefore the frequency of transformation events is low, on the contrary, thin cell layers explants with the high ability of the subepidermal layer to regenerate vegetative buds (up to 800 bud per explant) or flowers (up to 50 flowers per explant) in a short delay of time (10-12 days) are more suitable experimental model for systematic study of i) the transformation events, (via Agrobacterium, DNA uptake or microprojectiles coated DNA) ii) the inheritance of the inserted trait, and iii) the expression organ specific genes.

In tobacco and Arabidopsis TCL, the study of spatial / temporal cell / tissue / organ speicific gene expression is in progress (A. Gendy, C. Spriet) using the glucuronidase synthase (GUS) gene as marker gene.

POTENTIAL

Controlled and pure developmental patterns (flowers, vegetative buds, roots, callus, smatic embryos) have been obtained using thin cell layer method. The higher sensitivity and accessibility to signal molecules and to foreign DNA, the knowledges of morphological / cytological changes and of some molecular markers has led to new developments : fondamental study of mechanisms of morphogenetic differentiation as well as applied aspect in biotecnhology. Using TCL method, transgenic plants and seeds can be obtained in a short delay (Ammirato, 1987).

Genes expressed in tobbaco TCL are also expressed in entire plants. By combining two controlled morphogenetic programs, TCL method allows to "mimic" the entire plant by progressively inceasing the complexity of inter organ network, for example : direct root formation can be programmed before of after flower formation without or with stem / leaf formation.

The reversibility or irreversibility stage allows to select among the genes identified, the genes specific in the induction of different morphogenetic programs.

ACKNOWLEDGEMENTS

We deeply thank all our collaborators and our colleagues in all those research groups who have contributed to the development and improvment of the TCL method.

REFERENCES

Tran Thanh Van M., 1973, Direct flower neoformation from superficial tissue of small explant of *Nicotiana tabacum* L. Planta 115:87.

Tran Thanh Van M., Thi Dien N., and Chlyah A., 1974, Regulation of organogenesis in small explants of superficial tissue of *Nicotiana tabacum* L. Planta 119:149.

.Tran Thanh Van M., 1980, Thin cell layers : Control of morphogeneis by inherent factors and exogenously applied factors. Int. Rev. Cytol. 11 A: 175 Academic Press.

Albersheim P., Darvill A. G., McNeil M., Valent B. S., Sharp J. K., Nothnagel E. A., Davis K. R., Yamazaki N., Gollin D. J., York W. S., Dudman W. F., Darvill J. E., and Dell A., 1983, Oligosaccharins, naturally occurring carbohydrates with biological regulatory functions. In Structure and Function of Plant Genomes, eds. O. Ciferri and L. Dure III, pp. 293-312, Plenum N. Y.

Tran Thanh Van K., Toubart P., Cousson A., Darvill A. G., Gollin D. J., Chelf P., and Albersheim P., 1985[a], Manipulation of the morphogenetic pathway of tobacco explants by oligosaccharins. Nature 314:615.

Tran Thanh Van K., Trinh H. T., 1985[b], Organogenesis and cell differentiation in in vitro and in vivo systems. In Handbook of plant cell culture. eds. Evans, P. Ammirato, MacMilland N. Y.

Tran Thanh Van K., 1980[b], Control of morphogenesis or what shapes a group cells ? in Plant tissue culture. ed. Fiechter A., Springer Verlag, Berlin / Heidelberg / New York.

Tran Thanh Van K., 1981, Control of Morphogeneis. Ann. Rev. Plant. Physiol. 32 : 291.

Mutafstchiev S., Cousson A., and Tran Thanh Van K., 1987, Modulation of Cell growth and differentiation by pH and oligosaccharides. In British Plant Growth Regulator Group, Monograph 16. Advances in the Chemical Manipulation of Plant Tissue Cultures.

Richard L., 1987, Etude de l'influence des oligosaccharides pariétaux sur la croissance et l'organogenèse. Mise au point de methodes d'isolement et de culture d'assises cellulaires et de protoplastes de sous-épiderme de Nicotiana tabacum L. Université P. et M. Curie. DEA de Biologie et Physiologie végétales.

Thorpe T., Tran Thanh Van K., Gaspar T., 1978, Isoperoxidases in Epidermal Layers of Tobacco and Changes during Organ Formation in Vitro Physiol Plant. 44:388.

Kay L. E., and Basile D. V., 1987, Specific peroxydase isoenzymes are correlated with organogenesis. Plant Physiol., 84:89.

Tiburcio A. F., Kaur-Sawhney R., and Gaslton A., 1988, Polyanime Biosynthesis during Vegetative and Floral Bud Differentiation in Thin Layer Tobacco Tissue Cultures. Plant Cell Physiol. 29(7) : 1241.

Tiburcio A. F., Gendy C. A., and Tran Thanh Van K., 1989, Morphogenesis in tobacco subepidermal cells : putrescine as marker of root differentiation. <u>Plant cell tissue and organ culture</u> (in press).

Appelbaum A., Canellakis Z. N., Applewhite Ph. B., R-Kaur-Sawhney R., and Gaston A. W., 1988, Binding of Spermidine to a Unique Protein in Thin Layer Tobacco Tissue Culture. <u>Plant Physiol</u>. 88:996.

Meeks Wagner D., Dennis E. S., Tran Thanh Van K., and Peacock W. J., 1989, Tobacco Genes Expressed during in Vitro Floral Initiation and Their Expression during Normal Plant Development. <u>The Plant Cell</u>, 1:25.

Ammirato Ph. V., 1987, Speeding Transgenic Plants. <u>Biotechnoloy</u> 5:1015.

NUTRIENT REMOBILIZATION, NITROGEN METABOLISM AND CHLOROPLAST GENE EXPRESSION

IN SENESCENT LEAVES

Bartolomé Sabater, Antonio Vera, Rafael Tomás and Mercedes
Martín

Departamento de Biología Vegetal, Universidad de Alcalá de
Henares, Apdo. 20, Alcalá de Henares, 28871-Madrid. Spain

INTRODUCTION

Along plant ontogeny a continuous traffic of nutrients ensures the recovery of most nutrients from old structures to young, developing, structures (Sabater, 1985). Leaves accumulate the major percentage of nutrients
thus, the remobilization of components during leaf senescence is essential
for nutrient economy in the plant. The remobilization implies the activation
of several hydrolytic enzymes (Sabater and Rodríguez, 1978; Thomas and
Stoddard, 1980) and other activities, mainly related to amino acid metabolism (Calle et al., 1986; Cuello and Sabater, 1982; Peoples et al., 1980),
which form exportable low size nutrient molecules.

Chloroplasts are the main source of exportable nitrogen in senescent
leaves (Morita, 1980). However, the induction of leaf senescence requires
the synthesis of some specific polypeptides in chloroplasts (Cuello et al.,
1984; Sabater and Rodríguez, 1978; Yu and Kao, 1981) as well as in cytoplasm
(Thomas and Stoddard, 1980). In fact, inmediately before the induction of
natural (Martín et al., 1986) and detachment and dark accelerated (García
et al., 1983) senescence, chloroplast protein synthesis increases and the
organelles synthesize (García et al. 1983; Martín and Sabater, 1989) a
characteristic set of polypeptides (senescence polypeptides). Our purpose
is the identification in chloroplast DNA of the genes for senescence polypeptides.

HYBRID SELECT TRANSLATION OF CHLOROPLAST RNA FROM SENESCENT LEAVES

Depending on their physiological state, chloroplasts include different
untranslated mRNAs (Mullet, 1988). Thus, senescent chloroplasts show translational control (Martín and Sabater, 1989). For this reason, to identify
the mRNAs translated (and hence, the chloroplast DNA genes expressed) in
senescent chloroplasts, we have assayed the hybrid select translation
(Ricciardi et al., 1983) of RNAs isolated from barley (*Hordeum vulgare* L.
cv. Hassan) senescent chloroplasts (from 14-day-old leaf segments incubated
during 18 h in the dark). Translation assays were carried out in a light-
-driven, nuclease-treated chloroplast system, derived from senescent chloroplasts, which retains translational specificity and only synthesized senescence polypeptides when supplied with different chloroplast RNAs (Martín
and Sabater, 1989). As hybridization probes (to identify the chloroplast

Plant Aging: Basic and Applied Approaches
Edited by R. Rodríguez *et al.*
Plenum Press, New York, 1990

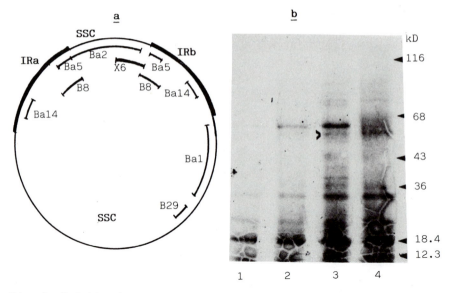

Fig. 1. Hydrid select translation of chloroplast RNA from senescent leaves. **a**: chloroplast DNA fragments assayed as probes in hybridization assays. **b**: SDS-PAGE and fluorography of the translation products with mRNAs selected by: Ba14 (lane 2), and Ba2 (lane 3). Lane 1 is a control of proteins synthesized without externally added RNA and lane 4 are the proteins synthesized with untreated whole chloroplast RNA from senescent leaves. Protein syntheses were carried out with [^{35}S]-methionine as described previously (Martín and Sabater, 1989).

genes expressed during senescence) we used several cloned fragments, of known sequence, of tobacco chloroplast DNA (Shinozaki et al. 1986; Sugiura et al., 1986) bound on a nylon membrane support.

Fig. 1a shows the location in tobacco chloroplast DNA of the different fragments used as probes in the hybrid selection assays. They account for about 50 % of the chloroplast genome and entirely include the single short chain (SSC) of chloroplast DNA. Fig. 1b shows the SDS-PAGE and fluorography of [^{35}S]-methionine labelled polypeptides synthesized after a typical hybrid select translation assay. In respect to the control (lane 1) of protein synthesis by nuclease-treated chloroplasts without externally added RNA, the addition of untreated whole chloroplast RNA of senescent leaves (lane 4) directed the synthesis of senescence polypeptides. Most of these polypeptides were also synthesized when the system of protein synthesis was supplemented with the mRNAs selected (lane 3) by the Ba2 probe which includes almost all the single short chain of chloroplast DNA. On the other hand, the mRNAs selected by the Ba14 probe, a 4.1 kbp fragment of the inverted repeats of chloroplast DNA, did not direct (lane 2) the synthesis of polypeptides clearly prominent over those synthesized in the control (lane 1) without external RNA. Taking into account mRNAs selected by other probes, only those selected by X6 probe directed the synthesis of 4 or 5 senescence polypeptides (results not shown). RNAs selected by probes: Ba5, B8, Ba1 and B29 (which includes the gene for the large subunit of ribulose-1,5-bisphosphate carboxylase/oxygenase and part of the gene for the ß sub-

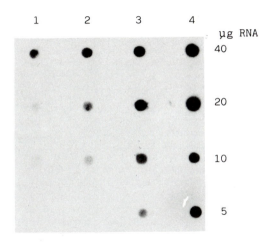

Fig. 2. Dot-blot assays of RNA complementary to BA2 fragment
of chloroplast DNA. Chloroplast RNA from: senescent leaves
(1) (14-day-old leaf segments incubated 18 h in the dark),
kinetin-treated leaves (2) (14-day-old leaf segments incubated 18 h in the dark in the presence of 40 μM kinetin),
expanding leaves (3) (7-day-old leaves from barley grown
under 16 h photoperiod) and young leaves (4) (from leaves
grown 6 days in the dark followed 1 day in light), were assayed at the indicated amounts for hybridization with [^{32}P]-
labelled Ba2 probe.

unit of thylakoid ATPase) yielded similar results as RNAs selected by probe
Ba14 (lane 2) when supplemented into the protein synthesis system derived
from senescent chloroplasts. In fact, the mRNAs selected by Ba2 account for
most of the senescence polypeptides (lanes 3 and 4).

Fig. 2 shows the results of dot-blot assays for the estimation of
levels of transcripts from genes of the Ba2 fragment. The level of Ba2
transcripts were lower in senescent leaves than at other stages of development, which agrees with the general diminution of chloroplast mRNAs during
senescence (Sabater, 1985). Young and expanding leaves showed high relative
levels of Ba2 transcripts, despite the fact that most of chloroplast DNA
genes highly expressed at these stages of development map in the single large
and inverted repeat chains. Probably, the translational control in chloroplasts is very important in determining which proteins are finally synthesized.

NADH DEHYDROGENASE ACTIVITY IN SENESCENT CHLOROPLASTS

All genes for senescence polypeptides identified so far map in the
single short chain of chloroplast DNA. This contrast with the genes mainly
expressed during plastid greening and leaf growth which mostly map in the
single large and inverted repeat chains. The single short chain includes
(Shinozaki et al., 1986) many open reading frames of unknown function and
six other showing high sequence homology with several components of mitochondrial NADH dehydrogenase. Thus, we assayed the presence in chloroplasts
of an activity similar to mitochondrial NADH dehydrogenase. Table 1 shows

Table 1. NADH: ferricyanide oxido-reductase activity, nmole.(min.mg prot)$^{-1}$, in chloroplasts

measured by	Chloroplasts isolated from barley leaves:			
	young	expanding	senescent	kinetin-treated
Ferricyanide reduction	291 ± 45	66 ± 8	94 ± 6	60 ± 9
NADH oxidation	142 ± 17	34 ± 6	48 ± 3	36 ± 4

Young: barley grown 6 days in the dark followed 1 day at light; expanding: barley grown 7 days under a 16 h photoperiod; senescent: 14-day-old barley leaves incubated 25 h in the dark; kinetin-tretaed: 14-day-old barley leaves incubated 25 h in the dark in the presence of 40 μM kinetin. Values are means ± SE of 3-8 independent experiments. NADH: ferricyanide oxido-reductase activity was measured according to Galante and Hatesi (1979) in 450 μg chlorophyll of chloroplasts resuspended in 50 mM Tris-HCl, pH 8.0.

that chloroplasts have a significant NADH dehydrogenase activity (measured as NADH: ferricyanide oxido-reductase) which depends on the physiological state of the leaves. The activity is higher in senescent than in kinetin-treated leaves (25 h after the induction of accelerated senescence) or in expanding leaves. It must be remembered that kinetin retards senescence and inhibits the synthesis of senescence polypeptides in chloroplasts (Martín and Sabater, 1989). Surprisingly, young chloroplasts show the highest activity NADH dehydrogenase, probably because the leaves have been grown 6 days in the dark and then 24 h under continuous light and the synthesis of NADH dehydrogenase components is activated in the dark. After 24 h under continuous light, a high level of NADH dehydrogenase formed in etioplasts should remain in young chloroplasts. Apparently, dark incubation (in etioplasts or during the induction of accelerated senescence) could stimulate, through activation of genes for NADH dehydrogenase, respiration in chloroplasts.

The function of other genes, encoded in the single short chain of chloroplast DNA and expressed during senescence, remains unknown, but it may be related to changes in chloroplasts allowing their degradation by hydrolytic enzymes probably synthesized in cytoplasm. By using DNA probes derived from chloroplast DNA of barley, we are now investigating the corresponding partner genes in barley.

The results shown here suggest that chloroplasts become a respiratory organelle in the dark. Most of the polypeptides synthesized by chloroplasts during senescence are incorporated in thylakoids (Guéra et al., 1989) and one possibility is that the respiratory chloroplast chain has some component in common with the photosynthetic electron transport chain.

ACKNOWLEDGEMENTS

This work was supported by the Spanish Comisión Interministerial de Ciencia y Tecnologia (Grants BIOT 23/85 and BIO 88-0002). We wish to thank Prof. Sugiura (Nagoya, Japan) for providing us tobacco chloroplast DNA probes.

REFERENCES

Calle, F., Martín, M., and Sabater, B., 1986, Cytoplasmic and mitochondrial localization of the glutamate dehydrogenase induced by senescence in barley (Hordeum vulgare), Physiol. Plant., 66:451.

Cuello, J., Quiles, M. J., and Sabater, B., 1984, Role of protein synthesis and light in the regulation of senescence in detached barley leaves, Physiol. Plant., 60:133.

Cuello, J., and Sabater, B., 1982, Control of some enzymes of nitrogen metabolism during senescence of detached barley (Hordeum vulgare L.) leaves, Plant Cell Physiol., 23:561.

Galante, Y. M., and Hatesi, Y., 1979, Purification and molecular enzymic properties of mitochondrial NADH dehydrogenase, Archiv. Biochem. Biophys., 192:559.

García, S., Martín, M., and Sabater, B., 1983, Protein synthesis by chloroplasts during the senescence of barley leaves, Physiol. Plant., 57: 260.

Guéra, A., Martín, M., and Sabater, B., 1989, Subchloroplast localization of polypeptides synthesized by chloroplasts during senescence, Physiol. Plant., 75, 382.

Martín, M., and Sabater, B., 1989, Translational control of chloroplast protein synthesis during senescence of barley leaves, Physiol. Plant., 75, 374.

Martín, M., Urteaga, B., and Sabater, B., 1986, Chloroplast protein synthesis during barley leaf growth and senescence: effect of leaf excision, J. Exp. Bot., 37:230.

Morita, K., 1980, Release of nitrogen from chloroplast during leaf senescence in rice (Oryza sativa L.), Ann. Bot., 46:297.

Mullet, J. E., 1988, Chloroplast development and gene expression, Annu. Rev. Plant Physiol., 39:475.

Peoples, M. B., Beilharz, V. C., Waters, S. P., Simpson, R. I., and Dalling, M. J., 1980, Nitrogen redistribution during grain growth in wheat (Triticum aestivun L.) II. Chloroplast senescence and degradation of ribulose-1,5-bisphosphate carboxylase, Planta, 149:241.

Ricciardi, R. F., Roberts, B. E., Paterson, B. M., and Miller, J. S., 1983, Transcription mapping and gene identification using hybridization selection and hybrid-arrest of translation, in: "Techniques in life sciences, Biochemistry, vol 85", pp 1-18, R. A. Flawell, ed., Elsevier, Amsterdam.

Sabater, B., 1985, Hormonal regulation of senescence, in: "Hormonal regulation of plant growth and development", S. S. Purohit, ed., p 169, Martinus-Nijhoff-Dr. W. Junk Pu., Dordrecht.

Sabater, B., and Rodríguez, M. T., 1978, Control of chlorophyll degradation in detached leaves of barley and oat through effect of kinetin on chlorophyllase levels, Physiol. Plant., 43:274.

Shinozaki, K., Ohme, M., Tanaka, M., Wakasugi, T., Hayashida, N., Matsubayashi, T., Zaita, N., Chunwongse, J., Obokata, J., Yamaguchi-Shinozaki, K., Ohto, C., Torazawa, K., Meng, B. Y., Sigita, M., Deno, H., Kamogashira, T., Yamada, K., Kusuda, J., Takaiwa, F., Kato, A., Tohdoh, N., Shimada, H., and Sugiura, M., 1986, The complete nucleotide sequence of the tobacco chloroplast genome: its gene organization and expression, EMBO J., 5:2043.

Sugiura, M., Shinozaki, K., Zaita, N., Kusuda, M., and Kumano, M., 1986, Clone bank of the tobacco (Nicotiana tabacum) chloroplast genome as a set of overlaping restriction endonuclease fragments: mapping of eleven ribosomal protein genes, Plant Science, 4:211.

Thomas, H., and Stoddart, J. L., 1980, Leaf senescence, Annu. Rev. Plant Physiol., 31:83.

Yu, S. M., and Kao, C. H., 1981, Retardation of leaf senescence by inhibitors of RNA and protein synthesis, Physiol. Plant., 52:207.

OXIDATIVE STRESS AND SENESCENCE IN OAT LEAVES

Victorio S. Trippi

Laboratorio de Fisiología Vegetal, Univ. Nac. de Córdoba

Casilla de Correo 395. 5000-Córdoba, Argentina

INTRODUCTION

Senescence is always associated with determinate growth and leaves can not escape from this behavior. In some cases determinate growth in plants seems result from genetic determination but in other cases it is a consequence of the environmental control (Trippi and Brulfert, 1973 a, b). However, people through years looked for an internal factor responsible for senescence. This communication deals with the effect of O_2 concentration on senescence in detached oat first-leaves.

EFFECT OF O_2 CONCENTRATION ON N_2 METABOLISM

During leaf growth nitrate reductase (NR) activity increases during the first 7 days and then decreases even though the leaves are still growing (Kenis and Trippi, 1987). In detached leaves placed under various O_2-concentrations, NR activity decreases as the O_2-concentration increases in both light and dark conditions (Kenis and Trippi, 1986) (Fig. 1). In attached leaves 100% O_2 prevents the normal increase in NR activity that takes place in the controls (under 21% O_2) between 5 and 7 days after sowing and 4% O_2 delays the decrease found after 7 days of growth. Addition of cystein, a protector of -SH group also delayed the decrease of enzyme activity (Kenis and Trippi, 1987). Recent work by Kenis and Campbell (unpublished) suggests that O_2 affects protein synthesis probably at a translational level. An inactivation provoked by O_2-concentration on ATPase has also been described (Chernyak and Kozlov, 1986).

EFFECT OF O_2 ON SENESCENCE

The effect of O_2 is somewhat different in the light than in the dark. Under light conditions O_2-concentrations above 0.3% accelerate the development of senescence parameters such as the decrease in chlorophyll

Plant Aging: Basic and Applied Approaches
Edited by R. Rodríguez *et al.*
Plenum Press, New York, 1990

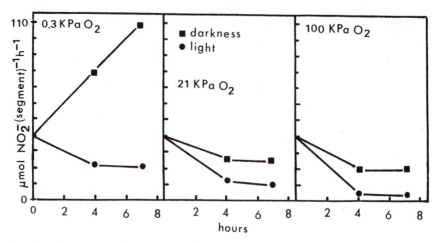

Fig. 1. Nitrate reductase activity under different O_2 concentration in darkness and light conditions.

content as well as the increase in soluble amino-acid (AA) and malondial-dehyde (MDA) contents and in membrane permeability (as measured by changes in conductivity of the incubation medium). The higher the O_2-concentration the faster the development of senescence (Trippi and De Luca d'Oro, 1985). In darkness 21% O_2 saturates O_2 requirements for senescence and 93% O_2 has no effect. Besides, 0.3% O_2 at 25°C provokes a rapid increase in MDA content (which indicates an increase in lipid peroxidation) and in conductivity after 24 h treatment, while chlorophyll remains at about its initial value, and AA content increases only slightly. Complementary observations show that any increase in permeability induces an increase in photo-oxidative degradation of chlorophyll in light conditions (Fig. 2) while it halts chlorophyll degradation in the dark. Later on, we observed that any condition affecting energy production like electron transport inhibitors such as DCMU, DNP, CNK, etc or environmental conditions such as temperature and water stresses capable of increasing permeability, also induced rapid photo-oxidative degradation of chlorophylls (De Luca d'Oro and Trippi, 1987; Luna and Trippi, unpublished). Present results suggest that membrane oxidations would be the first event taking place during senescence, and that energy metabolism could be involved in the development of senescence.

EFFECT OF O_2 ON ENERGY METABOLISM

Concerning energy metabolism, leaf-segments also react in a different manner to O_2 either in the light or in the dark. The difference is probably connected with the function of mitochondria or chloroplasts.

In general, O_2 concentrations above 0.5% in leaves and 4-10% O_2 in petals accelerate senescence and deterioration of energy metabolism (Trippi et al., 1988, 1989).

In leaf-segments placed in darkness, after growing in the light, 0.5%

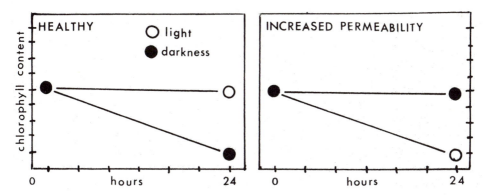

Fig. 2. Membrane permeability modulates chlorophyll breakdown in oat
leaves. Healthy membranes showed higher chlorophyll breakdown in
darkness than in light. Increased permeability prevented chlorophyll breakdown in darkness and induced photo-oxidative degradation in light conditions.

O_2 decreased ATP and total adenine nucleotides contents. This decrease
could result from the passage from light to darkness itself, and energy
production limited to mitochondrial activity. However, as tissue adaptation progressed parameters developed towards normal values.

In detached leaves nucleotides increase during the first 48 h and
then decrease in total nucleotides and ATP. This could result from the decrease in the use of energy by -SH enzymes like NR (Kenis and Trippi,
1987) and ATPase (Borochov and Faimar-Weinberg, 1984), but it could also
be due the increase in RNase activity (Trippi et al., 1989). However, it
is significant that the nucleotide pool decreases in the first phase of
senescence.

The second catastrophic phase of senescence might be a consequence
of the energy using system (EUS) inactivation and the decrease in the size
of the nucleotide pool. The higher the concentration of O_2, the faster the
decrease in size of nucleotide pool. The proposition is consistent with
decrease observed in ATP, ADP and AMP but also in NAD, NADP and NADPH during senescence of tobacco leaves (Yamamoto, 1963; Meyer and Wagner,
1986); with the ability of photosystem I and ferredoxin to produce O_2^- in
the absence of electron carrier (Thompson et al., 1987); generation of
$^1O_2$ by isolated thylakoids in absence of electron acceptors (Takahama and
Nishimura, 1975; Knox and Dodge, 1985) and generation of H_2O_2 in CO_2-free
atmosphere (Patterson and Myers, 1973), all facts capable of increasing
cell oxidative damages.

Summing up, oxidative damage to EUS and decrease in the size of
nucleotide pool should be responsible for destruction of energy producing
system (EPS) by enhancing effect on the production of active oxygen. It
is also significant that membrane oxidation and decrease in energy parameters is accompanied by the increase in membrane permeability.

OTHER STRESS CONDITIONS

Different stress conditions i.e. high or low temperatures and water deficit are capable of affecting senescence, through various mechanisms, but an increase in the hydroperoxide content and membrane permeability (De Luca d'Oro and Trippi, 1987) is always observed.

Recent work on the effect of water deficit induced on leaves by mannitol showed that chlorophyll loss and hydroperoxides and membrane permeability increases are closely related to the stress provoked by different concentrations of osmolite. Water stress effects are potentiated by high O_2-concentration (100% O_2), as suggested by the results on energy metabolism. Results also suggest that water stress affects not only the EUS but also the EPS function as supported by a decrease in energy charge. This effect was also observed when leaf-segments or petals are treated with highly concentrated H_2O_2. It seems probable that water protects tissues against oxidations because of the low solubility of O_2 in this medium (Luna and Trippi, unpublished).

Exogenous sugars also promote senescence of leaves under light conditions, determining an increase in lipoxygenase activity and hydroperoxides formation. However, in darkness sugars delayed senescence (Luna and Trippi, 1986). All these results suggest that damage induced by different stresses are mediated by oxidative phenomena.

EFFECT OF LEAF GROWTH ON CELL OXIDATIONS AND SENESCENCE

While some senescence parameters like chlorophyll and soluble AA contents increase during growth and then decrease during senescence, lipid hydroperoxide content (as measured by MDA content) and membrane permeability continuously increase during growth until the last period of senescence. Normal leaf growth is accompanied by an increase in the surface exposed to photo-oxidations due to cell enlargement, and by an increase in O_2 production in the chloroplast associated to photosynthetic differentiation and function. Therefore we present the hypothesis that growth itself could be a factor capable of increasing cell oxidations and senescence. Different treatments such as low temperature (4°C) and water stress provoked by 0.2 M mannitol in the light were used to produce leaves having smaller size than the controls (28°C and water respectively) as a result of different cell enlargement. After additional treatment of 24 h in the light, MDA was measured. To avoid any problem derived from cell number results were calculated on DNA basis. Other senescence parameters were measured after 96 h of treatment.

Results showed that oxidative treatments resulted in a degree of oxidation (MDA/DNA ratio) higher in those tissues with higher surface/DNA ratio than in those with a lower one, suggesting that the higher the cell surface the higher the hydroperoxide content and the faster the development of senescence. Therefore, results suggest that growth itself could be a senescence factor in leaves by increasing cell oxidative processes.

Results suggest that leaf growth and probably photosynthesis may increase oxidative processes in a well oxygenated atmosphere. Oxygen combined with photo-oxidative light intensities inhibit NR and probably others SH-enzymes like ATPase; stimulates lipid peroxidation leading to a progressive increase in membrane permeability; and provoks inhibition of the EUS and decrease in the nucleotides content, which in turn determines degradation of the EPS and concomitant increase in permeability.

Lipid peroxidation seems to be the earliest event of leaf senescence and responsible of the progressive increase in permeability. Latter increase in permeability depends on the nucleotide content and energy charge.

The increase in permeability is always linked to inactivation of degradative activity (probably enzymatic) of chlorophyll and therefore it prevents chlorophyll breakdown in darkness. Besides, increased permeability determines photo-oxidative degradation of chlorophyll and proteins under light conditions.

Sugars content that increases during leaf senescence, promotes oxidative processes in light condition (via activation lipoxygenase) and increases membrane permeability and senescence. In contrast, it delays senescence in darkness. Free radicals seem to be responsible for all oxidative damages inducing senescence (Fig. 3).

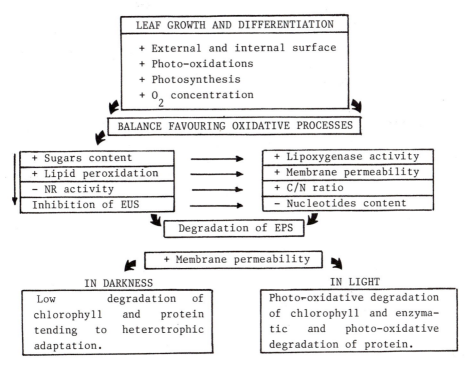

Fig. 3. Schematic representation of changes during senescence
induced by balance favouring oxidative processes.

CONCLUDING REMARKS

Besides the fact that O_2-concentrations higher than 0.5% are toxic for leaves and capable of accelerating senescence by enhancing oxidative phenomena, the question of whether senescence results from a genetic program (endogenous origin) or from external environment, still remain. Do we have to think that hyperoxia modulates a genetic program for senescence or that hyperoxia as well as other stress conditions destroys by just degrading cell components?

REFERENCES

Borochov, A., and Faiman-Weinberg, R., 1984, Biochemical and biophysical changes in plant protoplasmic membranes during senescence, Plant Physiol. 15: 1.

Chernyak, B. V., and Kozlov, I. A., 1986, Regulation of H^+-ATPases in oxidative and photophosphorilation, TIBS 11: 32.

De Luca d'Oro, G. M., and Trippi, V. S., 1987, Effect of stress conditions induced by temperature, water and rain on senescence development, Plant Cell Physiol. 28: 1389.

Kenis, J. D., and Trippi, V. S., 1986, Regulation of nitrate reductase in detached oat leaves by light and oxygen, Physiol. Plant. 68: 387.

—— 1987, Involvement of oxidation, proteolysis and reductant availability in the regulation of in vivo nitrate reductase in attached oat leaves during growth and senescence, Plant Cell Physiol. 28: 1307.

Knox, J. P., and Dodge, A. D., 1985, Singlet oxygen in plants, Phytochemistry 24: 889.

Luna, C., and Trippi, V. S., 1986, Membrane permeability regulation by exogenous sugars during senescence of oat leaf in light and darkness, Plant Cell Physiol. 27: 1051.

Meyer, R., and Wagner, K. G., 1986, Nucleotide pools in leaf and root tissue of tobacco plants: influence of leaf senescence, Physiol. Plant. 67: 666.

Patterson, C. O., and Myers, J., 1973, Photosynthetic production of hydrogen peroxide by Anacystis nidulans, Plant Physiol. 51: 104.

Takahama, U., and Nishimura, M., 1975, Formation of singlet molecular oxygen in illuminated chloroplasts. Effects on photoinactivation and lipid peroxidation, Plant Cell Physiol. 16: 737.

Thompson, J. E., Ledge, R. L., and Barber, R. F., 1987, The role of free radicals in senescence and wounding, New Phytol. 105: 317.

Trippi, V. S., and Brulfert, J., 1973a, Organization of the morphophysiologic unit in Anagallis arvensis L. and its relation with the perpetuation mechanism and senescence, Amer. J. Bot. 60: 641.

—— 1973b, Photoperiodic aging in Anagallis arvensis L. clones: its relation to RNA content, rooting capacity and flowering, Amer. J. Bot 60: 951.

—— and De Luca d'Oro, G. M., 1985, The senescence process in oat leaves and its regulation by oxygen concentration and light irradiance, Plant Cell Physiol. 26: 1303.

Trippi, V. S., Paulin, A., and Pradet, A., 1988, Effect of oxygen concentration on the senescence and energy metabolism of cut carnation flowers, <u>Physiol. Plant</u>. 73: 374.

—— , Gidrol, X., and Pradet, A., 1989, Effects of oxidative stress caused by oxygen and hydrogen peroxide on energy metabolism and senescence in oat leaves, <u>Plant Cell Physiol</u>. 30: 157.

Yamamoto, Y., 1963, Pyridine nucleotide content in the higher plant. Effect of age of tissue, <u>Plant Physiol</u>. 38: 45.

STRUCTURE AND FUNCTION OF GLUTAMATE DEHYDROGENASE DURING

TRANSIENT SENESCENCE OF MUSTARD (*SINAPIS ALBA* L.) COTYLEDONS

W. Lettgen, L. Britsch, H.I. Kasemir

Biologisches Institut II
Schänzlestr. 1
D-7800 Freiburg, FRG

INTRODUCTION

Glutamate dehydrogenase (GDH, EC 1.4.1.3) catalyses the reductive amination of 2-oxoglutarate to glutamate (NADH-GDH) especially under conditions of ample ammonia supply (Yamaya and Oaks, 1987). This enzyme seems to be ubiquitous in almost all species of higher plants and is recognized to have a regulatory position as a link between carbohydrate and amino acid metabolism. It is known to exhibit a complex isoform pattern (Nauen and Hartmann, 1980) which differs in various parts of the plant (Cammaerts and Jacobs, 1985) and depends on the developmental stage (Laurière, 1983) and on light treatment (Postius and Jacobi, 1976). An increase in GDH activity during leaf senescence (SEN) appears to be a general feature (Srivastava and Singh, 1987; Laurière, 1983) and it has been suggested that ammonia produced by proteolysis during SEN could be a regulatory factor for *de novo* synthesis of the enzyme. Kar and Feierabend (1984) demonstrated with detached wheat leaves that an increase in GDH activity followed a control mechanism which governed also the time course of SEN. Although the aminating reaction is often considered to be the primary role of GDH, the reverse reaction, i.e. the oxidative deamination of glutamate (NAD^+-GDH) may be important during the period of SEN which is characterized by an increased demand of energy and the need for mobile forms of nitrogen derived from amino acids released by proteolysis. In this case GDH may preferentially display its deaminating function acting as a link between amino acid degradation and the tricarboxylic acid cycle. Up to now little is known about the subcellular localization of GDH during senescence (Calle et al., 1986) which has been determined to be mainly in the cytoplasm and in mitochondria.
        In this report we present data concerning GDH in attached cotyledons of mustard seedling which underwent a transient SEN induced by light stress and nitrogen deficiency. By investigating the molecular structure, the subcellular localization and the time course of activity of mustard GDH we have tried to clarify its physiological function.

Plant Aging: Basic and Applied Approaches
Edited by R. Rodríguez *et al.*
Plenum Press, New York, 1990

MATERIAL AND METHODS

Plant material, growth conditions, determination of GDH activity and Western blotting were described by Lettgen et al., 1989). Labeling of GDH with dansylhydrazine after polyacrylamide gelelectrophoresis (PAGE) was performed according to Estep and Miller (1986). Specific antibodies against mustard GDH were induced in rabbits. Immunogold labeling of GDH and the preparation of electron micrographs were performed as described by V. Speth (Speth et al., 1987). The data shown in the figures are means of at least 6 independent experiments. SEM values range between 5 and 10%.

RESULTS

Recently Lettgen et al. (1989) have shown that during the first period of mustard seedling development (up to 4 d after sowing) the activity of GDH - which was already present in the seeds - decreased considerably. The decline could be slightly variated by light/dark changes and by different ion solutions (NH_4^+, NO_3^-). The behaviour of NAD^+- and NADH-GDH was similar. The enzyme (GDH1) consisted of 7 isoforms with an identical M_r of 440,000. Each isoform was composed of 4 different subunits ranging between 19 and 25 kDa and reacted with an antibody raised against the most prominent isoform of the seeds.

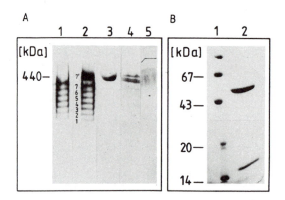

Fig. 1. Isoforms of GDH extracted from mustard cotyledons and separated on PAGE. A - (non-denaturing PAGE) - 1, 2: isoforms 1 - 7 (GDH1) and 7' (GDH2) of 3 d old seedlings; 3-5: isoform 7' (GDH2) of 14 d old seedlings; 1, 4: enzyme activity stain; 2: enzyme activity stain and in addition Coomassie stain; 3: Coomassie stain only; 5: Western blotting probed with anti-GDH1. B (SDS-PAGE) 1: M_r marker proteins; 2: Coomassie stain of the subunits of GDH2 isoform 7'.

Between 4 and 14 d after sowing a new isoform 7' (GDH2) was activated. In the young seedlings this form could only be detected as an inactive protein after native PAGE-separation (Fig. 1A). GDH2 consisted of a single subunit with 60 kDa which seemed to share common antigenic determinants with GDH1

It should be mentioned that in addition to the prominent GDH2 subunit a smaller band with M_r 15,000 appeared as soon as the enzyme became active.

The activity of GDH2 showed a different pattern when correlated with specific developmental stages: (1) Under short day conditions (8 h light/16 h dark) - no sign of SEN in cotyledons - GDH2 activity was relatively low whereby NADH-GDH2 activity was higher than NAD^+-GDH2 activity (Fig. 2). Both enzyme activities exhibited an inverse rhythmicity of approximately 48 h. (2) Under long day conditions (16 h light/8 h dark) - cotyledons underwent SEN - the activity of NAD^+-GDH2 increased considerably (Fig. 2)

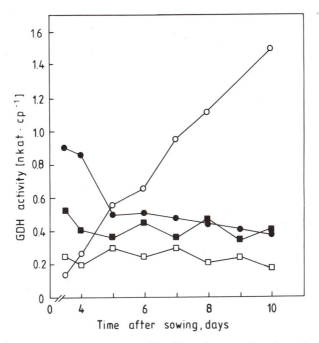

Fig. 2. Time course of GDH activity in mustard cotyledons of seedlings kept under long day (16h light/8 h dark, o●) or short day (8 h light/16 h dark, □■) conditions. Open symbols: deaminating activity, black symbols: aminating activity. cp: pair of cotyledons.

while the level of NAD^{++}-GDH was comparable to that of short-day seedlings. (3) When the seedlings were treated with N-compounds (solutions of 15 mM KNO_3 or NH_4Cl) the two GDH2-activities reacted quite differently. Whereas a rapid increase in NAD^+-GDH2 activity appeared after the application of KNO_3 only a weak response existed after the application of NH_4Cl (Fig. 3A) which was comparable to that of NADH-GDH2 (Fig. 3B).

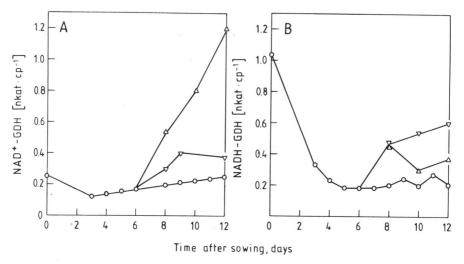

Fig. 3. Time course of GDH activity in mustard cotyledons of seedlings kept in continuous white light, treated up to 6 d with distilled water and thereafter with 15 mM of either KNO_3 ($\triangle$) or NH_4Cl ($\triangledown$). A: deaminating activity, B: aminating activity. cp: pair of cotyledons.

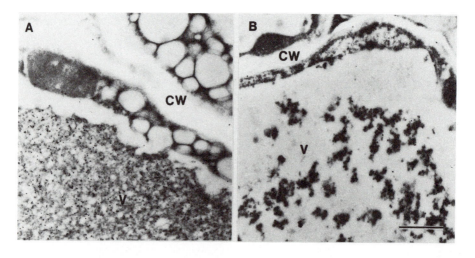

Fig. 4. Electromicrographs of sections of mustard cotyledons immunolabeled with anti-GDH1. Bound antibodies were visualized by immunogold labeling. A: 3 d old seedlings. B: 14 d old seedlings. Seedlings grown in continuous white light. Micrographs produced by V. Speth. CW: cell wall, V: vacuole, bar = 1μm.

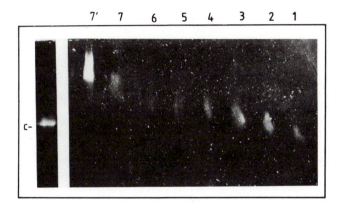

Fig. 5. Labeling of GDH-isoforms with dansylhydrazine after separation on non-denaturing PAGE. The isolated isoforms 1 - 7 of 3 d old and 7' of 14 d old seedlings were identified as glycoproteins by this method. C: ovalbumin (control).

To compare the localization of GDH in young and older mustard seedlings, immunogold labeling was performed on serial sections picked up on different grids. In cotyledons from 3 d old seedlings (mainly GDH1) the vacuoles of certain cells contain electron-dense granular material which was densely labeled with immunogold, (Fig. 4A) indicating the presence of GDH. These specific cells were packed with storage proteins (Fischer et al., 1988). In contrast, in cotyledons of 14 d old seedlings (mainly GDH2) labeling was found only in aggregated areas in the vacuoles of aging cells which show signs of lytic processes (Fig. 4B). No labeling could be detected in the mitochondria or chloroplasts. Since localization of GDH strongly suggested that the enzyme inside the vacuole is surrounded by proteolytic activities (Wittenbach et al., 1982) it was challenging to find out whether mustard GDH might be protected against proteolytic attack. As indicated in Fig. 5 both enzyme sets (GDH1 and GDH2) could be identified as glycoproteins which are known to be relatively resistent to proteolysis (Faye et al., 1989).

DISCUSSION

As indicated by Fig. 1 the molecular characteristics of mustard GDH change remarkably during the time course of seedling development. Whereas the young seedlings (up to 4 d after sowing) contain an enzyme (GDH1) which comprises 7 isoforms each composed of 4 subunits (Lettgen et al., 1989) the older seedlings (4-14 d after sowing) display a single protein (GDH2) which can already be detected as an inactive precursor in the younger seedlings (Fig. 1A). GDH2 is composed of only two subunits (Fig. 2A) and shows cross-reaction with antibodies to GDH1. As was recently concluded from results with young mustard seedlings (Lettgen et al., 1989) GDH1 mainly reacts in the deaminating direction and seems to display an anaplerotic function in connection with the degradation of storage proteins. Accordingly electron micrographs (Fig. 4A) show that immunolabeling of GDH1 is exclusively associated with storage protein which is known to decline gradually during the time course of seedling development (Fischer and Schopfer, 1988). Hence, GDH1 seems to be involved in the metabolism of mobile inorganic nitrogen during the transformation of cotyledons from storages into photosynthetic active organs.

A very similar function can be predicted for GDH2 from older mustard seedlings (older than 4d). When the seedlings were kept between 4 and 10 d under long-day conditions or in cWL the deaminating reaction of GDH2 increased several times (Fig. 2) whereas the aminating reaction remained fairly constant. Under these growth conditions the seedlings became senescent (indicated by the degradation of ribulose-bisphosphatecarboxylase and chlorophyll, see Kasemir et al., 1988) and very probably developed a need for the canalisation of transportable N-compounds. Transamination after the release of NH_4^+ from glutamate by GDH2 may be an important feature during this process. The increase of GDH2 seems to be restricted to certain cells in the cotyledons which were "sacrificed" by the cotyledons and eventually lysed completely (Fig. 4B). In this extreme situation the protection by glycosidation as indicated in Fig. 5 may be very reasonable. When inorganic nitrogen (NH_4^+, NO_3^-) was supplied to the seedlings (Fig. 3) senescence was reversed (data not shown) and GDH2 responded differently to the nitrogen solutions. Only after application of KNO_3 the deaminating GDH2 increased (Fig. 3A). This may be due to the fact that NO_3^- metabolism requires a high energy expenditure (Salsac et al., 1987). In this special case GDH activity may be decoupled from protein degradation.

Acknowledgement - The investigation was supported by grants from the Project Europäisches Forschungszentrum für Maßnahmen zur Luftreinerhaltung (PEF), Kernforschungszentrum Karlsruhe, FRG. The technical assistance of C. Nolte was gratefully acknowledged.

REFERENCES

Calle, F., Martin, M and Sabater, B., 1986, cytoplasmic and mitochondrial localization of the glutamate dehydrogenase induced by senescence in barley (*Hordeum vulgare*), Physiol. Plant., 66:451
Cammaerts, D. and Jacobs, M., 1985, A study of the role of glutamate dehydrogenase in the nitrogen metabolism of *Arabidopsis thaliana*, Planta, 163:517
Estep, T.N. and Miller T.J., 1986, Optimization of erythrocyte membrane glycoprotein fluorescent labeling with dansylhydrazine after polyacrylamide gel electrophoresis, Anal. Biochem., 157:100
Faye, L., Johnson, K.D., Sturm, A. and Chrispeels, M.J., 1989, Structure, biosynthesis, and function of asparagine-linked glycans on plant glycoproteins, Physiol. Plant., 75:309

Fischer, W., Bergfeld, R., Plachy, C., Schäfer, E. and
    Schopfer, P., 1988, Accumulation of storage materials,
    precosious germination and development of desiccation
    tolerance during seed maturation in mustard (*Sinapis
    alba* L.) Bot. Acta, 101:344
Kar, M. and Feierabend, J., 1984, Changes in the activities
    involved in amino acid metabolism during the senescence
    of detached wheat leaves, Physiol. Plant., 62:39
Laurière, C., 1983, Enzymes and leaf senescence, Physiol.
    Vég., 21:1159
Lettgen, W., Britsch, L. and Kasemir, H., 1989, The effect of
    light and exogenously supplied ammonium on glutamate
    dehydrogenase activity and isoforms in young mustard
    (*Sinapis alba* L.) seedlings, Botanica Acta, in press
Nauen, W. and Hartmann, T., 1980, Glutamate dehydrogenase
    from *Pisum sativum* L. Localization of the multiple forms
    and of glutamate formation in isolated mitochondria,
    Planta, 148:7
Postius, C. and Jacobi, G., 1976, Dark starvation and plant
    metabolism. VI. Biosynthesis of glutamic acid
    dehydrogenase in detached leaves from *Curcurbita maxima*.
    Z. Pflanzenphysiol., 78:133
Rosemann, D., Kasemir, H. and Oelmüller, R., 1988, Changes in
    ribulose-1,5-bisphosphate carboxylase and its
    translatable small subunit mRNA levels during senescence
    of mustard (*Sinapis alba*) cotyledons, Physiol. Plant.,
    73:257
Salsac, L., Chaillou, S., Morot-Gaudry, J.-F. and Lesaint,
    C., 1987, Nitrate and ammonium nutrition in plants,
    Plant Physiol. Biochem., 25:805
Speth, V., Otto, V. and Schäfer, E., 1987, Intracellular
    localization of phytochrome and ubiquitin in red-light-
    irradiated oat coleoptiles by electron microscopy,
    Planta, 171:332
Srivastava, H.S. and Singh, R.P., 1987, Role and regulation
    of L-glutamate dehydrogenase activity in higher plants,
    Phytochemistry, 25:597
Wittenbach, V.A., Lin, W. and Herbert, R.R., 1982, Vacuolar
    localization of proteases and degradation of
    chloroplasts in mesophyll protoplasts from senescing
    primary wheat leaves, Plant Physiol., 69:98
Yamaya, T. and Oaks, A., 1987, Synthesis of glutamate by
    mitochondria - An anaplerotic function for glutamate
    dehydrogenase, Physiol. Plant., 70:749

SECTION IV

MODULATION OF AGING AND MATURATION

EFFECTS OF VASCULAR TISSUE CONTAINING A MERISTEM ON ADJACENT PITH: HORMONE-LIKE FACTORS

Jiří Luštinec

Institute of Experimental Botany
Czechoslovak Academy of Sciences
Prague, Czechoslovakia

INTRODUCTION

The growth potential of meristems and young organized tissues is shown in their production of special growth stimulating metabolites. The capacity to produce these metabolites is low in senescent tissues and may be insufficient in tissue cultures which grow and/or develop poorly in spite of being supplied with different combinations of nutrients and growth regulators. These metabolites may involve, in addition to well known compounds such as cytokinins and auxins, some unknown regulatory factors. The existence of these factors can be proved using primary explants cultured on appropriate media and consisting of either a massive complex of senescent (old) tissue only or of senescent tissue with adjacent young and/or metabolically active tissue (e.g. cambial meristem and pith rays). The less active tissue in the complex explant can serve as a detector of regulatory impulses coming from the tissue(s) with high metabolic activity, while the explant consisting of a single tissue serves as a control. Explants of this type can be derived from different organs, especially stems and thick roots.

PLANT MATERIAL AND METHODS

Kale plants (*Brassica oleracea* L. var. *medullosa* cv. Krasa) 100-150 cm high with stems up to 12 cm in diameter were used in experiments described below. Their stems were cut to 4 mm sections from which blocks were excised using a double blade cutter. The blocks consisted of either the pith only--control explants (10 mm long x 10 mm wide x 4 mm high) - or of pith with the adjacent vascular tissue including a meristem - complex explants (15 or 25 mm in lenght, other dimensions the same). In some experiments the explants were cut from the stem perpendicularly to its longitudinal axis using a 8 mm cork borer. The explants were cultured in 100 ml conical flasks with 40 ml of sucrose-agar basal medium supplemented with auxins and cytokinins as indicated, or 40 ml of complete growth promoting medium (Horák et al., 1975). Cultivation took place in darkness at 26 °C. At the end of the cultivation period, the vascular tissue of the complex explants was separated from the pith. The later was compared with control explants derived from the corresponding position in the stem. Starch and soluble protein were determined as described (Luštinec et al., 1984a,b). Extraction, fractionation and bioassay of hormone-like factors was partially described before (Luštinec et al., 1985). Some further details are given below.

EFFECT OF VASCULAR TISSUE ON CALLUS FORMATION AND ORGAN REGENERATION

The presence of vascular tissue in explants cultured on complete media manifested itself in an increase in callus formation and organ regeneration, which occurred first of all in the region of vascular tissue but also in adjacent pith (Fig. 1). The stimulating effect of vascular tissue could not be achieved by any modification of culture media. The media used were selected as optimal for growth and differentiation of kale explants (Luštinec et al., 1984b. Luštinec, 1988).

EFFECT OF VASCULAR TISSUE ON SOLUBLE PROTEIN ACCUMULATION

If simple pith explants were cultured on sucrose-agar medium their content of soluble protein increased from 5% to 8% (dr.w.) or more during few days. The protein synthesis was apparently made possible by the intracellular supply of low molecular nitrogen compounds. If cytokinins and auxins were added to the sucrose medium, protein accumulation increased up to 4 fold. Protein accumulation in pith of complex explants considerably increased under the influence of vascular tissue, both on the simple sucrose medium and on this medium containing cytokinins and auxins (Luštinec et al., 1984b). Stimulation of protein accumulation by vascular tissue was strong up

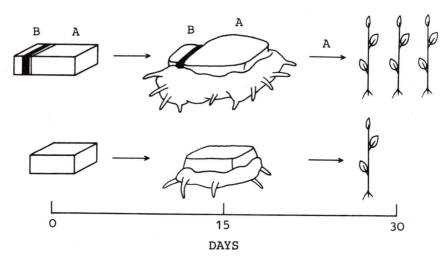

Fig. 1. The effect of vascular tissue on callus and organ formation in explants cultured on complete growth medium.

to a distance of at least 20 mm (Fig. 2). It occurred also in explants excised from the stem perpendicularly to its longitudinal axis and placed on the medium vertically so that the vascular tissue was not in direct contact with it.

EFFECT OF VASCULAR TISSUE ON STARCH ACCUMULATION

Simple pith explants cultured for several days on agar medium containing sucrose accumulated starch. Application of streptomycin, 5-fluorouracil and other inhibitors indicated that starch accumulation depended upon protein synthesis on 80S ribosomes (Luštinec et al., 1984a). If explants derived from young plants grown under natural long-day conditions contained vascular tissue, the amount of starch formed in adjacent pith parenchyma increased up to seven fold when compared with explants without vascular tissue (Fig. 3). Similar increase of starch content as caused by vascular tissue was achieved by the addition to medium of cytokinins in combination with auxins. A furher increase in starch accumulation could be achieved by application of cytokinin and auxin to explants containing vascular tissue. When explants were derived from adult plants grown under natural short-day conditions cytokinins and auxins had little or no effect, but vascular tissue enhanced starch formation considerably (Fig. 3).

Analysis of 5 mm segments of the pith taken at various distances from the vascular tissue showed that the effect of vascular tissue was significant up to a distance of at least 20 mm, the same as in the case of protein accumulation. The increase in starch synthesis did not depend upon direct con-

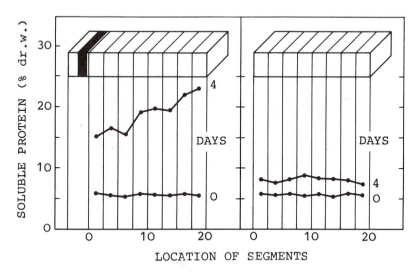

Fig. 2. The effect of vascular tissue on soluble protein content in segments of pith explants. Complex (25 mm x 10 mm x 4 mm) and control (20 mm x 10 mm x 4 mm) explants were cultivated for 0 to 4 days on medium containing 10 % sucrose. The locations (in mm) of particular segments are given with respect to the vascular tissue.

tact between the vascular tissue and the medium. When explants were excised from the stem perpendicularly to its longitudinal axis, the vascular tissue was separated from the medium by several mm of pith or cortex parenchyma. Under these conditions starch accumulation was still stimulated.

Microscopic observation of starch localization revealed that the meristematic region of vascular tissue was the centre of starch accumulation and hence perhaps of starch-inducing factor production. However, the substantial part of the starch stimulus translocated to the pith originated in the region of pith rays and xylem neighboring on the pith (Fig. 4).

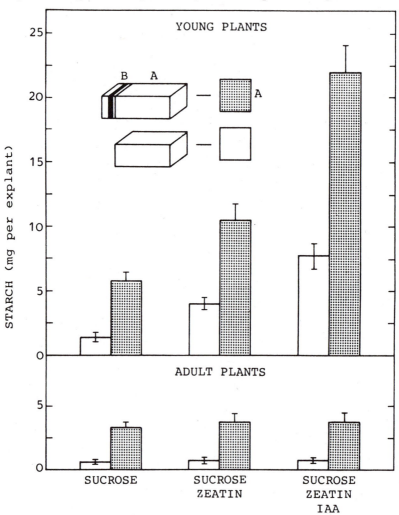

Fig. 3. The effects of the composition of the medium (10 % sucrose, trans-zeatin 10 µM, IAA 5 µM) and vascular tissue (black strip) on starch content in pith parenchyma of the explants, estimated after ten days of culture. The bars indicate SE of the means (n = 5), which correspond to five cultivation flasks with three explants in each (Luštinec et al., 1984a).

COMPARISON OF THE EFFECTS OF VASCULAR TISSUE ON CALLUS,PROTEIN
AND STARCH PRODUCTION AT LOW AND HIGH CONCENTRATION OF SUCROSE

Starch and protein accumulated in explants within a wide
range of sucrose concentrations in culture medium (Luštinec,
1989). Optimal concentration of sucrose for starch accumulation was 10-15 % (w/v), for the accumulation of soluble protein 27-32 % and for callus formation 2-7 %. At concentrations
above 20 % no cell proliferation in the explants was found.
The stimulatory effects of vascular tissue on starch and protein accumulation manifested itself both at low sucrose concentrations optimal for cell division and growth and at high
concentrations at which growth processes were suppressed (Luštinec et al., 1984b).

ISOLATION OF HORMONE-LIKE FACTORS

First Group of Factors

Explants containing vascular tissue and cultured for a
few days on sucrose medium were extracted with water and the
crude extract was fractioned chromatographically on the column
of Polyclar using water as an eluent. The ten-ml fractions of
1200 ml eluate were bioassayed for cytokinins using an Amaranthus bioassay and for starch inducing activity using a bioassay based on protein synthesis-dependent starch accumulation
in stem pith explants free of vascular tissue. Eight peaks of
high starch inducing activity occurred in fractions between
650 and 1125 ml of the eluate and two peaks occurred in fractions between 30 and 130 ml. The cytokinin values of these
fractions were zero. The activity of these fractions was compared with that of some known phytohormones and other metabolites using a starch bioassay (Luštinec et al., 1985).

The highest relative activities of compared factors were:
water control (1.0), most active eluate fraction (14.0),trans-
-zeatin (7.3), indol-3-ylacetic acid (2.0), aminocyclopropane
carboxylic acid (2.0), abscissic acid (0.8), 2,4-epi-bras-
sinolide (0.7), gibberellic acid (0.4), sucrose (2.3), glu-
cose-1-phosphate (2.0), adenosine diphosphoglucose (1.5).

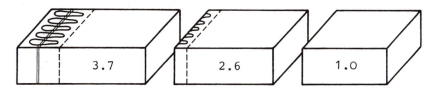

Fig. 4. The relative content of starch in pith parenchyma of complex explants containing vascular
tissue including the cambium (15 mm x 10 mm x
4 mm) and complex explants containing only a
part of vascular tissue as compared to that
of control explants (10 mm x 10 mm x 4 mm).
Explants were cultivated for 5 days on medium
containing 10 % sucrose. Initial relative
value of starch content was 0.07.

Second Group of Factors

Explants were extracted with water as before. The extract was mixed with Polyclar, which was then thoroughly washed with water. The adsorbed substances were released from Polyclar with 0.1 N NaOH. The Na^+ ions were removed by Dowex-50 and the extract was concentrated under vacuum at 40 ^{o}C. One ml of concentrated eluate corresponding to 60 g of fresh plant material was fractioned on the column of Sephadex LH-20 (2.5 x 120 cm) with water. High starch-inducing activity comparable with that of t-zeatin was detected in eluate fractions 930-960 ml, 1030-1060 and 1150-1190 ml.

DISCUSSION

Quantitative evaluation of the effects of vascular tissue could be influenced by several factors. Both tissues compared, i.e. pith part of complex explants and the control pith explants, differed from each other not only by their contact with vascular tissue, but also by the degree of wounding and exposure to the atmosphere, due to unequal surface-mass ratio (Luštinec et al., 1984b). The comparison of both pith tissues was charged with small errors which resulted in a certain overestimation of the effect of vascular tissue on protein and starch accumulation (+12 and +9 %, respectively) and in certain underestimation of the effect of vascular tissue on callus formation (-20 %).

Young cells of vascular tissue could act as a sink for regulatory metabolites, but also for mobile substances present in adjacent pith. This situation could occur, first of all, in explants growing on complete media. In explants cultured on simple sucrose media and especially at high sucrose concentrations, the sink effect could be slight or absent.

Experimental results in addition to published data suggested that the stimulatory effects of vascular tissue was the result of action of at least three factors translocated from this tissue to the pith parenchyma: 1) cytokinins, 2) auxin(s) and 3) unknown effective factor(s). Both cytokinins and auxins have been found in cambium and in xylem and phloem exudates (see Luštinec et al., 1984a). A low concentration of these factors may exist in the pith of young plants. These factors may be missing or their concentration may be very low in the pith of adult plants grown in a natural short day. The latter tissue was insensitive or nearly so to cytokinins and auxins, but not to the unknown effective factor(s) of vascular tissue.

The production of hormonal factors in the explants was induced or greatly intensified by exogenous sucrose which served at the same time as a source of energy and metabolic building blocks in processes induced by hormonal factors. The internal supply of sugars may thus limit the production, transport and action of these factors under natural conditions.

Anatomically the best detectable effect of vascular tissue was the accumulation of starch. The localization of starch granules in vascular tissue suggested that the cambial meristem with adjacent cells was the center of starch inducing factor(s) production. This seems to be in agreement with the

fact that in many plant organs starch accumulated in the proximity of or within the meristems. The same relationship was
found in tissue cultures, where in some cases starch accumulation preceeds the meristemization of the tissue leading to
shoot (Thorpe and Murashige, 1968, 1970) or flower formation
(Tran Thanh Van and Chlyah, 1976).

On the other hand it appeared that the hormonal stimulus
detected in the pith was produced not only in the meristematic
region of vascular tissue, but also in the region bordering
with the pith and composed of relatively old parts of xylem
and pith rays (Fig. 4). The production of hormonal factors in
meristematic and other regions of vascular tissue was independent of their growth activity, as followed from experiments
with high sucrose concentrations.

Hormone-like factors were isolated from the explants and
are supposed to be identical with those unknown effective
factors whose existence in vascular tissue containing explants
was proved experimentally. However, no direct evidence was obtained.

The chemical nature of the isolated factors is unknown.
It can only be assumed that at least some of them are phenolic
substances. Their biological activity, as far as determined,
is similar to that of cytokinins. Phenolic compounds with
cytokinin-like activity were already isolated from plant
sources. They were identified as dihydroconiferyl alcohol (Lee
et al., 1981) and dehydrodiconiferyl glucosides (Chen et al.,
1986; Bins et al., 1986). The relationship of studied hormone-
-like factors to these phenolic compounds, if any should be
elucidated.

ABSTRACT

Explants in form of blocks or cylinders were excised from
stem of kale (*Brassica oleracea* L., var. *medullosa*) in such a
way that they contained either pith parenchyma only or pith
parenchyma with adjacent vascular tissue including cambial
meristem, xylem and pith rays. During a few days cultivation
of explants on simple sucrose or complete media, hormone-like
factors were produced in vascular tissue and translocated to
the pith, where they affected metabolic and developmental
processes up to a distance of at least 20 mm. Depending on the
composition of culture media they stimulated soluble protein
and starch accumulation or callus and organ formation. The
production of hormone-like factors and their transport to the
pith were induced by exogenous sucrose. The hormone-like product of vascular tissue could partially be substituted by cytokinins and auxins if explants were derived from juvenile
plants, but not if the explants were derived from adult plants
grown under natural short-day conditions.

The hormone-like factors were extracted with water from
explants containing vascular tissue. The extract was fractioned using insoluble polyvinylpyrrolidone and Spehadex LH-20
Two groups of fractions with high activity in protein synthesis-dependent starch accumulation biotest were obtained. The
fractions were supposed to contain effective factors of vascular tissue differing from major classes of phytohormones.

REFERENCES

Binns, A. N., Chen, R. H., Wood, H. N., and Lynn, D. G., 1986, Cell division promoting activity of naturally occurring dehydrodiconiferyl glucosides: Do cell wall components control cell division?, Proc. <u>Natl. Acad. Sci. USA</u>, 84: 980.

Chen, R. H., Manning, K. S., Wood, H. N., and Lynn, D. G., 1986, The structural characterization of endogenous factors from *Vinca rosea* crown gall tumors that promote cell division of tobacco cells, <u>Proc. Natl. Acad. Sci. USA</u>, 84:615.

Horák, J., Luštinec, J., Měsíček, J., Kamínek, M., and Poláčková, D., 1975, Regeneration of diploid and polyploid plants from the stem pith explants of diploid marrow stem kale (*Brassica oleraces* L.), <u>Ann. Bot.</u>, 39:571.

Lee, T.S., Purse, J. G., Pryce, R. J., Horgan, R., and Wareing, P.F., 1981, Dihydroconiferyl alcohol - A cell division factor from *Acer* species, <u>Planta</u>, 152:571.

Luštinec, J., Kamínek, M., Beneš, K., and Conrad, K., 1984a, Hormone-like effect of vascular tissue on starch accumulation in stem explants of kale, *Brassica oleracea*, <u>Physiol. Plant.</u>, 61:224.

Luštinec, J., Kamínek, M., Kramell, H., and Sembdner, G., 1984b, Hormone-like effects of vascular tissue on development of stem explants of kale, <u>Biochem.Physiol.Pflanz.</u>, 179:227.

Luštinec, J., Conrad, K., Kamínek, M., Kramell, H., and Sembdner, G., 1985, Hormonal factors stimulating starch accumulation in stem explants of kale, <u>Biol.Plant.</u>, 27:281.

Luštinec, J., 1988, Kale (*Brassica oleracea* L. var. *acephala, medullosa, ramosa, sabellica*), <u>in</u>: "Biotechnology in Agriculture and Forestry, Vol. 6 Crops II", Y. P. S. Bajaj, ed., Springer-Verlag, Berlin Heidelberg.

Shibata, K., Kubota, K., and Kamisaka, S., 1974, Isolation and chemical identification of a lettuce cotyledon factor, a synergist of the gibberellin action in induced lettuce hypocotyl elongation, <u>Plant Cell Physiol.</u>, 15:191.

Thorpe, T. A., and Murashige, T., 1968, Starch accumulation in shoot-forming tobacco callus culture, Science, 160:421.

Thorpe, T. A., and Murashige, T., 1970, Some histochemical changes underlying shoot initiation in tobacco callus cultures, <u>Can. J. Bot.</u>, 48:277.

Tran Thanh Van, M., and Chlyah, A., 1976, Différenciation de boutons floraux, de bourgeons végétatifs, de racines et de cal à partir de l´assise sous-épidermique des ramifications florales de *Nicotiana tabacum* Wisc. 38, Étude infrastructurale, <u>Can. J. Bot.</u>, 54:1979.

HORMONAL CONTROL OF SENESCENCE

Bartolomé Sabater, Mercedes Martín, Francisco J. Sánchez and Antonio Vera

Departamento de Biología Vegetal, Universidad de Alcalá de Henares. Apdo. 20, Alcalá de Henares, 28871-Madrid. Spain

INTRODUCTION

Three approaches are currently used for the investigation of the hormonal control of senescence: 1) the assay of the effects on senescence of externally added growth regulators, 2) the measurement of hormonal concentration and/or sensitivity changes during senescence, and 3) the study of the mechanisms through which hormones affect some specific senescence sympton. Here, we mostly deal with the third approach in relation to the chloroplast breakdown symptom accompanying the senescence of the leaves of barley (*Hordeum vulgare* L. cv. Hassan).

Several difficulties, as interaction among hormones and dose-dependent effects (Sabater et al., 1981) complicate the investigations on the hormonal control of senescence and have been discussed in detail previously (Sabater, 1985). Differential hormone effects on diverse senescence symptoms (Sabater et al., 1981; Cuello and Sabater, 1982; Calle et al., 1986) suggest that the different senescence symptoms, although correlated among them, are not similarly controlled by hormones, indicating several levels for the action of hormones in senescence.

Since the pioneer discovery by Richmond and Lang (1957) that cytokinins retard the loss of chlorophyll and proteins in detached leaves of *Xanthium*, the study of the mechanism of these effects became a favoured topic. The enhancement of the senescence of the leaves of barley by detachment and incubation in the dark, and its retardation by cytokinins are preceded by increases in the activity of protein synthesis of chloroplasts (García et al., 1983; Martín et al., 1986). Detachment and incubation in the dark stimulate the synthesis of specific senescence polypeptides in chloroplasts, and kinetin treatment of leaf segments changes the type of polypeptides synthesized by chloroplasts. These effects of kinetin have been explored to investigate the mechanism of the control of senescence by cytokinins.

CYTOKININS INDIRECTLY AFFECT THE SYNTHESIS OF PROTEINS IN CHLOROPLASTS

One of the earliest (5-10 h after treatment) effects in chloroplasts of the treatment of detached leaves of barley with 40 μM kinetin is an increase in the activity of protein synthesis in chloroplasts (García et al., 1983). Table 1 shows, however, that kinetin added to isolated chloroplasts (III) has no sensible effect on the activity of protein synthesis (I). But

Plant Aging: Basic and Applied Approaches
Edited by R. Rodríguez *et al.*
Plenum Press, New York, 1990

Table 1. Effects of different treatments with 40 µM kinetin on barley chloroplast protein synthesis, Bq incorporated.(30 min.100 µg chl)$^{-1}$.

Treatment	Activity
I. Control. Chloroplasts isolated from 14-day-old leaves with the standard medium (García et al., 1983)........	44.1±3.6
II. Chloroplasts isolated from 14-day-old leaves incubated 18 h in the dark with kinetin.......................	339.0±77
III. As I, but the final KCl resuspension buffer including kinetin...	44.9±6.3
IV. As I, but all buffers from homogeneization including kinetin...	73.8±5.4
V. As I, except that chloroplasts (before final resuspension in KCl buffer) were treated for 20 min at 0 ºC with cytosolic fraction (Calle et al., 1986) from barley leaf incubated 18 h with kinetin (as in II)...........	362.0±95

Protein synthesis assays were carried out, essentially, as previously (García et al., 1983). Values are means ± SE of 2-5 independent experiments.

when kinetin is present in the isotonic buffer for the homogeneization of the leaves (IV), isolated chloroplasts show a clear increase in the activity of protein synthesis. In a different approach, if chloroplasts from leaves not treated with kinetin are resuspended in a cytosolic fraction of kinetin-treated leaves, and then isolated again (V), they show a high increase (720 %) in protein synthesis activity when compared with control I. Other controls of protein synthesis by the cytosolic fraction also show very low activity, 18.8 Bq.(30 min)$^{-1}$. Apparently, when kinetin interacts with an appropriate cytosolic fraction it stimulates chloroplast protein synthesis (IV). Moreover, kinetin induces a soluble cytosolic factor ("SK") in leaves which stimulates protein synthesis in chloroplasts (V).

The results indicated in Table 2 suggest that SK is a thermosensible protein which is inactivated by trypsin treatment. It retains almost 90 % activity after dialysis and loses 25 % activity after one week at -20 ºC. SK precipitates between 30 and 50 % saturation of ammonium sulfate, recovering its activity after resuspension and dialysis. Concentrated SK, after

Table 2. Effects of different treatments of SK on its stimulating activity on chloroplast protein synthesis.

	Stimulation (fold)
No treatment............................	8.2
5 min at 100 ºC.........................	2.8
7 days at -20 ºC........................	6.2
20 min with trypsin at 37 ºC............	2.7
" " + trypsin inhibitor.	5.0
4 h dialysis at 4 ºC....................	7.2

0.01 mg trypsin and 0.02 mg trypsin inhibitor were used, in the appropriate treatments, for 1 ml of "SK".

258

ammonium sulfate precipitation and dialysis, stimulates chloroplast protein synthesis when applied in the incubation mixture for protein synthesis (results not shown).

SK also affects the pattern of the polypeptides synthesized by chloroplasts (Fig. 1). Apparently, SK induces in senescent chloroplasts (lane 2) the synthesis of similar polypeptides to those synthesized by chloroplasts from kinetin-treated leaves (lane 1). Dots in Fig. 1 indicate the main differences between the polypeptides synthesized by chloroplasts treated (lane 2) and not treated (lane 3) with SK. As SK is applied to chloroplasts during 20 min at 0-5 ºC, it probably has no effect on transcription and it controls chloroplast protein synthesis at translational level. One possibility is that SK includes cytoplasmic precursors of polypeptides of nuclear--encoded chloroplast proteins induced by cytokinins.

CONTROL BY CYTOKININ OF TRANSLATABLE CHLOROPLAST mRNAs

Many evidences (for references, see Mullet, 1988) indicate the existence in chloroplasts of untranslated mRNAs and, hence, of a strong translational control of gene expression. In agreement with Fig. 1, and previous results (Martín and Sabater, 1989), the treatment of senescent leaf segments of barley with kinetin changes the type of several polypeptides synthesized by chloroplasts. This change seems, at least partially, to be responsible for the retardation of senescence by kinetin (Cuello et al., 1984). Translation of different chloroplast mRNAs in commercial reticulocyte lysate is not very efficient (Fig. 2) and suggests that the level of several translatable mRNAs for kinetin induced polypeptides (55, 32, 22,..kD, lane 3) increase in respect to those in senescent chloroplasts (lane 4). By comparing with the major chloroplast soluble polypeptide (lane 2), it seems that the 55 kD polypeptides synthesized in reticulocyte lysate (lane 3) is the large subunit of ribulose-1,5-bisphosphate carboxylase/oxygenase. Thus, probably, the treatment of senescent leaf segments with kinetin increases in chloroplasts the level of mRNAs for polypeptides usually synthesized during the earliest stages of chloroplast development.

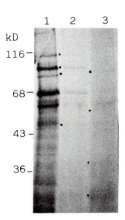

Fig. 1. SDS-PAGE and fluorography of the polypeptides synthesized by chloroplasts after different treatments with 40 µM kinetin. Light-driven protein syntheses were carried out (Martín and Sabater, 1989) in the presence of 1.1 MBq of [^{35}S]-methionine and 60 µg chl of chloroplasts from 14-day-old leaf segments incubated during 18 h in the presence of kinetin (1), chloroplasts (from 14-day-old leaves) treated with "SK" (2) and chloroplasts (from 14-day-old leaves) not treated with "SK" (3).

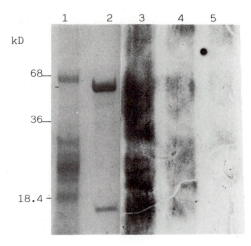

Fig. 2. SDS-PAGE of the chloroplast polypeptides and of the polypeptides synthesized by reticulocyte lysate with different chloroplast
mRNAs. Chloroplast RNA from 14-day-old leaf segments incubated 18 h
in the dark in the presence of kinetin (3) and in the absence of
kinetin (4) were translated in commercial reticulocyte lysate (Martín
and Sabater, 1989) with 1.1 MBq of [35S]-methionine. Lane 5 is a
control of protein synthesis without externally added RNA. For a
comparison, lanes 1 and 2 show a Coomasie stain of, respectively ,
membrane and soluble plastid polypeptides.

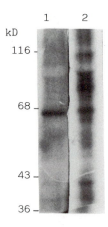

Fig. 3. SDS-PAGE and fluorography of the polypeptides synthesized in
nuclease-treated senescent chloroplasts by chloroplast RNA. Nuclease-
treated senescent chloroplasts were prepared as previously described
(Martín and Sabater, 1989) and they were supplemented with mRNA
isolated from 14-day-old leaf segments incubated during 25 h in the
dark in the presence (1) or in the absence of kinetin (2). Other
details as in Fig. 1 and the reference cited above.

A more efficient system for the translation of chloroplast mRNAs has been prepared (Martín and Sabater, 1989) from nuclease-treated senescent chloroplasts. However, this system did not translate mRNAs for the polypeptides induced by kinetin. It synthesized (Martín and Sabater, 1989) the same polypeptides (senescence polypeptides) when it is supplemented with mRNAs from senescent chloroplasts (isolated from detached leaf segments incubated during 18 h in the dark) or mRNAs from chloroplasts of leaf segments incubated during 18 h in the dark in the presence of kinetin. Clearly, chloroplasts from leaves treated 18 h with kinetin still contain mRNAs for senescence polypeptides. When chloroplast RNA was isolated after 25 h incubation of leaf segments with or without kinetin, almost all of the mRNAs for senescence polypeptides have disappeared in chloroplasts from kinetin--treated leaves. Fig. 3 shows that, when assayed by translation in nuclease--treated senescent chloroplasts, only the mRNA for the 66 kD polypeptide remains, at high translatable level, in chloroplasts from leaf segments incubated during 25 h in the presence of kinetin (lane 1). Many other senescence polypeptides were synthesized when the system of protein synthesis was supplemented with mRNAs isolated from chloroplasts of leaf segments incubated during 25 h in the absence of kinetin (lane 2).

Apparently, the control of gene expression is at least one of the mechanisms through which cytokinins retard the senescence of the leaves. The results shown in Fig. 2, indicate that cytokinins increase the levels of chloroplast mRNAs usually expressed during the earliest stages of chloroplast development. Probably also, cytokinins inhibit the synthesis in chloroplasts of the mRNAs for senescence polypeptides, but these mRNAs persist during the first 18 h after cytokinin treatment (Martín and Sabater, 1989). Twenty five h after cytokinin treatment, most of the mRNAs for senescence polypeptides disappear (Fig. 3). Although mRNAs for senescence polypeptides remain in chloroplasts during the first 18 h of cytokinin treatment, they are not used as templates in protein synthesis because a strong translational control in chloroplasts prevents their translation in chloroplasts of cytokinin-treated leaves (Martín and Sabater, 1989). Cytokinins control chloroplast gene expression during senescence at transcription and translation (and possibly at the level of mRNA stability).

In other chapter in this book, we show that most of the chloroplast genes for senescence polypeptides map in the single short chain of chloroplast DNA; in contrast with most of genes for the polypeptides synthesized by chloroplasts during their early development, which map in the inverted repeats and single large chains of chloroplast DNA. Thus, cytokinins retard senescence (at least in part) by inhibiting, at transcription and translation, the expression of genes of the single short chain of chloroplast DNA and by activating, also at transcription and translation, the expression of genes of the inverted repeats and single large chains of chloroplast DNA. At least the chloroplast gene translational control by cytokinins is mediated through cytoplasmic protein factor(s) (Table 1).

ACKNOWLEDGEMENTS

This work was supported by the Spanish Comisión Interministerial de Ciencia y Tecnología (Grants BIOT 23/85 and BIO 88-0002).

REFERENCES

Calle, F., Martín, M., and Sabater, B., 1986, Cytoplasmic and mitochondrial localization of the glutamate dehydrogenase induced by senescence in barley (Hordeum vulgare), Physiol. Plant., 66:451.

Cuello, J., Quiles, M.J., and Sabater, B., 1984, Role of protein synthesis and light in the regulation of senescence in detached barley leaves, Physiol. Plant., 60:133.

Cuello, J., and Sabater, B., 1982, Control of some enzymes of nitrogen metabolism during senescence of detached barley (Hordeum vulgare L.) leaves, Plant Cell Physiol., 23:561.

García, S., Martín, M., and Sabater, B., 1983, Protein synthesis by chloroplasts during the senescence of barley leaves, Physiol. Plant., 57:260.

Martín, M., and Sabater, B., 1989, Translational control of chloroplast protein synthesis during senescence of barley leaves, Physiol. Plant., 75:374.

Martín, M., Urteaga, B., and Sabater, B., 1986, Chloroplast protein synthesis during barley leaf growth and senescence: effect of leaf excision, J. Exp. Bot., 37:230.

Mullet, J. E., 1988, Chloroplast development and gene expression, Annu. Rev. Plant Physiol., 39:475.

Richmond, A. E., and Lang, A., 1957, Effect of kinetin on protein content and survival of detached Xanthium leaves, Science, 125:650.

Sabater, B., 1985, Hormonal regulation of senescence, in: "Hormonal regulation of plant growth and development", S. S. Purohit, ed., p 169, Martinus Nijhoff-Dr.W. Junk Pu., Dordrecht.

Sabater, B., Rodríguez, M. T., and Zamorano, A., 1981, Effects and interactions of gibberellic acid and cytokinins on the retention of chlorophyll and phosphate in barley leaf segments, Physiol. Plant., 51:361.

MODULATION, PURIFICATION AND FUNCTION OF A SOLUBLE AUXIN RECEPTOR

P.C.G. van der Linde[1] and A.M. Mennes[2]

[1]Bulb Research Centre, P.O. Box 85, 2160 AB Lisse, The
Netherlands; [2]Dept. Plant Molecular Biology, University of
Leiden, Nonnensteeg 3, 2311 VJ Leiden, The Netherlands

Introduction

Multicellular organisms use chemical signals, such as hormones, for the
regulation of their growth and development. Cells that are sensitive to
hormones show a specific response. In analogy with animal hormones it is
assumed that hormone-sensitive cells contain a protein that is able to
bind the hormone and, subsequently, to activate a biochemical process,
which eventually leads to the observed response.
The MolBas research group of the Dept. Plant Molecular Biology studies
auxin-binding proteins in cultured tobacco tissues. In these tissues, a
low concentration of auxin induces cell division whereas a high concentration induces root formation. Three types of auxin-binding proteins were
identified. One is membrane-bound and binds auxin and inhibitors of polar
auxin transport (Maan et al., 1985a). The second one is also membrane-
bound and binds IAA with a relative low affinity (Vreugdenhil et al., 1988).
This protein is probably involved in root formation (Maan et al., 1985b;
Nakamura et al., 1988). The third binding protein is soluble and binds IAA
with high affinity (Oostrom et al., 1975, 1980). On the basis of this
characteristic an involvement in IAA-induced cell division is suggested.

Detection

Binding of auxin to the soluble protein reaches equilibrium after ca. 30
min incubation at 25 C and is detected after separation of bound and free
hormone by means of the dextran-coated charcoal method. Optimal binding is
observed at pH 7.8 (Oostrom et al., 1975, 1980). The K_d for IAA is 16 nM.
The molecular weight of the protein is ca. 175 kD (Van der Linde et al.,
1984). The protein is not present in freshly excised stem pith but can be
detected after 1 day of culture (Bogers et al., 1980). The presence of this
protein was also demonstrated in soluble fractions of callus, shoot tips
and cell suspensions and in salt extracts of nuclei isolated from these
tissues. The protein is not present in salt extracts of nuclei isolated
from cell-suspension cultures at the stationary phase, but appears in this
fraction 5 min after the addition of 2,4-D (Libbenga et al., 1987).
The detection of this protein is hampered by its very low concentration in
soluble fractions (0-200 fmol/mg protein). The detection is enhanced by
preventing complex formation between polyphenols and proteins. When callus
tissue is grinded in Tris/2-ME buffer, pH 7.9, a very high contamination
of polyphenols is observed in fractions obtained by gel filtration of
soluble proteins. This contamination is reduced to ca. 6 % (w/w) by using

boric acid buffers, pH 6.8, supplemented with an inhibitor of polyphenol oxidase (Van der Linde et al., 1984). However, for shoot tips of the tobacco plant we found that Tris/DTT buffers, pH 7.5, give the best results. Sometimes, non-hyperbolic Scatchard plots are obtained from the binding assays. This could only be explained by assuming the presence of either positive co-operative binding or receptor inactivation during the binding assay. Therefore, we tested the effect of protease inhibitors on the concentration of the soluble receptor. Only antipain and leupeptin give a slight increase in the number of binding sites (Van der Linde, 1986). However, inhibition of phosphatase activity towards the receptor by adding a substrate for this enzyme and activation of protein kinases with magnesium and ATP increase the detectable amounts of the receptor in concentrated soluble fractions Van der Linde et al., 1985; Van der Linde,1986). Addition of the purified cAMP-dependent protein kinase from bovine heart and 1 mM cAMP decreases the concentration of the receptor (Van der Linde, 1986). Hence, it is concluded that auxin binding to this protein is regulated by phosphorylation and dephosphorylation.

Purification

Conventional protein purification methods were used in trying to purify the receptor. However, either a very low recovery is obtained, e.g., in ammonium-sulphate precipitation and in DEAE-cellulose ion-exchange chromatography, or the receptor elutes in several fractions, e.g., in gel filtration and in Concanavalin-A-bound Sepharose chromatography. The most reliable results are obtained in gel filtration of concentrated soluble fractions on Sepharose G-200 and AcA34 Ultrogel. The apparent molecular weight observed in these methods is 168 kD (Van der Linde et al., 1984; Van der Linde, 1986).

We also tried to purify the receptor by affinity chromatography. 5-OH-IAA was coupled to epoxy-activated Sepharose via the hydroxyl group. Concentrated soluble fractions of tobacco shoot tips were loaded onto the IAA-Sepharose after the addition of magnesium, a substrate for phosphatases and ^{32}P-ATP. Only 1 phosphorylated protein was obtained by elution with 0.01 mM IAA. This protein exhibits a molecular weight of 51.7 kD in SDS-PAGE (Van der Linde, 1986; Libbenga et al., 1987). Therefore, we conclude that the receptor in its active form is a phosphoprotein, consisting of several subunits.

Function

In order to examine the function of the soluble receptor, Bailey et al. (1985) studied the modulation of the receptor during the growth of suspension cultures. They found that the receptor is present in the soluble fraction at day 0, i.e. directly after transfer of the cells to fresh medium, and disappears from this fraction during the exponentional growth phase. At this stage, the receptor is present in salt extracts of isolated nuclei. The receptor reappears in the soluble fraction when the cells reach the stationary phase. By that time, the receptor cannot be detected anymore in salt extracts of isolated nuclei. They concluded that the nucleus was the primary site of action for the receptor. A similar conclusion was drawn by Libbenga et al. (1987), who observed an accumulation of the receptor in nuclei, isolated 30 min after the addition of auxins to 2,4-D starved cells. This accumulation shows hormone specificity. Within the same time period, auxins induce a rapid and transient increase in the activity of RNA polymerases (Van der Linde, 1986). Therefore, we investigated the activity of the soluble receptor on transcription in isolated nuclei.

Partially purified fractions containing the soluble auxin receptor increase the RNA-polymerase-II activity of isolated callus nuclei in the presence of IAA. This increase is correlated with the calculated occupancy of the receptor (figure 1) and shows hormone specificity (figure 2)(Van der Linde et al., 1984; Bailey, 1983). The IAA-Sepharose-purified receptor shows a similar response in nuclei, isolated from the stationary phase of cell-suspension cultures (Van der Linde, 1986; Libbenga et al., 1987). It is concluded that the soluble auxin receptor is involved in auxin-induced transcription preceding cell division.

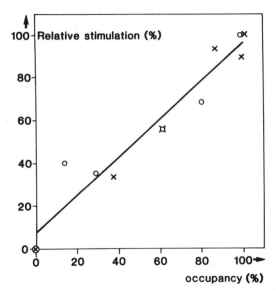

Figure 1. Relation between the relative stimulation of transcription and the relative occupancy of the soluble auxin receptor, observed by Bailey (1983) (x) and by Van der Linde et al. (1984) (o). The maximal stimulation observed by Bailey was 30 % and by Van der Linde et al. 43 %. Y=0.9X + 6.3; Cr = 0.97.

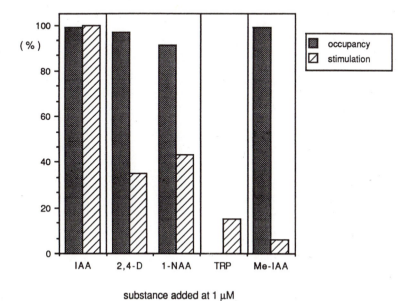

substance added at 1 µM

Figure 2. Hormone-specificity of the stimulation of transcription of the soluble auxin receptor. IAA gives the highest stimulation. 2,4-D and 1-NAA act as partial agonists. Me-IAA act as an anti-auxin. An indole-like non-auxin as tryptophan does not give a response.

In the same cells auxins induce the transcription of specific genes in vivo
(Van der Zaal et al., 1987a). Some changes in gene transcription are already
observed within 15 min after addition of auxins (Van der Zaal et al., 1987b).
In nuclei isolated at the stationary phase of the cultures, the induced genes
are not transcribed in vitro, but their transcription is already observed
in nuclei isolated 15 min after the addition of 2,4-D. Whether the soluble
auxin receptor is able to induce the transcription of these genes in
stationary phase nuclei in vitro is still under investigation.

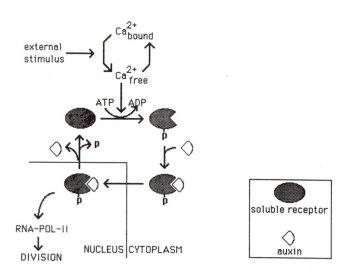

Figure 3. Proposed action mechanism for auxin-induced cell division. For
explanation see text.

A model for auxin-induced cell division

On basis of these results, the following model for auxin-induced cell
division is proposed (figure 3). Auxin binds to a soluble, phosphorylated
receptor in the cytoplasm. The auxin-receptor complex migrates to the nucleus
and activates the RNA-polymerase-II activity. This increases the transcrition
of genes that are involved in the regulation of cell division and, consequently,
the cells divide. The receptor is inactivated by phosphatases in the nucleus.
It looses the phosphate group and the auxin and, subsequently, it is
recycled to the cytoplasm in an inactive conformation. In the cytoplasm, the
receptor can be activated by phosphorylation if the cytoplasmic free calcium
concentration is high enough to activate the responsible protein kinase.
Since the endogenous auxin concentration of suspension-cultured cells at any
stage of growth is high enough to saturate the soluble receptor (Elliott et
al., 1987), it is hypothesized that the main regulatory signal for cell
division is a stimulation of the internal free calcium concentration, which
may be achieved by many (environmental) stimuli.

References

Bailey, H.M. (1983) Receptors for plant growth substances. Ph.D. Thesis.
   Leicester Polytechnic, United Kingdom.
Bailey, H.M., Barker, R.D.J., Libbenga, K.R., Van der Linde, P.C.G., Mennes,
   A.M., and Elliott, M.C. (1985) Auxin binding site in tobacco cells.
   Biol. Plant. 27, 105-109.
Bogers, R.J., Kulescha, Z., Quint, A., Van Vliet, T.B., and Libbenga, K.R.
   (1980) The presence of a soluble auxin receptor and the metabolism of
   3-indole-acetic acid in tobacco-pith explants. Plant Sci. Lett. 19, 311
   -317.

Elliott, M.C., O'Sullivan, A.M., Hall, J.F., Robinson, G.M., Lewis, J.A.,
Armitage, D.A., Bailey, H.M., Barker, R.D.J., Libbenga, K.R., and Mennes,
A.M. (1987) Plant cell division. The role of IAA and IAA-binding proteins.
pp. 245-255 in: Molecular biology and plant growth control. Alan R. Liss,
Inc.

Libbenga, K.R., Van Telgen, H.-J., Mennes, A.M., Van der Linde, P.C.G., and
Van der Zaal, E.J. (1987) Characterization and function analysis of a
high-affinity cytoplasmic auxin-binding protein. pp. 229-243 in: Molecular
biology and plant growth control. Alan R. Liss, Inc.

Maan, A.C., Kühnel, B., Beukers, J.J.B., and Libbenga, K.R. (1985a) Naphthyl-
phthalamic acid-binding sites in cultured cells of Nicotiana tabacum.
Planta 164, 69-74.

Maan, A.C., Van der Linde, P.C.G., Harkes, P.A.A., and Libbenga, K.R. (1985b)
Correlation between the presence of membrane-bound auxin binding and
root regeneration in cultured tobacco cells. Planta 164, 376-378.

Nakamura, C., Van Telgen, H.-J., Mennes, A.M., Ono, H., and Libbenga, A.M.
(1988) Correlation between auxin resistance and the lack of a membrane-
bound auxin-binding protein and a root-specific peroxidase in Nicotiana
tabacum. Plant Physiol. 88, 845-849.

Oostrom, H., Kulescha, Z., Van Vliet, T.B., and Libbenga, K.R. (1980)
Characterization of a cytoplasmic auxin receptor from tobacco-pith
callus. Planta 149, 44-47.

Oostrom, H., Van Loopik-Detmers, M.A., and Libbenga, K.R. (1975) A high-
affinity auxin receptor for indoleacetic acid in cultured tobacco-pith
explants. FEBS Lett. 59, 194-197.

Van der Linde, P.C.G., Maan, A.C., Mennes, A.M., and Libbenga, K.R. (1985)
Auxin receptors in tobacco. pp. 397-403 in: Proc. 16th FEBS meeting, part
C, VNU Science Press, Utrecht.

Van der Linde, P.C.G. (1986) Modulation, purification and function of a
soluble auxin receptor. Ph.D. Thesis, University of Leiden, The Netherlands.

Van der Linde, P.C.G., Bouman, H., Mennes, A.M., and Libbenga, K.R. (1984)
A soluble auxin-binding protein from cultured tobacco tissues stimulates
RNA synthesis in vitro. Planta 160, 102-108.

Van der Zaal, E.J., Memelink, J., Mennes, A.M., Quint, A., and Libbenga, K.R.
(1987a) Auxin-induced mRNA species in tobacco cell cultures. Plant Mol.
Biol. 10, 145-157.

Van der Zaal, E.J., Mennes, A.M., and Libbenga, K.R. (1987b) Auxin-induced
rapid changes in translatable mRNA's in tobacco cell suspension. Planta
172, 514-519.

Vreugdenhil, D., Burgers, A., and Libbenga, K.R. (1979) A particle-bound
auxin receptor from tobacco-pith callus. Plant Sci. Lett. 16, 115-121.

FREE RADICALS IN STRESSED AND AGING PLANT TISSUE CULTURES

Erica E. Benson

Department of Agriculture and Horticulture
Nottingham University School of Agriculture
Sutton Bonington, Loughborough, Leics. UK. LE12 5RD

INTRODUCTION

Free radicals are highly reactive molecular species which possess an unpaired electron. Oxy free radicals are especially important in biological tissues, since metabolism is dependent on the transfer of electrons, oxidation/reduction reactions and molecular oxygen. Free radical activity is therefore a normal feature in both plant and animal cells (e.g. in electron transport, lipid metabolism, detoxification and phagocytosis). However, free radicals can also initiate harmful reactions in the cell and their activity is tightly controlled. Cells are equipped with anti-oxidants to ensure that any free radicals that "leak" from normal metabolic processes are removed. Unfortunately such control is challenged if tissues undergo pathological disease, severe stress and physical injury. Under these circumstances, reduced antioxidant status and metabolic impairment can soon lead to free radical attack of macro molecules (lipids, proteins and DNA). In the long term these events can lead to further metabolic disorder, necrosis and cell and tissue death.

Free radicals are implicated in aging phenomena and plant senescence. However, very little is understood about age and stress-related free radical activity in plant tissues that have been grown in vitro. The objective of this review is to report on both the direct and indirect evidence for free radical activity in plant tissue cultures and discuss the implications of oxidative stress on culture performance and stability.

Nooden (1988) has proposed that senescence decribes the endogenously controlled degenerative processes that lead to death. In the whole plant this can be the death of an organ or organs (e.g. leaves, petals) or the entire organism. In contrast aging is described as passive degeneration due to non-regulated, exogenous factors. Aging may not lead to death directly but increase the probability of death occurring by decreasing resistance to stress and disease. The growth of plants in vitro is largely controlled by exogenous factors, specifically through the application of growth regulators and the maintenance of nutrients in the culture medium.

PHYTOHORMONES AND FREE RADICAL RELATED PLANT SENESCENCE

The hormonal control of plant senescence can directly and indirectly involve free radical activity. Since plant growth regulators are widely used to manipulate growth and differentiation in plant tissue cultures an understanding of their role in proposes complex models which describe the involvement of lipoxygenase, plant hormones (notably cytokinins) and calcium in controlling senescence. Lipoxygenase incorporates molecular oxygen in linoleic and linolenic acids (Zimmerman and Vick, 1988). Lipid peroxides are produced and these provide substrates for a hydroperoxide cyclase which produces an 18-carbon, cyclic oxo-acid which is a precursor of jasmonic acid, a compound known to promote plant senescence. Lipoxygenase can produce several free radical species (the superoxide radical, $O_2^{\cdot-}$ and peroxyl and alkoxyl radicals $ROO^{\cdot}$, $RO^{\cdot}$), which, in the absence of free radical scavengers can attack cellular components. Cytokinin lowers lipoxygenase activity in senescing tissues and is thought to reduce free radical damage in two ways (Leshem, 1984, 1988):

1. By direct scavenging in which a free radical abstracts a hydrogen from the amine bond which is present in all members of the cytokinin group of hormones.

2. By incipient prevention on inhibiting xanthine oxidase the enzyme which catalyses the breakdown of xanthine to uric acid and oxy free radicals.

Auxins are also involved in oxidative phenomena and may indirectly contribute to free radical activity. Indole acetic acid (IAA) breakdown occurs via the IAA oxidase enzyme, and H_2O_2 is involved in this reaction. Peroxidase and catalase activity is associated with the dissipation of H_2O_2 which, if allowed to accumulate in the cell, can lead to the formation of the highly toxic hydroxyl radical ($OH^{\cdot}$) via the Haber-Weiss/Fenton reaction (Haber and Weiss, 1934).

$$O_2^{\cdot-} + H_2O_2 \xrightarrow{\quad Fe^{3+}/Fe^{2+} \quad} O_2 + OH^{\cdot} + OH^{-}$$

A general increase in peroxidase activity has been noted in senescent tissues (Thompson et al., 1987).

Ethylene evolution is thought to be free radical mediated via 1-amino cyclopropane-1-carboxylic acid (ACC). Recent studies on cell-free systems of barley roots by Nilsen et al. (1988) suggest that the enzymic peroxidation of fatty acids generates $O_2^{\cdot-}$ which then oxidises ACC to ethylene. However, they caution that this mechanism may not be universal. Ethylene is also linked to polyamine production and the antisenescent polyamines share a common metabolite with S-adenosylmethionine (SAM). Mattoo and Aharoni (1988) show that the production of ethylene is inhibited by polyamines in order of cationic progression. Thus, spermine the strongest base is the most effective inhibitor.

EVIDENCE FOR FREE RADICAL PRODUCTION IN PLANT TISSUE CULTURES

1. Phenol oxidation

This is the most documented aspect of oxidative stress. Symptoms involve a blackening or browning of explants and the leakage of oxidised products into the media. Plants naturally high in phenolics (woody species) are the most susceptible. Phenol oxidation products can seriously inhibit growth in vitro and can cause tissue death. Although phenols play an important role in plant lignification and protection, their wound response in tissue culture systems is highly undesirable. Thus, oxidised phenols combine with proteins either reversibly through hydrogen bonding or irreversibly by oxidation (Compton and Preece, 1986). Free radical activity is associated with lignification and a scheme involving $O_2^{\cdot-}$ has been proposed by Elstner (1984). Studies on mammalian tissues have also demonstrated that phenols can be converted to protein binding species by the superoxide radical (Wolff and Youngman, 1984). The production of H_2O_2 during peroxidase/phenol age-related oxidative phenomena may be useful. Leshem (1984, 1987 and 1988) oxidation may also provide an indirect source of free radicals since H_2O_2 can react with metal cations in the Fenton reaction to produce the toxic OH' radical (Haber and Weiss, 1934).

2.   Lipid peroxidation

Lipid peroxidation can occur via lipoxygenase activity or through the direct attack of free radicals on membrane lipids. Many breakdown products result from lipid peroxidation and these can be used as convenient markers of free radical injury. The volatile lipid peroxidation product, ethane (largely derived from the oxidation of linolenic acid) has been used to assess stress and damage in a range of plant tissue culture systems:

Protoplasts. Schnabl et al. (1983a and b) used ethane as a marker of aging in isolated protoplasts of Vicia faba. The rate of aging was reduced when the protoplasts were immobilised in calcium/alginate matrices. Ethylene evolution occurred after four days in protoplasts that were maintained in liquid mannitol medium. In contrast even after seven days the immobilised protoplasts produced very low levels of the hydrocarbon. This reduction in aging was confirmed by the finding that photosynthetic activity was maintained in the immobilised cells but was reduced in the suspended protoplasts. Subsequent studies by Schnabl and Youngman (1985) confirm that the immobilisation of plant protoplasts inhibits enzymic lipid peroxidation. Using the above systems they measured: (1) lipid acyl hydrolase, the enzyme responsible for the excision of fatty acids from membranes, (2) lipoxygenase, the enzyme which oxidises the free fatty acids by means of an oxy free radical mechanism, (3) hydroperoxides, the products of lipid peroxidation, and (4) ethane. All the "markers" associated with lipid peroxidation were reduced in the immobilised cells compared to protoplasts which were maintained in liquid suspension. It was concluded that immobilisation significantly increases stability and prolongs the life of plant protoplasts. The reasons for this were unclear, however it was hypothesised that immobilisation maintains cell compartmentation and somehow protects the fragile cell wall lacking cells.

Cassells and Tamma (1986) used ethane to determine the effects of pre-isolation stress on protoplast release and survival in Nicotiana tabacum. Donor plants were stressed by water limitation or waterlogging, both treatments deleteriously affecting post-isolation survival but not immediate viability. Ethane evolution was found to be a more reliable parameter of stress than ethylene. The presence of the lipid peroxidation marker indicates that oxidative damage may be important in determining successful protoplast isolation. This is supported by the finding that protoplasts are stabilised by the application of antioxidants (Saleem and Cutler, 1986).

Cells and callus tissue. Yamaguchi and Fukuzumi (1982) detected free fatty acids, aldedhydes, alcohols and ketones in volatile extracts from callus cultures of Eucalytus polybractea, Pinus radiata and Eucalyptus robusta. Although these workers do not comment on the origin of these compounds the volatile profiles are typical of those of lipid peroxidation breakdown products (e.g. hexanols and hexanals). A range of short chain volatile hydrocarbons associated with lipid peroxidation have been detected in cell and callus cultures of Daucus carota (Benson and Withers, 1987). This study shows that the hydroxyl radical may also be produced. The evolution of methane from dimethyl sulphoxide which was incorporated into cryoprotectant and culture media was correlated with the OH' scavenging properties of the solvent. On cryopreservation it was noted that the carrot cells produced increased amounts of methane, ethane and pentane. Garcia and Einset (1983) have shown that tobacco callus stressed with high concentrations of 2,4-D and NACL produce ethane. They suggest that this volatile may be used to monitor severe stress in plant tissue cultures.

<u>Organised cultures</u>. Van Aartrijk et al. (1985a and b) monitored ethane and ethylene in tissue culture bulb scale explants of <u>Lilium speciosum</u>. Ethane was produced during early culture initiation, and this was thought to be a symptom of lipid peroxidation induced by wounding. NAA and high temperature treatment seemed to exacerbate the production of the volatile. Ethane evolution decreased during the later stages of culture.

3.   Singlet oxygen (O_2^1)

Singlet oxygen is not a free radical but an electronically excited state of molecular oxygen. Because of its reactive nature it behaves in a similar way to a free radical and can initiate free radical formation. Using the method of chemiluminescence, Benson and Noronha-Dutra (1988) have monitored O_2^1 production in cryopreserved and non-cryopreserved callus cultures of <u>D. carota</u> and shoot-tips of <u>Brassica napus</u>. The excited state of oxygen was highest in freshly thawed tissues, however it was also detected in cultures which had been wounded in the presence of light. This suggests that free radicals and singlet oxygen may be implicated in the bleaching of plant cultures, a common symptom of photooxidative stress.

Ishi (1986) has also demonstrated the production of singlet oxygen during the isolation of protoplasts from oat leaves. Since lipids, DNA and proteins are targets of O_2^1 attack it is suggested that protoplast damage during isolation may be due in part to the reactive oxygen species.

THE CONTROL OF OXIDATIVE DAMAGE IN PLANT TISSUE CULTURES

Compton and Preece (1986) give a comprehensive list of the means by which exudation, a symptom of oxidative stress, can be reduced in plant tissue cultures. These can be summarised:

1.   Minimise explant wounding.
2.   Wash explants before culture initiation.
3.   Apply antioxidants.
4.   Use reduced salts in the media.
5.   Careful application of plant growth regulators.
6.   Incorporation of activated charcoal.
7.   Culture under low light.
8.   Use rapid subculture regimes.

These approaches can indirectly support the previous evidence that free radicals have a significant role in tissue culture stress. Antioxidants such as glutathione and ascorbate are natural components of cellular free radical scavenging mechanisms. However, it is important to caution their general application to plant tissue cultures. Cellular antioxidant mechanisms are tightly controlled and interdependent on complex recycling mechanisms (Benson, 1989). Thus their topical application to cultures may not produce the desired effects, and may even initiate more damage. An example of this is the application of ascorbate when iron salts are present. The ascorbate Fe^{2+}/Fe^{3+} couple is a system known to produce the toxic hydroxyl radical (Halliwell and Gutteridge, 1985a and b). The latter is also formed when iron salts react with peroxides. It is not therefore surprising that under certain circumstances free Fe^{3+} in culture media causes browning and culture aging. Vagera (1984) shows that the growth and development of anthers from <u>N. tabacum</u> is dependent on the free iron content of the medium. Anthers cultured on low free iron media retained their light colour and did not undergo browning. However, a compromise had to be reached in which enough free iron was available, but levels were such that browning was reduced.

Activated charcoal has been used in the culture of oxygen sensitive micro-organisms (Hoffman et al., 1983). When incorporated into bacteriological media it prevented photochemical oxidation of media components, reduced H_2O_2 levels and detoxified accumulating free fatty acids. The precise role of activated charcoal in plant tissue culture systems is unknown but there is evidence to suggest that it is important in preventing oxidation. Polyamines have been applied to tissue cultures in an attempt to prevent phenolic oxidation (Compton and Preece, 1986). This corroborates their function as free radical scavengers, and their role in preventing plant senescence (Drolet et al., 1986).

THE CONSEQUENCES OF FREE RADICAL DAMAGE IN PLANT TISSUE CULTURES

Oxidative stress can severely affect the establishment, development and proliferation of explants in tissue culture and this is particularly the case for woody species. Once established, cultures prone to oxidative stress can lose their potential to proliferate and culture browning may be a constant problem. A less researched aspect of free radical damage in plant tissue cultures concerns the involvement of oxidative stress in molecular stability. Plant tissue culture largely involves the production of new and improved cultivars and varieties (genetic manipulation), the increased proliferation of existing genotypes (micropropagation) and the establishment of cultures that produce economically important compounds (secondary products). Genotypic and phenotypic stability is therefore paramount for the successful exploitation of in vitro techniques and factors which increase the potential for tissue culture instability must be avoided.

Studies on animal and microbial systems have shown that free radicals cause both direct and indirect damage of the genome and they have been implicated in carcinogenesis and mutagenesis. Lipid peroxidation products are mutagenic to bacteria and it has been suggested that it is the hydroperoxide group which is responsible for the genomic damage (Frankel, 1987; Frankel et al., 1987). Peroxidation products can interact directly with DNA; the resulting compounds are fluorescent and can be readily detected (Frankel et al., 1987). Fluorescent products were detected when calf thymus DNA was incubated with peroxidising arachidonic acid. Structural changes in the DNA were observed together with a reduced template activity for rat liver RNA polymerase (Reiss and Tappell, 1973). Singlet oxygen is extremely reactive and can damage deoxynucleosides. OH$^{.}$ also attacks DNA, forming thymine glycols (Ames et al., 1987).

Whilst there are presently many studies linking free radical activity with genetic damage in mammalian systems there remains a paucity of information on their effects in the plant genome. Genetic instability may be linked with free radical damage in stored seeds (Benson, 1989), but to date the evidence is either indirect or circumstantial. Plant tissue cultures may be particularly prone to oxidative stress and in the long term this may be expressed within the genome. If the basis of in vitro genetic instability is to be understood fully the contribution of free radical injury must not be overlooked.

REFERENCES

Ames, B. C., Hollstein, M. C. and Cathcart, R., 1982. Lipid peroxidation and oxidative damage to DNA, in: "Lipid Peroxides in Biology and Medicine", pp 339-351. K. O. Yagi, ed., Academic Press Inc., London.

Benson, E. E. and Withers, L. A., 1987. Gas chromatographic analysis of volatile hydrocarbon production by cryopreserved plant tissue cultures: A non-destructive method for assessing stability. Cryo-letters, 8:35-46.

Benson, E. E. and Noronha-Dutra, A. A., 1988. Chemiluminescence in cryo-preserved plant tissue cultures: The possible involvement of singlet oxygen in cryoinjury. Cryo-letters, 9:120-131.

Benson, E. E., 1989. Free radical damage in stored plant germplasm. I.B.P.G.R. Status report, Rome (in press).

Cassells, A. C. and Tamma, L., 1986. Ethylene and ethane release during tobacco protoplast isolation and protoplast survival potential in vitro. Physiol. Plant., 66:303-308.

Compton, M. E. and Preece, J. E., 1986. Exudation and plant establishment. I.A.P.T.C. Newsletter (50):9-18.

Drolet, G., Dumbroff, E. B., Legge, R. L. and Thompson, J. E., 1986. Radical scavenging properties of polyamines. Phytochem., 25:367-372.

Elstner, E. F., 1984. Comparison of inflammation in pine needles and humans, in: Oxygen Radicals in Chemistry and Biology, pp 969-978. W. Bor, M. Saran and D. Tait, eds, Walter de Gruyter and Co., Berlin.

Frankel, E. N., 1987. Biological significance of secondary lipid oxidation products. Free Rad. Res. Comm., 3:213-225.

Frankel, E. N., Neff, W. E., Brooks, D. D. and Fujimoto, K., 1987. Fluorescence formation from the interaction of DNA with lipid oxidation degradation products. Biochim. et Biophys. Acta, 919:239-244.

Garcia, F. G. and Einset, J. W., 1983. Ethylene and ethane production in 2,4-D treated and salt treated tobacco tissue cultures. Ann. Bot., 51:287-295.

Haber, F. and Weiss, J., 1934. The catalytic decomposition of H_2O_2 by iron salts. Proc. R. Soc. Land., A147:332.

Halliwell, B. and Gutteridge, J. M. C., 1985a. The role of transition metals in superoxide-mediated toxicity, Chapter 2, pp 46-82, in: "Superoxide Dismutase", Vol. II. L. W. Oberley, ed., CRC press Inc., Florida.

Halliwell, B. and Gutteridge, J. M. C., 1985b. Free radicals in biology and medicine. Clarendon Press, Oxford.

Hoffman, P. S., Pine, L. and Bell, S., 1983. Production of superoxide and hydrogen peroxide in media used to culture Legionella pneumophila: Catalytic decomposition by charcoal. Applied and Environmental Microbiology, March 1983, 784-791.

Ishi, S., 1986. Generation of singlet oxygen during isolation of protoplasts from oat leaves by enzymes, Abstr. 382, in: "VI International Congress of Plant Tissue and Cell Culture". I.A.P.T.C. Minnesota. Abstracts.

Leshem, Y. Y., 1984. Interactions of cytokinins with lipid-associated oxy free radicals during senescence: a prospective mode of cytokinin action. Can. J. Bot., 62:2943-2949.

Leshem, Y. Y., 1987. Membrane phospholipid catabolism and Ca^{2+} activity in control of senescence. Physiol. Plant., 69:551-559.

Leshem, Y. Y., 1988. Plant senescence processes and free radicals. Free Radical Biology and Medicine, 5:39-49.

Mattoo, A. K. and Aharoni, N., 1988. Ethylene and plant senescence, pp 242-269, in: "Senescence and Aging in Plants". L. D. Nooden and A. C. Leopold, eds, Academic Press Inc., London.

Nilsen, H-G, Sagstuen, E. and Aarnes, H., 1988. A cell-free ethylene forming system from barley roots. J. Pl. Phys., 133:73-78.

Nooden, L. D., 1988. The phenomena of senescence and aging, pp 2-51, in: "Senescence and Aging in Plants". L. D. Nooden and A. C. Leopold, eds, Academic Press Inc., London.

Reiss, U. and Tappel, A. L., 1973. Fluorescent product formation and changes in structure of DNA reacted with peroxidizing arachidonic acid. Lipids, 8:199-202.

Saleem, M. and Cutler, A. J., 1986. Stabilising cereal protoplasts with antioxidants, Abstr. 66, in: "International Congress of Plant Tissue and Cell Cultures". I.A.P.T.C. Minnesota. Abstracts.

Schnabl, H., Youngman, R. J. and Zimmerman, U., 1983a. Maintenance of plant cell membrane integrity and function by the immobilisation of protoplasts in alginate matrices. Planta, 158:392-397.

Schnabl, H., Elbert, C. and Youngman, R. J., 1983b. Release of ethane from immobilised plant cell protoplasts in response to chemical treatment. Physiol. Plant, 59:46-49.

Schnabl, H. and Youngman, R. J., 1985. Immobilisation of plant cell protoplasts inhibits enzymic lipid peroxidation. Pl. Science, 40:65-69.

Thompson, J. E., Legge, R. L. and Barber, R. F., 1987. The role of free radicals in senescence and wounding. New Phytol., 105:317-344.

Vagera, J., 1984. Prolonged vitality of tissue cultures on media without free trivalent iron. Plant tissue and cell culture application to crop improvement. F. J. Novak, L. Halvel and J. Dolezel, eds, Proc. Int. Symp. Inst. Exp. Bot. Czechoslovak Academy of Sciences, Prague.

Van Aartrijk, J., Blom-Barnhoorn, G. J. and Bruinsma, J., 1985a. Adventitious bud formation from bulb-scale explants of Lilium speciosum Thunb. in vitro production of ethane and ethylene. J. Pl. Physiol., 117:411-422.

Van Aartrijk, J., Blom-Barnhoorn, G. J. and Bruinsma, J., 1985b. Adventitious bud formation from bulb-scale explants of Lilium speciosum Thunb, in vitro. Effects of aminoethoxyvinyl-glycine, 1-amino cyclopropane-1-carboxylic acid, and ethylene. J. Pl. Physiol., 117:401-410.

Wolff, T. and Youngman, R. I., 1984. Oxidative metabolism of phenols: conversion of bromogatechol to protein binding species by $O_2^{\cdot-}$, pp 165-170, in: "Oxygen Radicals in Chemistry and Biology". W. Bors, M. Saran and D. Tait, eds, Walter de Gruyter and Co., Berlin.

Yamaguchi, T. and Fukuzumi, T., 1982. Volatile compounds from callus cells of woody plants. Proc. 5th Intl. Congr. Plant Tissue and Cell Culture, Tokyo.

Zimmerman, D. C. and Vick, B. A., 1988. Lipid peroxidation in plants - products and physiological roles, pp 196-212, in: "Lipid Peroxidation in Biological Systems". A. Sevarian, ed., Proc. of the American Oil Chemists Society.

POLYAMINES AND AGING: EFFECT OF POLYAMINE BIOSYNTHETIC INHIBITORS ON PLANT

REGENERATION IN MAIZE CALLUS CULTURED <u>IN VITRO</u>

A.F. Tiburcio[1], X. Figueras[1], I. Claparols[2], M. Santos[2],
and J.Mª. Torné[2]

[1]Laboratori Fisiologia Vegetal, Facultat de Farmàcia
 Universitat Barcelona, 08028, Barcelona
[2]C.I.D.-C.S.I.C. Jordi Girona Salgado 18-26
 08034 Barcelona, Spain

INTRODUCTION

Culture age is an important factor in expressing the genetic potential of plant cells and tissues <u>in vitro</u>, since it is well known that many cultures lose their morphogenetic capacity as they aged (Vasil et al., 1984). Although the precise biochemical and molecular mechanisms underlying the gradual loss of totipotency are not known, some compounds such as polyamines (PAs) are involved in these phenomena. Thus, the diamine putrescine (Put), the triamine spermidine (Spd), the tetraamine spermine (Spm), and their biosynthetic enzyme arginine decarboxylase (ADC) are biochemical markers of both plant cell aging and plant cell differentiation. In senescing leaves of cereals incubated in darkness both the ADC activity and the endogenous PA levels progressively decrease, while exogenous application of PAs, especially Spd and Spm, inhibits or retards the symptoms of senescence (Kaur-Sawhney et al., 1979; 1982). On the other hand, using tobacco thin cell layer cultures (TCL; Tran Thanh Van, 1973) we have demonstrated that Spd is a marker of floral differentiation (Tiburcio et al., 1987; 1988; Kaur-Sawhney et al., 1988), while the Put formed via ADC is a marker of root differentiation (Tiburcio et al., 1989a).

In recent years, the irreversible suicide inhibitors, DL-alpha-difluoro-methylarginine (DFMA; Kallio et al., 1981) and DL-alpha-difluoromethyl-ornithine (DFMO; Metcalf et al., 1978) have been shown specifically to inhibit plant ADC and ODC activities, respectively (Slocum and Galston, 1987). The availability of these inhibitors has permitted the demonstration of which pathway (ADC and/or ODC) is involved in a particular developmental process (Tiburcio et al., 1989b). In addition, the use of these inhibitors may also have some biotechnological applications. For instance, we have previously shown that the pretreatment of oat seedlings with the inhibitor DFMA was able to reduce the symptoms of senescence of the leaves subjected to osmotic stress. It was suggested that this antisenescence effect of the pretreatment with DFMA was due to the increase of endogenous levels of Spd and Spm (Tiburcio et al., 1986a,b). These results encouraged us to examine whether protoplasts isolated from DFMA-pretreated leaves will grow better in culture. As we expected, protoplasts isolated from such leaves appeared greener, with greater cytoplasmic streaming, more uniform regeneration of cell wall, and showing less lysis than protoplasts extracted from control leaves (Tiburcio et al., 1986b). Under a biotechnological point of view, the leaf pretreatment with DFMA resulted to be not only a new method to retard the osmotically-induced senescence of oat leaves, but also a new method to improve the viability and cell division of protoplasts isolated from such leaves.

Plant Aging: Basic and Applied Approaches
Edited by R. Rodríguez et al.
Plenum Press, New York, 1990

277

Furthermore, the use of the PA-biosynthetic inhibitor DFMA was more effective than the exogenous application of PAs.

In the present investigation, we have studied the effect of prolonged treatments with the PA-biosynthetic inhibitors DFMA and DFMO on bud differentiation and plant regeneration in maize calli cultured in vitro.

MATERIALS AND METHODS

Plant Material and Callus Culture

Plants of W64A $_{02}$ maize inbred line were grown in natural conditions in the field and inmature embryos were used as plant material for in vitro culture. The meristematic calli were obtained by the atrophic tissue method and cultured on a maintenance MS2 medium (M-medium) containing 2 mg/l of 2,4-D as previously described by Torné et al. (1984).

Treatment with Inhibitors

The meristematic calli cultured on the M-medium were treated with 0.5 mM DFMA or 0.5 mM DFMO for 3 consecutive months. Then, the calli were transferred to the same M-medium lacking the inhibitor or to a differentiation MS0.25 medium (D-medium) containing 0.25 mg/l of 2,4-D and lacking the inhibitor. The 2,4-D was removed from the D-medium at the second month of culture. The pH of all culture media was always adjusted to 5.8 before autoclaving.

Polyamine Analysis

The PAs were analyzed following procedures detailed in an earlier report (Tiburcio et al., 1985). Briefly, samples were extracted with 5% PCA and centrifuged at 27,000 g for 20 min, after which aliquots from the pellet and supernatant fractions were hydrolyzed with 12 N HCl. The nonhydrolyzed PCA supernatant containing the free PAs(S-fraction), as well as the hydrolyzed PCA supernatant (SH-fraction) and the hydrolyzed pellet (PH-fraction) containing the PAs liberated from conjugates were then dansylated, solvent purified, separated by TLC and quantified using a spectrophotofluorimeter.

Some dansylated samples were also analyzed by HPLC as described by Carbonell and Navarro (1989).

Protein Analysis

Protein was determined according to the method of Bradford (1976) in the insoluble PCA-pellet resuspended in 1 N NaOH. Bovine-gamma-globulin (Sigma) was used as a standard.

Hydroxycinnamoyl Amides (HCAs) Analysis

Conjugated PAs to hydroxycinnamic acids were analyzed as described by Wyss-Benz et al. (1988). The samples were extracted with water/methanol (1:1,v/v) using a Polytron homogenizer. Extracts were centrifuged at 20,000 g for 20 min and loaded onto a weakly acidic cation-exchanger column. The HCAs were eluted with 8 M acetic acid/methanol (1:1,v/v) and filtered prior to injection into the HPLC system.

Samples were injected onto a reverse phase column Nucleosil C-18 column (250 x 4.6 mm, particle size 5 um) and eluted with a programmed gradient of 25 mM sodium acetate pH 4.5 to acetonitrile at a flow rate of 1ml/min and at 30ºC. The acetonitrile gradient was as previously described by Wyss-Benz et al. (1988).An attached fluorescence spectrophotometer was used as a detector. The HCAs were quantitated by using a HP 9000 computer (Wyss-Benz et al., 1988).

RESULTS

Effect of the Addition of Inhibitors on Bud Differentiation

Figure 1a shows the effect of the treatment with 0.5 mM DFMA or DFMO for 3 consecutive months on bud differentiation in meristematic maize calli cultured on a M-medium. The DFMA-treatment produces a 6-fold decrease in the number of differentiated buds, whereas similar concentration of DFMO does not significantly reduce the number of buds in relation to the controls (Fig. 1a). These observations are similar to previous results reported in tobacco TCL explants (Tiburcio et al., 1988). It should be stressed that the DFMO-treated calli show a similar appearance to the controls, whereas the DFMA-treated calli show a tendency to decrease the green areas and consequently to increase its yellowish color (not shown).

Effect of the Removal of Inhibitors on Plant Regeneration

Figure 1b shows the effect of the removal of DFMA or DFMO on plant regeneration in meristematic maize calli cultured on a D-medium for 1 month. Compared to the controls, in the calli previously treated with DFMO there is a 2-fold decrease in the number of regenerated shoots, whereas in the calli previously treated with DFMA there is an increase (more than 3-fold) in the number of regenerated shoots. Furthermore, the plants regenerated from the cultures previously treated with DFMA are more developed than the controls. In contrast, the plants regenerated from the cultures previously treated with DFMO are less developed than the controls (Fig. 2).

Effect of the Addition of Inhibitors and their Removal on Polyamine Levels

The meristematic maize callus contains Put (51%) as the main PA followed by Dap (22%), Spd (18%) and Spm (9%), in this order (Fig. 3). The main PA-fraction is the SH-fraction (51%), followed by the S-fraction (29%) and the PH-fraction (20%). Upon analysis of the HCAs, only one (feruloyl-Put; FP) of the four known HCAs (p-coumaroyl-Put, caffeoyl-Put, FP, and diferuloyl-Put) was detected in the meristematic callus. Therefore, FP is not the only HCA-conjugate of Put, but also the major PA-bound form in this callus system.

Figure 4 shows the effect of the treatment with DFMA or DFMO as well as their removal on total PA (free + bound) levels in maize calli cultured on either a M-medium or a D-medium. In the presence of DFMA total PA levels progressively decrease reaching the lowest value at the third month of culture on the M-medium, in which PA levels are 2-fold lower than the controls. In the presence of DFMO total PA levels also decrease during the first 3 months of culture on the M-medium, but in this case the values are always higher than in the controls (1.3-fold at the third month) and in the DFMA-treated calli (2.6-fold at the third month).

Total PA levels increase when the control and DFMA-pretreated calli are transferred from a M-medium to a D-medium (fourth month; Fig. 4) lacking the inhibitor. On the contrary, total PA levels markedly decrease (2.5-fold) when the DFMO-pretreated calli are transferred from a M-medium to a D-medium lacking the inhibitor. Similar effects occur when the calli pretreated with inhibitors are transferred to a M-medium lacking the inhibitors (fourth and fifth months of culture on the M-medium; Fig. 4).

Effect of the Addition of Inhibitors and their Removal on Protein Levels

The general protein profile is very similar to that observed for total PA levels (data not shown). In the presence of inhibitors the protein levels progressively decrease in the calli during the first 3 months of culture. Upon tranfer to M- or D-media lacking the inhibitors, the protein levels decrease in the calli pretreated with DFMO, whereas the protein levels increase in the calli pretreated with DFMA (data not shown).

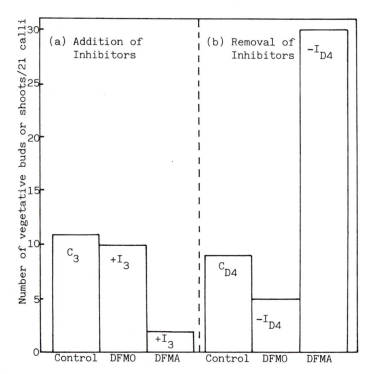

Fig. 1. (a) Effect of the addition of inhibitors (+I) on differentiation of buds and (b) of the removal of inhibitors (-I) on regeneration of shoots. In subscript months of culture on M- or D-media.

Fig. 2. Effect of the removal of inhibitors on plant regeneration after 2 months of culture on D-medium.

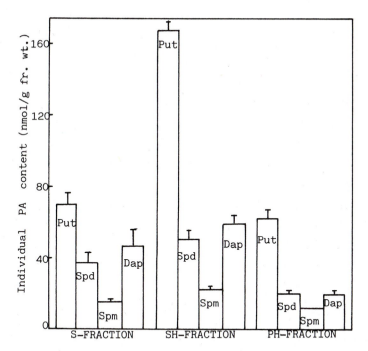

Fig. 3. Pattern of different free and bound PAs in the meristematic maize
callus. Dap=diaminopropane. Values are means of 3 replicates. The
standard errors are shown by bars.

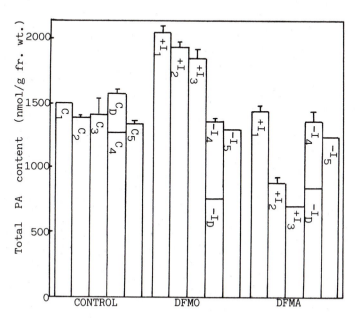

Fig. 4. Effect of the addition of inhibitors (+I) and their removal (-I) on
total PA levels. In subscript months of culture on M- or D-media.

DISCUSSION

        In this study we show that a prolonged treatment with DFMA for 3
consecutive months results in a dramatic decrease in the number of buds
differentiated from maize callus,with a concomitant marked decrease of total
PA levels. Since decreased PA content is a characteristic step during cell
aging (Kaur-Sawhney et al., 1982; 1985), we suggest that a senescence effect
is exerted on the calli by the DFMA-treatment which may explain the reduction
of morphogenetic capacity. This is further supported by the decrease of both
callus greening and protein levels induced by the DFMA-treatment. Upon
transfer the calli pretreated with DFMA to a D-medium lacking the inhibitor
there is a marked increase in the number of regenerated shoots, which in
addition are more developed than the controls. Concomitantly, the total PA
levels increase in such cultures. Since increased PA content is a
characteristic step during cell division and growth in young tissues (Kaur-
Sawhney et al., 1982; 1985), we suggest that a rejuvenation effect is exerted
on the calli upon DFMA-removal which may explain the improvement of
morphogenetic capacity. This is further supported by the marked increase on
protein content observed upon removal of DFMA from the cultures.
        The results obtained with DFMO in the maize callus are in concordance
with previous observations in tobacco TCL. Thus, a prolonged treatment with
DFMO for 3 consecutive months does not significantly decrease the number of
differentiated buds, supporting thus the view that the activity of ODC is not
involved in organ differentiation (Tiburcio et al., 1988; 1989). The
paradoxical effect of DFMO in increasing PA levels is probably related to its
unexplained, but consistently observed enhancement of plant ADC activity
(Flores and Galston, 1982; Slocum and Galston, 1987). Upon transfer the DFMO-
pretreated calli to a D-medium lacking the inhibitor there is a marked
decrease in the number of regenerated shoots, which in addition are less
developed than the controls. Concomitantly, the total PA levels decrease in
such cultures. Since as indicated above this is a characteristic step during
cell aging, we suggest that a senescence effect is exerted on the calli upon
DFMO-removal which may explain the reduction of morphogenetic capacity. This
is further supported by the decrease on protein content observed upon removal
of DFMO from the cultures.
        The antisenescence effect observed in the maize calli after the DFMA-
removal is probably due to the increased rates of endogenous PA levels, such
as it was previously demonstrated in oat protoplasts pretreated with DFMA
(Tiburcio et al., 1986b). Perhaps, a direct mechanism may be related to the
PA effects on plant proteolytic enzymes. Thus, Kaur-Sawhney et al.(1982') showed
that exogenous PAs strongly inhibited the activity of plant proteases. More
recently, Balestreri et al. (1987) have demonstrated that such inhibition is
achieved by a conformational change of the enzyme induced by the binding of
the PA to a site which is not identical to the substrate binding site.
According to these authors, during the onset of leaf senescence the shortage
of water and the resulting increase in ionic strength, prevents the binding of
PAs to the proteolytic enzymes thus maintaining the latter in the fully active
conformation (Balestreri et al., 1987). This mechanism is in accordance with
the view that leaf senescence is regulated at a post-transcriptional level
(Thomas and Stoddart, 1980).
        In conclusion, we describe a new method to improve plant regeneration
from maize callus. Through specific inhibition of ADC activity by DFMA we
first block the formation of PAs required for bud differentiation. Then, the
removal of DFMA from the cultures results in an increase on shoot regeneration
with a concomitant rise of PA levels. Previously, Kaur-Sawhney et al. (1985)
reported a similar effect in cultures of Vigna aconitifolia. However, they
used a combination of three inhibitors DFMA, DFMO and dicyclohexylamine
(inhibitor of Spd biosynthesis) to block PA biosynthesis, and therefore their
finding is less specific than the method reported here.

ACKNOWLEDGEMENTS

We wish to thank Merrell-Dow Research Institute for a generous gift of
DFMA and DFMO, Markus Wyss-Benz for the HCAs-analysis and Dr. Juan Carbonell
for the HPLC analysis of some PA samples. The research was supported by a
grant from CICYT : BIO88-0215.

REFERENCES

Balestreri, E., Cioni, P., Romagnoli, A., Bernini, S., Fissi, A., Felicioli,
    R. , 1987, Mechanism of polyamine inhibition of a leaf protease,
    Arch. Biochem. Biophys., 255: 460.
Bradford, M.M., 1976, A rapid method for the quantitation of microgram
    quantities of protein utilizing the principle of protein-dye binding,
    Anal. Biochem., 72:248.
Carbonell, J., and Navarro, J.L., 1989, Correlation of spermine level with
    ovary senescence and fruit set and development in Pisum sativum,
    Planta, (in press).
Flores, H.E., and Galston, A.W., 1982, Polyamines and plant stress:
    Activation of putrescine biosynthesis by osmotic shock, Science,
    217:1259.
Kallio, A., McCann, P.P., and Bey, P., 1981, DL-alpha-difluoromethylarginine:
    A potent enzyme-activated irreversible inhibitor of bacterial arginine
    decarboxylases, Biochemistry, 20:3163.
Kaur-Sawhney, R., and Galston, A.W., 1979, Interaction of polyamines and
    light on biochemical processes involved in leaf senescence, Plant Cell
    Environ., 2:189.
Kaur-Sawhney, R., Shih, L.M., Flores, H.E., and Galston, A.W., 1982, Relation
    of polyamine synthesis and titer to aging and senescence in oat
    leaves, Plant Physiol., 69:405.
Kaur-Sawhney, R., Shih, L.M., Cegielska, T., and Galston, A.W., 1982',
    Inhibition of protease activity by polyamines. Relevance for control
    of leaf senescence, FEBS Lett., 145:345.
Kaur-Sawhney, R., Shekhawat, N.S., and Galston, A.W., 1985, Polyamine levels
    as related to growth, differentiation and senescence in protoplast-
    cultures of Vigna aconitifolia and Avena sativa, Plant Growth Regul.
    3:329.
Kaur-Sawhney, R., Tiburcio, A.F., and Galston, A.W., 1988, Spermidine and
    flower bud differentiation in thin layer tobacco tissue cultures,
    Planta, 173:282.
Metcalf, B.W., Bey, P., Danzin, C., Jung, M.J., Casara, P., and Vevert, J.P.,
    1978, Catalytic irreversible inhibition of mammalian ornithine
    decarboxylase (EC 4.1.1.17) by substrate and product analogues,
    J. Amer. Chem. Soc., 100:2551.
Slocum, R.D., and Galston, A.W., 1987, Inhibition of polyamine biosynthesis
    in plants and plant pathogenic fungi, in: "Inhibition of Polyamine
    Metabolism: Biological Significance and Basis for New Therapies,"
    P.P. McCann, A.E. Pegg and A. Sjoerdsma, eds., Academic Press, New
    York.
Thomas, H., and Stoddart, J.L., 1980, Leaf senescence, Annu, Rev. Plant
    Physiol., 31:83.
Tiburcio, A.F., Kaur-Sawhney, R., Ingersoll, R., and Galston, A.W., 1985,
    Correlation between polyamines and pyrrolidine alkaloids in developing
    tobacco callus, Plant Physiol., 78:323.
Tiburcio, A.F., Masdéu, M.A., Dumortier, F.M., and Galston, A.W., 1986a,
    Polyamine metabolism and osmotic stress: I. Relation to protoplast
    viability, Plant Physiol., 82:369.

Tiburcio, A.F., Kaur-Sawhney, R., and Galston, A.W., 1986b, Polyamine
metabolism and osmotic stress: Improvement of oat protoplast by an
inhibitor of putrescine biosynthesis, <u>Plant Physiol</u>., 82:375.

Tiburcio, A.F., Kaur-Sawhney, R., and Galston, A.W., 1987, Regulation by
polyamines of plant tissue culture development, <u>in</u>: "Advances in the
Chemical Manipulation of Plant Tissue Cultures," M.B. Jackson, S.H.
Mantell and J. Blake, eds., British Plant Growth Regulator Group,
Bristol.

Tiburcio, A.F., Kaur-Sawhney, R., and Galston, A.W., 1988, Polyamine
biosynthesis during vegetative and floral bud differentiation in thin
layer tobacco tissue cultures, <u>Plant Cell Physiol</u>., 29:1241.

Tiburcio, A.F., Gendy, C.A., and Tran Thanh Van, K., 1989a, Morphogenesis in
tobacco subepidermal cells: putrescine as marker of root
differentiation, <u>Plant Cell Tiss. Org. Cult</u>., (in press).

Tiburcio, A.F., Kaur-Sawhney, R., and Galston, A.W., 1989b, Polyamine
metabolism, <u>in</u>:"The Biochemistry of Plants. A Comprehensive Treatise,"
B.J. Miflin, ed., Academic Press, New York, (in press).

Torné, J.M., Santos, M.A., and Blanco, J.L., 1984, Methods of obtaining maize
totipotent tissues. II. Atrophic tissue culture, <u>Plant Sci. Lett</u>.,
33:317.

Tran Thanh Van, K., 1973, Direct flower neoformation from superficial tissues
of small explant of <u>Nicotiana tabacum</u> L., <u>Planta</u>, 115:87.

Vasil, V., Vasil, I.K., and Chin-yi Lu, 1984, Somatic embryogenesis in long-
term callus cultures of <u>Zea mays</u> L. (Gramineae), <u>Am. J. Bot</u>., 71:158.

Wyss-Benz, M., Streit, L., and Ebert, E., 1988, Hydroxycinnamoyl amides in
stem explants from flowering and non-flowering <u>Nicotiana tabacum</u>,
<u>Physiol. Plant</u>., 74:294.

HORMONE ACTION AND SENSITIVITY:

POSSIBLE RELATION TO AGING

P.C.G. van der Linde

Bulb Research Centre, P.O. Box 85, 2160 AB Lisse

The Netherlands

Introduction

Hormones are organic chemicals produced by specialised cells in specific
glands and transported to target cells, where they evoke a specific response.
The target cells are characterized by the presence of receptors, which
bind the hormone and, subsequently, initiate the response of the cell. This
classical hormone concept is fading. Hormones are also produced by cells
outside specialised glands and non-target tissue also contains receptors for
hormones. Moreover, typical mammalian hormones and neurotransmitters have
been isolated from protozoa and plants and have been shown to influence
development processes in non-mammalian tissues (Roth et al., 1982) Conversely,
typical plant hormones were found to be produced by mammalian cells and
micro-organisms and to influence development processes in non-plant tissue
(Amagai, 1984; Ebright and Beckwith, 1985; LePage-Degivry et al., 1986).
Finally, peptide and lipid hormone-like substances have been isolated from
plants (Watson and Waaland, 1986; Scherer et al., 1988).
Therefore, it is hypothesized that the different signal molecules, identified
today, already had a regulatory function before the evolutionary segregation
of plants and animals took place, and that plants "chose" other signal
molecules to perform a hormonal function than animals. On the basis of this
hypothesis it is not inconceivable to presume the presence of a conserved
basic scheme for the action mechanism of signal molecules throughout nature.
In this respect mammalian hormones have been studied most extensively.

Perception and transduction of hormones

Hormones are percepted by receptors, which are located at the membranes, in
the cytoplasm and in the nucleus. Some membrane-bound receptors are part of
an ion-channel and hormone-binding directly affects the activity of the
channel (Changeux et al., 1984). Other receptors are transmembrane proteins
and aggregate upon hormone binding. Subsequently, their protein-kinase
domain in the cytoplasm is activated (Schlessinger, 1988). A third kind of
membrane-bound receptors activates one of the different GTP-binding
proteins (G-proteins), which are involved in the regulation of the activity
of adenylate cyclase, guanylate cyclase, and phospholipase C. These enzymes
synthesize important second messengers, viz., cAMP, cGMP, inositol-1,4,5-
triphosphate (IP3), and diacylglycerol (DG). Changing their concentration
affects the cytoplasmic pH and calcium concentration, the membrane potential,
protein phosphorylation, carbohydrate metabolism and respiration (Berridge,
1985,1987; Siess, 1989). Increasing the calcium concentration also activates
phospholipase A2. This enzyme releases arachidonic acid from the membranes,

Plant Aging: Basic and Applied Approaches
Edited by R. Rodríguez *et al.*
Plenum Press, New York, 1990

which is metabolized in the presence of free radicals by lipoxygenase and cyclo-oxygenase to leukotrienes, thromboxanes, and prostaglandines. These compounds are important intercellular signals. Some act as a calcium-ionophore and enhance the phospholipase-A2 activity. This autocatalytic process is known as the arachidonic-acid cascade. How membrane-percepted hormones influence gene transcription is not understood.

Hormone-binding to a cytoplasmic or nuclear receptor induces a state of high affinity for certain nuclear acceptor sites in the receptor. As a result, the activity of RNA polymerase II is increased by enhancing the rate of initiation. The underlying mechanism for the response is not clear. It is believed that in the presence of the hormone the receptors are continuously recycled between an active and an inactive state by phosphorylation and dephosphorylation, and that the rate of turnover determines the level of the response (Rossini, 1984).

Evidently, perception of hormones leads to a rapid disturbance of the homeostasis in the cell. To be able to survive, the cell has to restore the homeostasis.

Restoration of the homeostasis

Hormone-sensitive cells metabolize hormones and second messengers rapidly. This ensures no further change in the activity of effectors by phosphorylation. Phosphatase action returns the activity of the phosphorylated enzymes to the original level. Hormone-sensitive cells desentisize after hormone perception, either by changing the affinity of the receptors by phosphory-lation (Lefkowitz and Caron, 1984) and methylation (Alberts et al., 1983) or by decreasing the receptor concentration by internalization (Hollenberg, 1985). When the response-chain has reached the arachidonic-acid cascade, an autocatalytic process is triggered, which can only be stopped by scavenging the formed free radicals and hydroperoxides. Some enzymes involved in the scavenging processes are superoxide dismutase, catalase, glutathion reductase, and ascorbate dehydrogenase. Furthermore, the cells stabilize the membranes by polyamines. Due to active pumping of calcium into the extracellular space and into cell organels, the original low level of cytoplasmic free calcium is restored.

The processes leading to disturbance and restoration of the homeostasis are finely tuned. This is illustrated by the oscillating behaviour of the cytoplasmic free calcium concentration after hormone application. The type of oscillation depends on the hormones in question (Grapengiesser et al., 1989).

How about plants ?

For all classes of plant hormones receptors have been described (Chadwick and Garrod, 1986). They occur as membrane-bound, cytoplasmic and nuclear proteins. Some receptors have been localized only in specific cells (Löbler and Klämbt, 1985; Jacobs and Gilbert, 1983), which emphasizes the excistence of target cells in plants. Although for none of the receptors an involvement in hormone-induced responses has been demonstrated unequivocally, several reports suggest that they are involved (Van der Linde et al., 1984; Thompson et al., 1983; Maan et al., 1985).

Little is known about the transduction mechanisms in plant cells. Many enzymes involved in the signal-transduction pathways in animal cells are also present in plant cells, e.g., G-proteins, adenylate cyclase, cAMP-binding proteins, phosphodiesterase, phospholipase C and a calcium- and phospholipid-activated protein kinase (Brown and Newton, 1981). Fluctuations in the concentration of typical second messengers, e.g., calcium, IP3, and DG, after hormone application has also been detected (Zbell and Walter-Back, 1988). Furthermore, IP3 mobilizes calcium (Ranjeva et al., 1988) and inhibits cell-to-cell transport (Tucker, 1988). An autocatalytic peroxidation cycle is also present in plant cells. Calcium activation of phospholipase A2 releases polyunsaturated fatty acids (PUFA), e.g., linoleic and linolenic acid, which are metabolized by lipoxygenase in the presence of free radicals to PUFA-hydroperoxides. Subsequently, malondialdehyde, jasmonic acid and ethylene

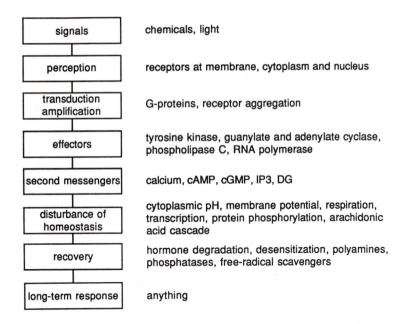

Figure 1. The sequence of events involved in signal perception, transduction and response (left) and the molecular processes (right).

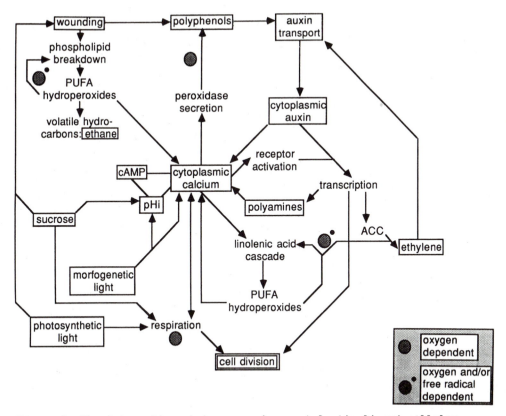

Figure 2. The interactions between environmental stimuli and cellular processes in adventitious bud formation on scale explants of lily cultured in vitro. For explanation see text.

are formed in the presence of oxygen, ACC and free radicals. Some PUFA-hydroperoxides act as a calcium ionophore and the cycle is closed. The activity of the so-called linolenic-acid cascade is high after wounding and during senescence (Leshem, 1987).

After disturbance of the homeostasis the plant cell tries to restore the original situation by similar mechanisms as animal cells. Hormones and second messengers are rapidly metabolized. Calcium is pumped into the extracellular space and into cell organels. Although down-regulation of hormone sensitivity is not a major subject in plant hormone research, the phenomenon was already observed by Reid and Pratt in 1972 for the ethylene-induced respiration in potato tubers. Plant cells also contain free-radical scavenging systems, e.g., superoxide dismutase, catalase, glutathion reductase, ascorbate dehydrogenase, and alpha-tocopherol, cytokinins and ascorbic acid. In this way, the activity of the autocatalytic linolenic-acid cascade can be controlled. Finally, Felle (1988) clearly showed the presence of oscillations in cytoplasmic free calcium, pH and membrane potential after auxin administration.

Although we are still far from a complete understanding of the transduction pathways in plants, it is clear that similar processes are involved as in animal cells. This may confirm the hypothesis of the presence of a conserved scheme for signal perception and transduction in nature (figure 1).

What is so unique about plants ?

To be able to survive in an ever changing environment, plants adapt their hormone sensitivity. In other words; environmental stimuli are translated inside the cell into changes in the concentration of receptors, second messengers, hormones, and substrates for transducers and effectors. An example from tissue culture illustrates the ability of plant cells to react to environmental stimuli in the same way as to hormonal stimuli. The number of adventitious buds formed on scale explants of lily bulbs can be regarded as an indication of the number of cell divisions induced in tissue culture, since almost no callus is formed. The number of adventitious buds is increased by a high auxin concentration, the application of an inhibitor of polar auxin transport, increasing the culture temperature, increasing the sucrose concentration of the medium, increasing the light intensity, and by additional wounding of the tissue (Van Aartrijk et al., 1989). The similarity in the effects of the various stimuli clearly suggests, that they are translated by the cells into the same processes.

On the basis of these results and the mechanisms of action of plant hormones, the following interactions are proposed to explain the observed effects (figure 2): Wounding induces degradation of membrane lipids. This process is mediated by free radicals and gives rise to the evolution of volatile hydrocarbons, e.g., ethane (Elstner and Konze, 1975). Some of the formed PUFA-hydroperoxides act as calcium ionophores (Serhan et al., 1981) on neighbouring cells. In this way, the effect of auxins on increasing the cytoplasmic calcium concentration is enhanced, and, subsequently, the pH of the cytoplasm (Felle, 1988; Rasi-Caldogno et al., 1987) and the cAMP level is changed. The increased calcium levels activate protein phosphorylation and the linolenic-acid cascade. ACC scavenges the formed free radicals and ethylene is formed, which inhibits polar auxin transport (Suttle, 1988). Wounding-induced polyphenols may cause the same effect (Jacobs and Rubery, 1988). The increase in the calcium concentration also induces the secretion of peroxidases, an increase in carbohydrate metabolism and respiration. It also activates the soluble auxin receptor, which binds auxin and stimulates transcription (Van der Linde et al., 1984). This increases the level of ACC synthase, arginine decarboxylase (Roustan et al., 1988) and cellulases. As a consequence, the polyamine levels are rapidly increased (Burtin et al., 1989; Fallon and Phillips, 1988) and the membranes are stabilized (Roberts et al., 1986; Agazio et al., 1988). Calcium is pumped out of the cytoplasm and the homeostasis is restored. Due to the increase in the respiration and transcription, cell division takes place and bud formation can occur. The scheme also explains why a linear correlation is observed in the ratio of

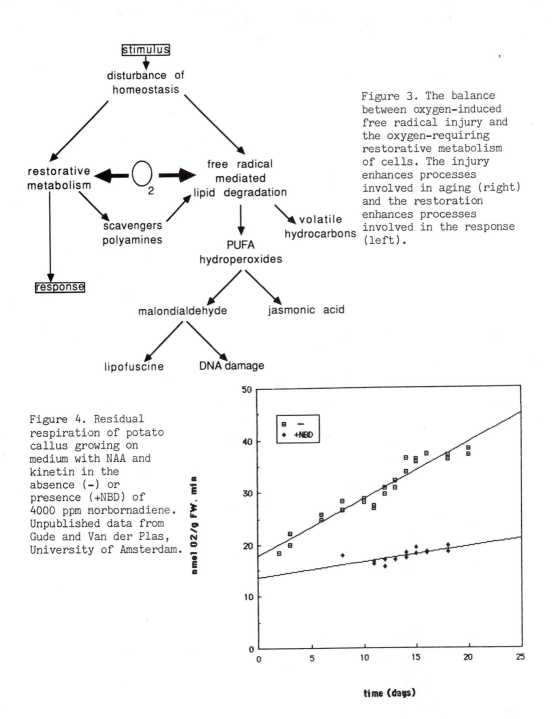

stimulus

disturbance of homeostasis

restorative metabolism

O₂

free radical mediated lipid degradation

scavengers polyamines

volatile hydrocarbons

PUFA hydroperoxides

response

malondialdehyde   jasmonic acid

lipofuscine   DNA damage

Figure 3. The balance between oxygen-induced free radical injury and the oxygen-requiring restorative metabolism of cells. The injury enhances processes involved in aging (right) and the restoration enhances processes involved in the response (left).

Figure 4. Residual respiration of potato callus growing on medium with NAA and kinetin in the absence (–) or presence (+NBD) of 4000 ppm norbornadiene. Unpublished data from Gude and Van der Plas, University of Amsterdam.

nmol O2/g FW. min

time (days)

Table 1. Effect of the presence of 4000 ppm norbornadiene on the average respiration and the growth of potato callus.

| Conditions | Respiration (nmol O_2/g F.W. min) | | Fresh weight (mg) |
	Total	Residual	
–	48.2	34.9	121
+NBD	49.3	17.1	103

Unpublished data from Gude and Van der Plas, University of Amsterdam.

289

ethane over ethylene evolution during the first days of culture and the number of adventitious buds observed at the end of the culture (Van Aartrijk et al., 1984). Ethane can be regarded as a marker for the initial damage and ethylene as a marker for the restorative metabolism. Evidently, the events induced during the first days of culture determine the scale of the regeneration process.

Environmental factors directly interact with some of the processes in figure 2. Red light (morfogenetic light) changes the calcium flux over the plasma-lemma (Roux et al., 1986; Takaga and Nagai, 1988), decreases the internal pH (Piwowarczik, 1988) and has been shown to decrease the number of membrane-bound auxin receptors in etiolated tissue. Since red-light effect on the swelling of protoplasts is mimiced by cAMP, a close interaction between the internal calcium concentration and cAMP level is proposed. Photosynthetic light induces free-radical damage and influences respiration. Sucrose uptake is accompanied by acidification of the cytoplasm (Felle, 1988) and enhances free-radical injury of photosynthetic light (Trippi, this volume). Extreme temperatures and drought cause structural changes in the membrane lipids, which makes the membranes more permeable for cations (Erlandson and Jensén, 1989), and release free fatty acids, which are able to modulate the activity of the H^+-ATPase (Palmgren et al., 1988). Therefore, it is not surprising that a cold treatment increases the sensitivity of plant tissue for auxin (Van Aartrijk and Blom-Barnhoorn, 1981) and gibberellin (Hanks, 1984).

Relation to aging

In humans the rate of aging is related to the accumulation of malondialdehyde-derived compounds and to the concentration of free radicals. Malondialdehyde is a toxic and undigestible byproduct of free-radical mediated lipid peroxidation. It forms complexes with DNA and with all kinds of organic molecules, which accumulate as lipofuscine; the fluorescent pigment of old people. The DNA damage is thought to be the cause of genetic aging. Lipofuscine-like compounds also accumulate in senescing plant tissue (Düggelin et al., 1988). As has been shown above, hormones and environmental stimuli induce free-radical mediated lipid peroxidation in plant cells. It is tempting to suggest that the rate of aging depends on the balance between the damage, caused by free-radical mediated lipid peroxidation, and the restorative metabolism of the cells (figure 3). Since both processes require oxygen, the rate of aging can be measured by determination of the residual oxygen consumption in relation to the total oxygen consumption by the cells. When the residual oxygen consumption increases in time, the cells age. This was found for growing potato callus (figure 4). Because every growing cell produces ethylene, they are always exposed to the aging hormone. When the action of ethylene on the cells is inhibited by application of norbornadiene, almost no increase in residual oxygen consumption in time was observed in the same callus. Table 1 shows, that neither the total respiration nor the growth of the callus was significantly affected. On the other hand, increasing the oxygen concentration should also increase the concentration of free radicals, and, subsequently, enhance lipid peroxidation and aging. When the oxygen concentration is raised from 21 (normal air) to 70 %, the mitochondrial respiration of potato callus is increased with 40 %, probably due to a better diffusion of oxygen to the centre of the callus. However, the residual respiration is increased with 270 % and the callus fresh weight with 100 % (Van der Plas and Wagner, 1986). This illustrates that a rapid growth, which is wanted in most applications of tissue culture, is not very favourable since it enhances aging: an observation which many plant physiologists will confirm.

References

Agazio, M.D., Giardina, M.C., and Grego, S. (1988) Plant Physiol. 87, 176-178.
Alberts, B., Bray, D., Lewis, J., Raff, M., Roberts, K., and Watson, J.D.

(1983) Molecular Biology of the Cell. Garland Publishing Inc., NY, London

Amagai, A. (1984) J. Gen. Microbiol. 130, 1961-1965.

Berridge, M.J. (1985) Scientific Am. 253(4),124-134.

Berridge, M.J. (1987) Ann. Rev. Biochem. 56, 159-193.

Brown, E.G. and Newton, R.P (1981) Phytochem. 20, 2453-2463.

Burtin, D., Martin-Tanguy, J., Paynot, M., and Rossin, N. (1989) Plant Physiol. 89, 104-110.

Chadwick, C.M. and Garrod, D.R. (1986) Hormones, receptors and cellular interactions in plants. University Press, Cambridge.

Changeux, J.-P., Devillers-Thiéry, A., and Chemouilli, P. (1984) Science 225, 1335-1345.

Düggelin, T., Bortlik, K., Gut, H., Matile, P., and Thomas, H. (1988) Physiol. Plant. 74, 131-136.

Ebright, R.H. and Beckwith, J. (1985) Mol. Gen. Genet. 201, 51-55.

Elstner, E.F. and Konze, J.R. (1976) Nature 263, 351-352.

Erlandson, A.G.I. and Jensén, P. (1989) Physiol. Plant. 75, 114-120.

Fallon, K.M. and Phillips, R. (1988) Plant Physiol. 88, 224-227.

Felle, H. (1988) Physiol. Plant. 74, 583-591; Planta 174, 495-499; Planta 176, 248-255.

Grapengiesser, E., Gylfe, E., and Hellman, B. (1989) Arch. Biochem. Bioph. 268, 404-407.

Hanks, G.R. (1984) Sci. Hortic. 23, 379-390.

Hollenberg, M.D. (1985) Trends in Pharm. Sci. 162, 242-245.

Jacobs, M. and Gilbert, S.F. (1983) Science 220, 1297-1300.

Jacobs, M. and Rubery, P.H. (1988) Science 241, 346-349.

LePage-Degivry, M.-T., Bidard, J.-N., Rouvier, E., Bulard, C., and Lazdunski, M. (1986) Proc. Natl. Acad. Sci. USA 83, 1155-1158.

Lefkowitz, R.J. and Caron, M.G. (1984) pp. 209-231 in:Current topics in cellular regulation vol. 28.

Leshem, Y.Y. (1987) Physiol. Plant. 69, 551-559.

Löbler, M. and Klämbt, D. (1985) J. Biol. Chem. 256, 9854-9859.

Maan, A.C., Van der Linde, P.C.G., Harkes, P.A.A., and Libbenga, K.R. (1985) Planta 164, 376-378.

Palmgren, M.G., Sommarin, M., Ulskov, P., and Jørgensen, P.L. (1988) Physiol. Plant. 74, 11-19.

Piwowarczyk, W. (1988) Planta 173, 42-45.

Ranjeva, R., Carrasco, A., and Boudet, A.M. (1988) FEBS Lett. 230, 137-141.

Rasi-Caldogno, F., Pugliarello, M.C., and De Michelis, M.I. (1987) Plant Physiol. 83, 994-1000.

Reid, M.S. and Pratt, H.K. (1972) Plant Physiol. 49, 252-255.

Roberts, D.R., Dumbroff, E.B., and Thompson, J.E. (1986) Planta 167, 395-401.

Rossini, G.P. (1984) J. Theor. Biol. 108, 39-53.

Roth, J., LeRoith, D., Shiloach, J., Rosenzweig, J.L., Lesniak, M.A., and Havrankova, J. (1982) New Eng. J. Med. 306, 523-527.

Roustan, J.-P., Henry, M., and Fallot, J. (1988) C.R. Acad. Sci. III, 307, 781-784

Roux, S.J., Wayne, R.O., and Datta, N. (1986) Physiol. Plant. 66, 344-348.

Scherer, G.F.E., Martiny-Baron, G., and Stoffel, B. (1988) Planta 175, 241-253.

Schlessinger, J. (1988) Trens in Biochem. Sci. 13, 443-447.

Serhan, C., Anderson, P., Goodman, E., Dunham, P., and Weissmann, G. (1981) J. Biol. Chem. 256, 2736-2741.

Siess, W. (1989) Physiol. Rev. 69, 58-178.

Suttle, J.C. (1988) Plant Physiol. 88, 795-799.

Takagi, S. and Nagai, R. (1988) Plant Physiol. 88, 228-232.

Thompson, M., Krull, U.J., and Venis, M.A. (1983) BBRC 110, 300-304.

Tucker, E.B. (1988) Planta 174, 358-363.

Van Aartrijk, J., Blom-Barnhoorn, G.J., and Van der Linde, P.C.G. (1989) Lilies. Handbook of Plant Cell Culture Vol. 5. (in press).

Van Aartrijk, J. and Blom-Barnhoorn, G.J. (1981) Sci. Hortic. 14, 261-268.
Van der Linde, P.C.G., Bouman, H., Mennes, A.M., and Libbenga, K.R. (1984)
  Planta 160, 102-108.
Van der Plas, L.H.W. and Wagner, M.J. (1986) Plant Cell Tissue Organ Cult.
  7, 217-225.
Watson, B.A. and Waaland, S.D. (1986) Plant Cell Physiol. 27, 1043-1050.
Zbell, B. and Walter-Back, C. (1988) J. Plant Physiol. 133, 353-360.

SECTION V

GENETIC MANIPULATION

CELLULAR AND MORPHOGENIC REORIENTATION INDUCED BY INSERTING

FOREIGN DNA

Albert Boronat

Unitat de Bioquímica
Facultat de Farmacia
Universitat de Barcelona
08028-Barcelona, Spain

THE ROLE OF PHYTOHORMONES IN PLANT DEVELOPMENT

Although the molecular bases underlaying plant development are poorly understood, it is widely accepted that plant
growth and differentiation are greatly influenced by several
classes of phytohormones. A classical example is the control
of organogenesis in tissue-cultured tobacco by auxins and
cytokinins (Skoog and Miller, 1957).

Phytohormones are widely used in plant tissue culture and
there is a large body of information concerning their effects in plant development and morphogenesis (Wareing and
Phillips, 1981; King, 1988). The current knowledge of the
role of phytohormones is based mainly on experiments in
which hormones have been exogenously applied. This approach
presents problems related with the uptake of the exogenously
applied compounds. In order to minimize these problems, many
studies have been performed with isolated parts of the
plant, although this also has negative effects because of
the disruption of the normal interaction between the different organs and tissues.

AGROBACTERIUM TUMEFACIENS AS A MODEL SYSTEM TO STUDY PHYTO-
HORMONE EFFECTS ON PLANT DEVELOPMENT

An alternative model system for studying the role of
phytohormones in plant development is provided by plant
cells genetically transformed by the soil bacterium Agrobacterium tumefaciens. This phytopathogen is the causative
agent of crown gall disease, a neoplastic growth that affects many dicotyledoneous plant species. The disease is the
direct result of the transfer of a particular DNA fragment
from the Ti plasmid (the T-DNA) within the bacterium to the
plant cell genome, where its integration and expression
induces the tumor formation (for recent reviews see Zambrys-

Plant Aging: Basic and Applied Approaches
Edited by R. Rodríguez *et al.*
Plenum Press, New York, 1990

ki, 1988; Zambryski et al., 1989). The integrated T-DNA
encodes novel enzymatic activities whose overproduction
results in the perturbation of normal plant cell development
(Morris, 1986). It is known that the T-DNA contains three
genes, generally referred to as onc genes, that specify the
biosynthesis of auxin and cytokinin in the transformed cells
(Fig. 1) Two of these genes, iaaM and iaaH, code for the
enzymes tryptophan monooxygenase and indolacetamide hidrola-
se, which constitute a new two-step pathway, leading to the
conversion of tryptophan into the active auxin indole-3-ace-
tic acid (IAA) (Fig. 2). Another gene, iptZ, codes for
isopentenyl trasferase which catalyzes the synthesis of the
cytokinin isopentenyl-5'-AMP by condensation of isopentenyl
pyrophosphate and 5'- AMP (Fig. 2) (Weiler and Schröder,
1987; Memelink et al., 1987). The endogenous synthesis of
auxin and cytokinin in the trasformed cells is responsible
for the tumorous morphology, and also for their ability to
grow "in vitro" without the addition of hormones to the
culture media.

EXPRESSION OF THE T-DNA PHYTOHORMONE GENES FROM AGROBACTE-
RIUM TUMEFACIENS IN TRANSGENIC PLANTS

The observation that T-DNA contains genes that encode for
enzymes of auxin and cytokinin synthesis in the transformed
plant cells is of great interest, since it provides unique
tools for studying the role of phytohormones in plant development. The function of these genes was first analyzed by
studying the effect of spontaneous and induced mutations in
tumor morphology and growth (reviewed by Memelink et al.,
1987). If all three onc genes are inactivated, transformation still occurrs although no tumors are formed, and phenotypically normal plants can be regenerated. Mutations in
either the iaaM or the iaaH gene induce the formation of
slow growing tumors with proliferation of shoots . Such
tissues contain increased level of cytokinins, but essentially normal levels of auxin. The exogenous addition of
auxin give rise to unorganized growth of the tumor "in
vitro". Mutations in the iptZ gene induce tumors that frequently produce roots. Thus, relatively high cytokinin/auxin
ratios induce shoot formation and suppress root formation,
while low ratios have opposite effects. Intermediate ratios
result in unorganized tumor growth.

Based in a different approach, the phytohormone genes
from the T-DNA have been fused to strong constitutively
expressed promoters and their effect examined in transgenic
plants. The endogenous overproduction of auxin and cytokinin
had remarkably effects on plant development and morphology.
Regenerated transformed petunias overproducing cytokinin
resulted an almost completely suppression of root formation
and a lack of apical dominance (quoted in Klee et al.,
1987a). Transformed petunias constitutively expressing the
iaaM gene under the control of the CaMV 19S promoter induced
an increase of approximately 10 fold in the plant IAA levels. The transformed petunia also displayed an extremely
abnormal morphology. In this case the main altered traits
were : reduced leaf size, extreme leaf curling, occasional

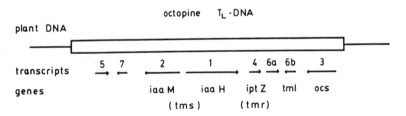

1 Kb

octopine T_L-DNA

plant DNA

transcripts   5 7   2   1   4 6a 6b   3

genes   iaa M   iaa H   ipt Z   tml   ocs

                    (tms)   (tmr)

Fig. 1. Transcription map of of the T_L-DNA of <u>Agrobacterium</u> <u>tumefaciens</u>. Transcripts 1 and 2 (<u>iaaH</u> and <u>iaaM</u>) encode auxin biosynthetic enzymes. Transcript 4 (<u>iptZ</u>) encodes a cytokinin biosynthetic enzyme. Mutations in <u>tml</u> results in an increase of tumor size. <u>Ocs</u> is the octopine synthase locus. Mutations in <u>iaaM</u> and <u>iaaH</u> induce overproduction of shoots (<u>tms</u>). Mutations in <u>iptZ</u> induce root formation (<u>tmr</u>).

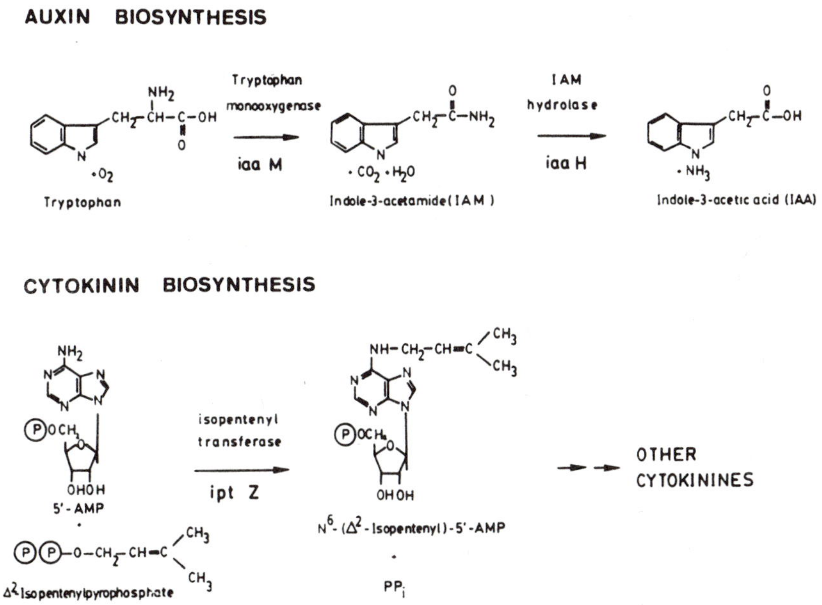

AUXIN BIOSYNTHESIS

Tryptophan monooxygenase
iaa M

IAM hydrolase
iaa H

$+ O_2$

$+ CO_2 + H_2O$

$+ NH_3$

Tryptophan   Indole-3-acetamide(IAM)   Indole-3-acetic acid (IAA)

CYTOKININ BIOSYNTHESIS

isopentenyl transferase
ipt Z

OTHER CYTOKININES

5'-AMP

N^6-(Δ^2-Isopentenyl)-5'-AMP

Δ^2Isopentenylpyrophosphate

PP_i

Fig. 2. Pathways for biosynthesis of auxin and cytokinin genes of the T-DNA of <u>Agrobacterium</u> <u>tumefaciens</u>.

formation of avdentitious roots from the stems, an almost complete apical dominance and increased amounts of xylem and phloem cells (Klee et al., 1987b). Despite the morphological abnormalities of these plants, they could undergo a complete life cycle and their reproductive capacity was not affected, suggesting that most cells in the plant are essentially unaffected by the high endogenous level of auxin (Klee et al., 1987).

While the constitutive overproduction of auxin and cytokinin genes has been interesting, the real promise of endogenous phytohormone overproduction will be in conjunction with promoter elements that allow tissue-specific or inducible expression of these genes. Initial approaches have been attempted by different groups, although no clear patterns have yet been established. Klee et al. (1987b) have reported the expression of the iaaM gene in petunia under the control of the soybean 7S promoter. This promoter is known to be properly regulated in this plant, showing a specific expression in immature embryos. The regenerated transformed plants presented a normal phenotype and the expression of the iaaM gene in the developing embryos appeared not to be lethal.

An inducible overexpression of cytokinin has been attempted by Medford et al. (1989) using a chimeric isopentenyl transferase gene. The A. tumefaciens iptZ gene was placed under the control of the maize heat-shock promoter (hsp70) and transferred to tobacco and Arabidopsis thaliana. The heat-induced plants showed high increased levels of cytokinins (50 to 20 fold over that of the nontransformed control plants), although the transformed uninduced control plants also showed significant increased level (3 to 7 fold) of cytokinins. The regenerated transgenic plants were affected in various aspects of development including release of axilary buds, reduced stem and leaf area and an underdeveloped root system. Unexpectedly, the altered phenotype was also observed in the uninduced plants and no further changes were apparent in the heat-shock induced plants. Since the heat-shock treatment was given at daily intervals of 2 hours, it is reasonable to think that the transient increases in the cytokinin levels were unable to modify the plant responses beyond of those produced by the basal uninduced levels. Alternatively, a lack of the corresponding cellular factor(s) responsible for mediating the cytokinin effect can also be considered. It is interesting to point out that the transformed plants were only affected in growth but not in differentiation, including flowering. These experiments suggest that cytokinin may play an important role in plant growth and that the implementation of the plant differentiation program can be mediated by factors other than the absolute cytokinin levels.

AGROBACTERIUM RHIZOGENES AND THE HAIRY ROOT PHENOTYPE

The expression of oncogenes carried by the Ri plasmid of Agrobacterium rhizogenes represents another interesting and attractive model for studying the mechanisms underlaying cell differentiation and growth control in plants. Infection of a plant by A. rhizogenes incites the formation of adventitious roots at the site of the infection. Such roots can be grown "in vitro" in the absence of the inciting bacterium. This morphogenic event is the result of the transfer of genetic information contained in the T-DNA of the Ri plasmid into the plant cell genome. Although the mechanisms of genetic transformation are similar between A. rhizogenes and A. tumefaciens, many features differentiate the morphogenic action of Ri and Ti plasmids (for a review see Birot et al., 1987). Plants can be regenerated from the transformed tissues although they often show developmental abnormalities (usually bushy growth and wrinkled leaves), known as the hairy-root phenotype, which depends on the plant species and the particular transformation event.

The characterization of the T-DNA genes from the Ri plasmids and their implications in root proliferation has been complicated by the variable response of different plants and certain plant organs to the infection. The Ri plasmid present in the A. rhizogenes strain A4 contains two noncontiguous T-DNA regions (T_R-DNA and T_L-DNA). Both T-DNA regions participate in root induction, either individually or toghether, depending on the plant species and tissue. The T_R-DNA contains two genes homologous to the iaaA and iaaH genes of the Ti plasmid, although auxin production seems to play a conditional and nonessential role in the root induction, depending on the plant species and tissue.

Transformed plants containing the T_L-DNA show a variety of developmental abnormalities as a consequence of the combined expression of the genes rolA, rolB and rolC. These abnormalities include loss of apical dominance, severely wrinkled leaves and reduced internode distances. Although the functions encoded by these genes are synergistic (Spena et al., 1987), each of them can independently influence plant morphogenesis. When expressed separately they induce distinct developmental abnormalities that are characteristic of each rol gene (Schmülling et al., 1988). The rolA gene has shown to cause the most severe affects and induces an aberrant phenotype similar to that of the entire T_L-DNA (Sinkar et al., 1988). The independent expression of rolB and rolC induce less severe growth alterations. Tobacco plants carrying only the rolB gene show altered leaf morphology, increased stigma and flower size, heterostylity and increased formation of adventitious root on the stem, while the expression of rolC results in branched plants with altered leaf morphology and reduced flower size and pollen production (Schmülling et al., 1988).

Interestingly, new growth abnormalities are generated when rolB and rolC genes are overpressed (Schmülling et al., 1988). When the rolB gene was expressed under the control of the strong CaMV 35S promoter early senescence was aparent, even prior to flowering, and cellular death ocurred both in callus and young leaves. Transgenic plants expressing the rolC genes under the control of the CaMV 35S promoter had a dwarf and bushy phenotype, with small lanceolated leaves, leading to very marked inhibition of the onset of senescence.

Concerning the involvement of rol genes in root induction, the expression of the rolB gene alone is able to induce root formation in tobacco and kalanchoe tissues. Nevertheless, the rolA and rolC genes by themselves are able to induce root formation in tobacco but not in kalanchoe (Vilaine et al., 1987; Cardarelli et al., 1987a; Spena et al., 1987). Expression of the rolB gene togheter with either rolA and rolC genes increased the induction of root formation on both tobacco and kalanchoe (Spena et al., 1987). The different response observed among plants species is not fully understood, but suggests that different plants must have specific factors controlling differentiation that respond to different stimuli.

At present, nothing is known about the mode of action of the rol genes, although the hairy-root phenotype has been related to an altered response to phytohormones in the transformed plants (Cardarelli et al., 1987b; Shen et al., 1988; Schmülling et al, 1988)

CONCLUDING REMARKS

It seems clear that expression in transgenic plants of the onc genes present in the T-DNAs of the Ti and Ri plasmids is bringing new insigths into the molecular mechanisms underlying cell differentiation in plants. At present, these approaches are limited only by the availability of appropriate transcriptional promoters and other genes involved in phytohormone biosynthesis or in mediating their effetcs.

References

Birot, A.M. Bouchez, D., Casse-Delbart, F., Durand-Tardif, M., Jouanin, L., Pautot, V., Robaglia, C., Tepfer, D., Tepfer, M., Tourneur, J. and Vilaine, F. (1987) Studies and uses of the Ri plasmids of Agrobacterium rhizogenes. Plant Physiol. Biochem. 25, 323-335.

Cardarelli, M., Mariotti, D., Pomponi, M., Spano, L., Capone, I and Constantino, P. (1987a). Agrobaterium rhizogenes T-DNA capable of inducing hairy root phenotype. Mol. Gen. Genet. 209, 475-480.

Cardarelli, M., Spano, L., Mariotti, D., Mauro, M.L., Van Sluys, M.A. and Constantino, P. (1987b). The role of auxin in hairy root induction. Mol. Gen. Genet. 208, 457-463.

King P.J. (1988). Plant hormone mutants. Trends in Genetics 4, 157-162.

Klee, H, Horsch, R. and Rogers, S. (1987a) Agrobacterium-mediated plant transformation and its further application to plant biology. Ann. Rev. Plant. Physiol. 38, 467-486.

Klee, H., Horsch, R., Hinchee, M., Hein, M. and Hoffmann, N. (1987b) The effects of overproduction of two Agrobacterium tumefaciens T-DNA auxin biosynthetic gene products in transgenic petunia plants. Genes Develop. 1, 86-96.

Medford, J., Horgan, R, El-Sawi Z. and Klee, H. (1989) Alteration of endogenous cytokinins in transgenic plants using a chimeric isopentenyl transferase gene. Plant Cell 1, 403-413.

Memelink J., de Pater, B.S., Hoge J.H.C. and Schilperoort, R.A. (1987) T-DNA hormone biosynthetic genes: Phytohormones and gene expression in plants. Develop. Genet. 8: 321-337.

Morris , R.O. (1986) Genes specifying auxin and cytokinin biosynthesis in phytopathogens. Ann. Rev. Plant. Physiol. 37, 509-538.

Skoog, F. and Miller, C.O. (1957) Chemical regulation of growth and organ formation in plant tissues cultured in vitro. Symp. Soc. Exp. Biol. 11, 118-131.

Schmülling, T., Schell, J. and Spena, A. (1988) Single genes from Agrobacterium rhizogenes influence plant development. EMBO J. 7, 2621-2629.

Shen, H.S., Petit, A., Guern, J. and Tempé, J. (1988) Hairy roots are more sensitive to auxin than normal roots. Proc. Natl. Acad. Sci. U.S.A. 85, 3417-3421.

Sinkar, V.P., Pythoud, P., White, F.F., Nester, E.W. and Gordon, M.P. (1988) Genes Develop. 2, 688-697.

Spena, A., Schmülling, T., Koncz, C. and Schell, J. (1987) Independent and synergistic activity of rol A, B and C loci in stimulating abnormal growth in plants. EMBO J. 6, 3891-3899.

Vilaine, F. and Casse-Delbart, F. (1987) Further insight concerning the T_L region of the Ri plasmid of Agrobacterium rhizogenes strain A4: Transfer of a 1.9 kb fragment is sufficient to induce transformation roots on tobacco leaf fragments. Mol. Gen. Genet. 210, 111-115.

Wareing, P.F. and Phillips, I.D.J.. (1981) Growth and differentiaTion in plants. 3rd edition. Pergamon Press (Oxford).

Zambryski, P.(1988) Basic processes underlying Agrobaterium-mediated DNA transfer to plant cells. Ann. Rev. Genet. 22, 1-30.

Zambryski, P., Tempe, J. and Schell, J. (1989) Transfer and function of T-DNA genes from Agrobacterium Ti and Ri plasmids in plants. Cell 56, 193-201.

GENETIC MANIPULATION OF FOREST TREES: IMPLICATIONS FOR

PHYSIOLOGICAL PROCESSES

Don J. Durzan

Department of Environmental Horticulture
University of California
Davis, CA 95616 Fax (916) 752-1819

Introduction

With the recent advances in biotechnology of woody perennials (Durzan, 1988ab, 1989, 1990; Hällgren, 1988; Haissig et al., 1987; Torrey, 1988), considerable interest has developed in gene expression, particularly as transgenic trees are being regenerated. Progress in physiological aspects of biotechnology is evaluated against current schemes for domestication and tree improvement (Haissig et al., 1987; Timmis et al., 1986). An attempt is made to illustrate how elite or foreign genes may be expressed in quantitative terms against different genetic backgrounds. Notions from metabolic control theory involving preconditioning factors become useful, especially when examining the rejuvenation of cells for clonal propagation and gene insertion. The general area involving gene manifestation, penetrance and expressivity has been termed molecular phenogenetics (Durzan, 1990).

Changes in Domestication and Improvement Strategies

Historically, the modern management practices of intensive cultivation have relied upon clonal genotypes derived by recurrent selection from breeding populations (Fig. 1). Advances in vegetative propagation involving somatic polyembryogenesis (Durzan, 1988ab), the recovery of embryos from protoplasts (Gupta and Durzan, 1987), and the insertion and expression of

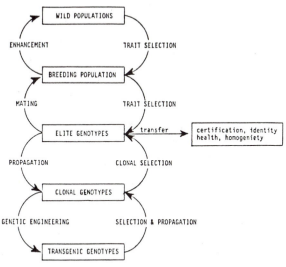

Figure 1. A general scheme for domestication and tree improvement modified to include recombinant DNA technologies. The scheme starts with wild populations and passes through classical breeding methods to biotechnologies (propagation, genetic engineering) resulting in the certification of germplasm (elite, clonal and transgenic genotypes) (Modified from Haissig et al., 1987).

Plant Aging: Basic and Applied Approaches
Edited by R. Rodríguez *et al.*
Plenum Press, New York, 1990

foreign genes in cells of rejuvenated trees (Dandekar et al., 1987) or in protoplasts (Gupta et al., 1988), all indicate that the next requirement will be criteria for the design and evaluation of transgenic genotypes.

Tree improvement strategies will have to consider population genetics and the appropriate biotechnological fixes, especially those that are based on recombinant DNA methods and on enzymes that improve genome diagnostics (e.g., restriction fragment length polymorphism (RFLP) -- Tanksley et al., 1989). Strategies need to combine classical genetics and modern genetic engineering methods (Fig. 2). When combined, the breeder has a wider array of options and tools to deal with the domestication of natural and transgenic ideotypes. Not all aspects implied in Figure 2 have been exploited. Whatever the eventual result, many aberrations, lethals and substandard trees will be recovered that need to be removed from the breeding and production populations. Detecting, searching and sorting procedures will be based on direct probes using RFLP fragments (e.g., Tanksley et al., 1989) and on various

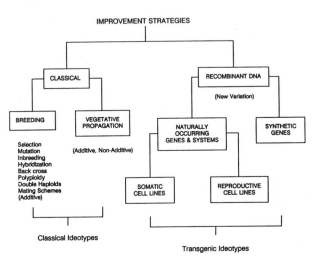

Figure 2. Tree improvement strategies are sexual and/or asexual and employ classic or modern biotechnological methods (e.g., recombinant DNA methodology).

phenotypic expressions over a range of environmental test conditions. For some trees, like Douglas fir, it may take 9 to 10 years to evaluate clonal performance. Considerable time-saving could be had by screening individual mother ideotypes for undesirable and elite fragments. Even so, the molecular bases for elite traits need to be understood and applied. The study of different genetic backgrounds against an array of environmental conditions to determine precise adaptational capabilities remains a fundamental aspect in tree improvement (Wright, 1968). Better methods are needed to diagnose phenotypic expressions.

Metabolic Phenotypes for Elite Traits

The most useful traits often are found in mature trees after decades of cultivation and adaptation. Individuals bearing elite traits are often difficult to propagate from the mature state (Timmis et al., 1986). Vegetative propagation and gene insertion currently become successful only after rejuvenation.

Since nitrogen (N) is a main limiting factor in forest soils (Bidwell and Durzan, 1975; Durzan, 1975), we have focused on free amino acids as one indicator of phenotype development. The study of amino acids integrates several strategies for the efficient use of nitrogen, carbon and water for protein and nucleic acid synthesis and for specific phenotype development under wide environmental conditions.

One example of how N flux in the free amino acid pool contributes to leaf growth and development is seen in Figure 3. From large data sets and behavioral displays of this type, the irritability and dynamics of the tissue systems can be characterized (Durzan, 1987ab). Results lead to the need for sharper concepts and new terminology relating gene manifestation to phenotypic expression.

Interpretations of Phenotypic Dynamics

For simplicity, we will examine the flux of an arbitrarily selected, single metabolite, that could be found on maps of metabolic phenotypes as in Figure 3. For each target tissue, we recognize that topological position, developmental stage, whole-tree correlations and microclimate affect phenotypic dynamics. Moreover, these historical factors precondition future behavior. Hence, Figures 4, 5 and 6 have introduced the notion of physiological preconditioning (Rowe, 1964) and illustrate how preconditioning can affect simple metabolite behavior. Figure 4 deals with the distributive behavior of a response surface that may vary from 0 to 100%, or from 0 to some other level. Figure 5 deals with the stability of metabolic flux and is based on the departure from mean behavior of the system, i.e. the average intensity (level or rate) is fixed at 0. Flux can vary in a positive or negative way. In Figure 6, the behaviors illustrated in Figure 5 show how the imposition of rejuvenation might or might not affect metabolic flux in

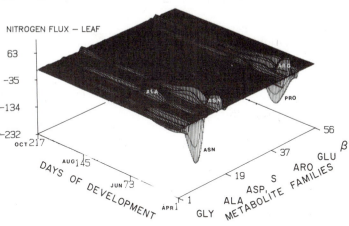

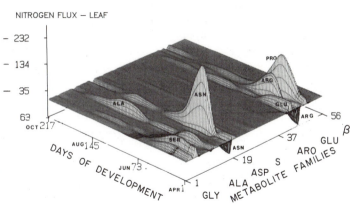

Figure 3. Elite trees have phenotypes that can be expressed as metabolic response surfaces. For leaves on shoots developing from bud break (April) to leaf fall (October, day 217), the flux of nitrogen for each metabolite in the free amino acid pool can be portrayed as a departure from mean behavior. The flux (μmoles N in each free amino acid per unit fresh weight per day) for individual amino acids is arranged in families and then as hierarchies of biosynthesis from the left to the right. Families are **gly** glycine-derived metabolites, **ala** alanine, **asp** aspartate, **S** sulfur amino acids, **aro** aromatic amino acids, **glu** glutamate, β beta-amino acids. Numbers refer to lanes along which fluxes are portrayed over time. The top graph shows the positive departures from the mean. The bottom graph shows the reverse, i.e. the negative departures from the mean. The z-axis shows the intensity of the flux. From surfaces of this type, various behaviors can be observed and identified as illustrated in Figures 4, 5 and 6.

the presence or absence of preconditions. The notions become quite complex, but illustrate the diagnostic power now available through computer-assisted technologies (*cf.* Durzan, 1990).

Figure 4. An interpretation of several roles that physiological preconditioning might -- or might not -- play in the expression of simple metabolite behaviour, as displayed by response surfaces with a set baseline. Some effects of preconditioning are shown graphically by broken lines. Metabolite levels or rates in the absence of physiological preconditions are shown by solid lines.

Preconditioning represents a variable potential that is established but not always immediately or fully expressed unless a signal or inducing event occurs. Induction by way of a signal[1] refers to the emergence of a capability that depends uniquely on past physiological states and their inherent preconditions. In this portrayal of metabolite behavior, we are dealing with a signal from the development state of the system.

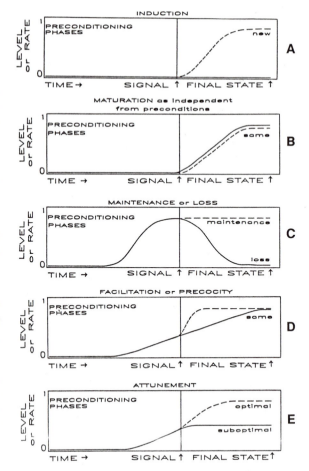

A. **Induction** refers to cases in which a new expression (level, rate) depends entirely on established preconditions.

B. **Maturation** refers to the development of a new capability independent of physiological preconditions, as distinct from earlier physiological states. The new capability is also reached at the same time, independent of preconditions.

C. **Maintenance of Loss** requires the development of a full capability during the preconditioning phase. Preconditioning is necessary to maintain or reduce the capability after a state signal or inducing event.

D. **Facilitation** affects only the rate of development of a capability after a signal based on the physiological state. Through preconditioning, the capability is realized earlier (precociously, as shown) or later (not shown) than normal.

E. **Attunement** refers to a situation in which preconditioning enables the full development of a capability, i.e. the metabolic phenotype is maximized or optimized.

---

[1] Threshold values may be material or temporal (S. Wright, 1968) and involve cyclical pathways.

Figure 5. An interpretation of several roles that physiological preconditions might -- or might not -- play in the expression of metabolite flux or system stability as seen as a ± departure from equilibrium or a mean. The effects of preconditioning are shown by broken lines. Levels or rates in the absence of physiological preconditions are shown by solid lines. With rejuvenation, we are dealing with a time-reversal in the behavior of metabolic flux.

Preconditions represent a variable potential that is established but not always immediately or fully expressed unless a signal or inducing event occurs. Induction by way of a signal[1] refers to the emergence of a capability that depends uniquely on past physiological states and their inherent preconditions. In this portrayal of metabolite behaviors, we are dealing with a signal from the state of the system.

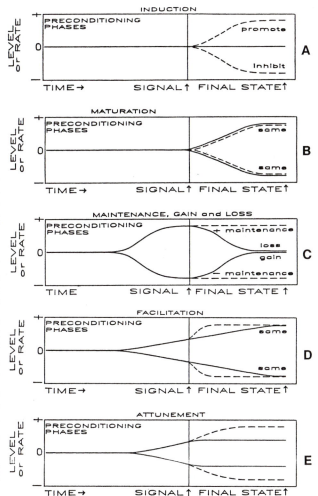

A. **Induction** refers to cases in which a new expression depends entirely on preconditions that may result in promotion or inhibition, leading to an arbitrary final state or discrete outcome.

B. **Maturation** refers to the development of a new capability or state ideally independent of physiological preconditions, as distinct from earlier but essential physiological states. Identical levels (±) or rates are reached at the same time.

C. **Maintenance or Loss** requires the development of full capability (±) during the preconditioning phase. Events responsible for preconditioning maintain (±) or alter (gain, loss) the capability after a state signal or inducing event.

D. **Facilitation** affects only the rate of development of a capability after a state signal. The capability is initiated during the phase of physiological preconditioning and is realized earlier (precociously) than in the absence of preconditions.

E. **Attunement** refers to situations in which preconditioning initiates and enables the full development of a capability or state.

---

[1]Threshold values may be material or temporal (S. Wright, 1968) and involve cyclical pathways.

Figure 6. Illustrations of how the "rejuvenation" of cells and tissues is translated into a metabolic flux. The time-reversals, cited in Figure 5, are spliced on the right end of each illustration (A to E). The effects of physiological preconditions leading to mature cells (at signal) are shown for a metabolic indicator by solid lines up to some arbitrary final state. Upon rejuvenation, the metabolic flux, unconditioned or preconditioned, returns to a level found in the preconditioning phases before the signal for maturation occurs.

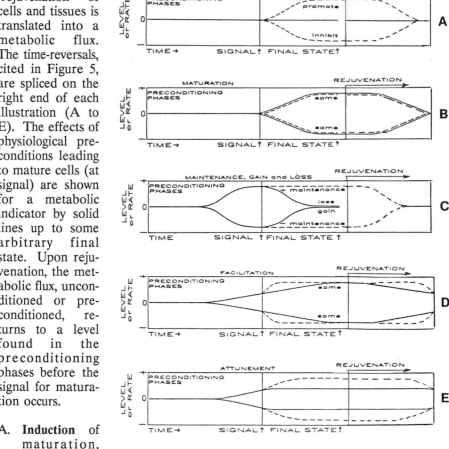

A. **Induction** of maturation, which depends on preconditions that promote or inhibit flux, is erased with the return of the flux to the presignal baseline.

B. Positive or negative metabolic flux leading to a final state in **maturation** returns to the presignal condition independently of the influence of preconditioning factors.

C. **Maintenance or loss** of flux signalled during the onset of maturation must return to the presignal levels during complete rejuvenation.

D. **Facilitation** of flux rates signalled during maturation return to the presignal rate during rejuvenation.

E. **Attunement** of metabolic flux to the mature state must be removed and returned to the presignal level or rate.

References

Bidwell, R.G.S., and Durzan, D.J., 1975, Some recent aspects of nitrogen metabolism, in: "Historical and Recent Aspects of Plant Physiology: A Symposium Honoring F.C. Steward," Cornell Univ. Press, Coll. Agric. and Life Sci., pp. 162-227.

Dandekar, A.M., Gupta, P.K., Durzan, D.J. and Knauf, V., 1987, Transformation and foreign gene expression in micropropagated Douglas-fir (*Pseudotsuga menziesii*), *Bio/Tech.*, 5:587-590.

Durzan, D.J., 1975, Nutrition and water relations of forest trees: a biochemical approach, Third North American Conf. Tree Biology, C.P.P. Reid and C.H. Fechner, eds., Colorado State Univ., Fort Collins, Colorado, pp. 15-63.

Durzan, D.J., 1987a, Physiological states and metabolic phenotypes in embryonic development, in: "Cell and Tissue Culture in Forestry," J.M. Bonga and D.J. Durzan, eds., Martinus Nijhoff, Dordrecht, Vol 2:405-439.

Durzan, D.J., 1987b, Plant growth regulators in cell and tissue culture of woody perennials, *Plant Growth Regulation*, 6:95-111.

Durzan, D.J., 1988a, Somatic polyembryogenesis for the multiplication of tree crops, *Biotech. Genetic Eng. Revs.*, 6:339-376.

Durzan, D.J., 1988b, Process control in somatic polyembryogenesis, in: "Molecular Genetics of Forest Trees," J.-E. Hällgren, ed., Swedish Univ. of Agric. Sci., Dept. Forest Genetics and Plant Physiol., Umeå, Sweden, Frans Kempe Symp., Report No. 8, 147-186.

Durzan, D.J., 1989, Performance criteria in response surfaces for metabolic phenotypes of clonally propagated woody perennials, in: "Applications of Biotechnology in Forestry and Horticulture, V. Dhawan, ed., (in press).

Durzan, D.J., 1990, Molecular phenogenetics in fruit breeding, *Acta Hort* (in press).

Gupta, P.K., and Durzan, D.J., 1987, Somatic embryos from protoplasts of loblolly pine proembryonal cells, *Bio/Tech.*, 5:710-712.

Gupta, P.K., Dandekar, A.M. and Durzan, D.J., 1988, Somatic proembryo formation and transient expression of a luciferase gene in Douglas fir and loblolly pine protoplasts. Plant Science 58:85-92.

Haissig, B.E., Nelson, N.D., and Kidd, G.H., 1987, Trends in the use of tissue culture in forest improvement, *Bio/Tech.*, 5:52-59.

Hällgren, J.-E., ed., 1988, "Molecular Genetics of Forest Trees," Frans Kempe Symp., Swedish Agric. Univ., Umeå, Sweden, June 14-16, 1988, Rept. No. 8, 147-186.

Rowe, J.S., 1964, Environmental preconditioning, with special reference to forestry, *Ecology*, 45:399-403.

Tanksley, S.D., Young, N.D., Paterson, A.H., and Bonierbale, M.W., 1989. RFLP mapping in plant breeding: New tools for an old science, *Bio/Tech.*, 7:257-264.

Timmis, R., Abo El-Nil, M.M. and Stonecypher, R.W., 1986, Potential gain through tissue culture, in: "Cell and Tissue Culture in Forestry," Vol. 1, General Principles and Biotechnology, J.M. Bonga and D.J. Durzan, eds., Martinus Nijhoff, Dordrecht, pp. 198-215.

Torrey, J.F., 1988, Biotechnology applied to tree improvement of underground systems of woody plants, in: "Genetic Manipulation of Woody Plants," J.W. Hanover and D.E. Keathley, eds., Plenum Press, N.Y., pp. 1-21.

Wright, S., 1968, "Evolution and the Genetics of Populations," Vols. 1 to 4, Univ. Chicago Press, Chicago.

BIOTECHNOLOGY IN FOREST TREE IMPROVEMENT: TREES OF THE FUTURE

Vladimír Chalupa

Forestry and Game Management Research Institute
156 04 Praha 5, Zbraslav-Strnady
Czechoslovakia

INTRODUCTION

Forest trees are the important source of renewable raw material and play a significant role in the formation of environment, with important functions in the field of soil protection, water retention, CO_2 absorption and carbon storage, recreation and health improvement. The growing importance of all these functions can be expected in the near future and society is likely to highly appreciate them. With the growing world population, the deforestation occurs, above all in developing countries. In some industrial countries, air pollution, acid rains, drought, wind, snow, pests and diseases cause serious damages on forest stands. Present situation calls for urgent intensification of forest tree breeding. Long reproductive cycle of forest trees is a serious obstacle for effective tree improvement. Conventional tree breeding techniques using controlled crossing for transfer of desirable traits are time-consuming. The application of parasexual genetic recombination methods (cytogenetic manipulations, recombinant DNA technology) in forestry will accelerate tree improvement programs.

Biotechnology in forest tree improvement will contribute to the formation of more resistant and productive genotypes. Forest trees of the future should possess high growth rates, good stem and branching form, high quality wood, and should be resistant to pests and diseases and tolerant of environmental stresses. Biotechnology will be used for the production of trees with the important valuable traits. Forest trees with valuable traits will be produced by propagation of existing superior genotypes, by propagation of hybrids obtained by sexual crossing or by cytogenetic manipulations and somatic hybridization and by propagation of selected somaclonal variants and transgenic trees obtained by recominant DNA technology. Organogenesis and somatic embryogenesis are the most promising methods in *in vitro* propagation of selected superior trees, hybrids and transgenic trees.

Plant Aging: Basic and Applied Approaches
Edited by R. Rodríguez *et al.*
Plenum Press, New York, 1990

FOREST TREE MICROPROPAGATION BY ORGANOGENESIS

Micropropagation contributes to the improvement of forest by multiplication of valuable genotypes. New improved forest trees arise by selection and propagation of existing superior genotypes and by combining their valuable genes by controlled crosses or by cytogenetic manipulations and somatic hybridization. Selection of elite genotypes, controlled crosses, production of hybrid seeds and *in vitro* tree propagation are of great importance in the forest tree improvement.

In our experiments with the production and multiplication of valuable genotypes, investigations were focused on improvement of multiplication methods. Field trials were established to test performance of micropropagated trees. Explants were taken either from seedlings and other juvenile material (embryos, cotyledons) or from adult trees. For micropropagation of adult trees, either explants taken from juvenile parts of tree (root suckers, stump sprouts, epicormic shoots) were used, or very small explants were cultured (apical meristems, needle primordia) or explants taken after rejuvenation (pruning, grafting, shoot-tip micrografting) were used as initial material.

The induction of adventitious and axillary shoots was greatly affected by the type and concentration of external cytokinin. The high cytokinin activity exhibited adenine-type cytokinins BAP and PBA and an active phenylurea derivative thidiazuron (Chalupa 1985b, 1987a, 1988, 1989b), and kinetin and zeatin. Adenine-type cytokinins BAP and PBA stimulated effectively cell division and formation of adventitious and axillary buds of all tested conifer and broadleaved forest tree species. The new adenine-type cytokinin PBA (6-benzylamino-9-(2-tetrahydropyranyl)-9H-purine) showed high activity and significantly stimulated bud and shoot formation and shoot proliferation. Thidiazuron (N-phenyl-N´-1,2,3-thidiazol-5-ylurea) stimulated cell division, callus growth and bud formation. Thidiazuron displayed high cytokinin activity at lower concentrations than adenine-type cytokinins. In our experiments thidiazuron was used either alone or together with adenine-type cytokinins. Low concentration of thidiazuron stimulated formation of adventitious buds in forest tree organ cultures, however, the shoot growth was slow and shoots produced on media containing thidiazuron were short.

Conifer tissue culture has the potential for multiplication of superior and resistant genotypes by exploiting the totipotency of conifer cells. Four important European conifer species (*Larix decidua, Pinus sylvestris, Picea abies* and *Pseudotsuga menziesii*) were cultured and multiplied in our experiments. Juvenile material (seeds or seedlings) was used for experiments. The multiplication methods were refined and the systematic study of the field performance of micropropagated trees was started. The survival, growth rates, root development and growth forms of micropropagated trees planted in the field were investigated.

Multiplication of larch (*Larix decidua*) was achieved using adventitious and axillary buds. Adventitious buds were induced on cotyledons, needles, vegetative buds and needle primordia cultured on WPM supplemented with BAP, PBA or thidiazuron (Chalupa 1983b, 1985b, 1989a). The formation of axillary buds was stimulated on shoot tips soaked in a cytokinin solution for 2-4 hours. Shoot elongation from induced adventitious and axillary buds was promoted on WPM containing a low auxin concentration. Root initiation was stimulated on excised microshoots cultured first on WPM containing auxins (NAA and IBA) and later transferred to auxin-free medium. The survival of micropropagated trees planted in the field was high. The height and diameter growth of trees produced *in vitro* was comparable to the growth of trees produced from seeds (Fig. 1). The development and number of roots formed at microshoots affected the growth of micropropagated trees. Trees with well developed roots exhibited fast growth, good stability and orthotropic growth habit.

Shoot multiplication of Scots pine (*Pinus sylvestris*) was stimulated on WPM, MS and QL medium supplemented with a low concentration of cytokinin (PBA or BAP) and auxin (Chalupa 1983b, 1985b, 1989a). Shoot segments cultured on bud inducing medium produced 2-6 axillary shoots within 4-5 weeks. WPM and QL medium containing low concentration of cytokinin (PBA or BAP) and auxin (IBA) stimulated also the elongation of needle fascicle buds. Shoots developed from needle fascicle buds were multiplied as shoots produced from axillary buds. Root initiation was stimulated on microshoots cultured first on medium containing auxins and later transferred to auxin-free medium. The height growth of micropropagated Scots pine trees planted in the field was comparable with that of seedlings (Fig. 1).

Multiplication of Norway spruce (*Picea abies*) was achieved using adventitious buds induced on embryos, cytoledons, hypocotyls, vegetative buds and needle primordia (Chalupa 1977, 1983b, 1985b). Micropropagated Norway spruce

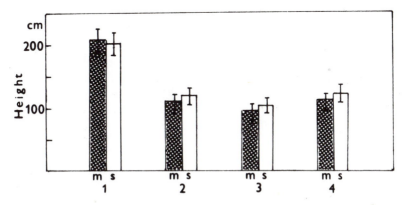

Fig. 1. Field performance of micropropagated conifers. Mean height ($\pm$ S.E.) of micropropagated trees (m) and trees produced from seeds (s) at the end of the fifth growing season. 1-*Larix decidua*, 2-*Pinus sylvestris*, 3-*Picea abies*, 4-*Pseudotsuga menziesii*

trees planted in the field exhibited normal growth and appearance.

For micropropagation of Douglas-fir (*Pseudotsuga menziesii*) we used adventitious and axillary buds induced on organ cultures. (Chalupa 1977, 1983b). Shoot multiplication using axillary buds was achieved on WPM supplemented with a low auxin (IBA) concentration. Shoots excised from cultures were rooted in non-sterile substrate. Some micropropagated plants had exhibited plagiotropic growth which disappeared after two or three years of growth. The height growth of micropropagated trees was comparable to that of trees produced from seeds (Fig. 1).

In experiments with European broadleaved forest trees we achieved multiplication of many species by axillary bud culture method (Fig. 2). The multiplication methods were sufficiently refined to yield large numbers of plants (Chalupa 1979, 1981a,b, 1983a,b, 1984, 1985b, 1987a,b, 1988, 1989b). Growth and development of explants was controlled by hormones present in nutrient medium. The shoot proliferation was promoted by a low auxin to cytokinin ratio. High multiplication rates were obtained on WPM, MS and DKW medium supplemented with a low concentration of cytokinin (PBA or BAP $0,1-1.0$ mg.1^{-1}) and auxin (IBA). Shoots excised from proliferating cultures were rooted on low salt agar medium containing low levels of auxin (IBA, NAA). Most microshoots (70-100 %) formed roots within 2-3 weeks. Rooted plantlets were transplanted into soil and were grown under high air humidity for 2-3 weeks. Survival rates of 70-90 % were ob-

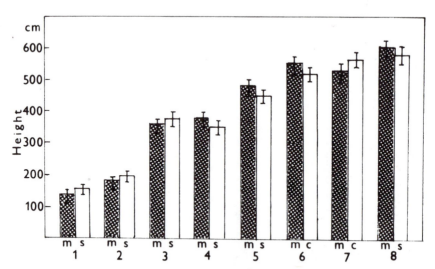

Fig. 2. Field performance of micropropagated broadleaves. Mean height ($\pm$ S.E.) of micropropagated trees (m) and trees produced from seeds (s) or cuttings (c) at the end of the fifth growing season. 1-*Quercus robur*, 2-*Tilia cordata*, 3-*Betula pendula*, 4-*Sorbus aucuparia*, 5-*Robinia pseudoacacia*, 6-*Salix alba*, 7-*Populus euramericana* cv. *robusta*, 8-*Populus tremula*

tained for tested species. Prior to field planting the plants were hardened off and grown in partial shade for 1-2 months. After 5-9 years of field trials, the height and diameter growth, root development and growth form of micropropagated trees was evaluated. Survival of micropropagated trees planted in the field was high and losses during the first winter were low (0-10 %). The height and diameter growth of micropropagated trees was comparable to the growth of trees produced from seeds (Fig. 2). Micropropagated trees planted in the field exhibited normal growth and appearance. The height growth of micropropagated trees was greatly affected by the development of root system. The main reason of height variation of micropropagated trees of the same origin consisted mainly in the different growth and development of roots after planting.

FOREST TREE REGENERATION BY SOMATIC EMBRYOGENESIS

Somatic embryogenesis is expected to play an important role in forest tree improvement. Embryogenesis is considered to be a suitable method of producing a large number of uniform trees in a short time. Recently somatic embryogenesis was reported for important coniferous species (Durzan 1982, 1987, 1988, Hakman et al. 1985, Chalupa 1985a, Nagmani and Bonga 1985, Gupta and Durzan 1986a,b, Durzan and Gupta 1987).

We induced somatic embryogenesis in tissue cultures of Norway spruce (*Picea abies*), oak (*Quercus robur*) and poplar (*Populus euramericana* cv. *robusta*). In the earlier experiments with regeneration of *Picea abies* from cell and tissue cultures (Chalupa and Durzan 1973) some of the cells cultured in liquid medium exhibited signs of organized growth (symmetry and polarity) and the formation of embryoidlike structures was observed. In recent experiments with *Picea abies* (Chalupa 1985a) some immature and mature embryos produced translucent callus which consisted of meristematic cells and elongated suspensor-like cells. Media with reduced hormone concentration stimulated the formation of bipolar embryos. The further development of somatic embryos was stimulated after their transfer to WPM containing a low concentration of auxin (IBA 0.1 mg.1^{-1}). Cytoledons, needles and roots developed after several transfers to the fresh medium. The plantlets were transplanted into potting mixture where their growth continued.

Somatic embryogenesis was also induced in *Quercus robur* cultures initiated from immature acorns (Chalupa 1985b). Some explants produced white callus with embryogenic potentiality. The calli retained their morphogenic potential through several subcultures and globular structures were formed on the surface. The calli with globular structures produced somatic embryos after transfer to media with reduced hormone concentration. Morphogenic callus was also initiated from anthers of *Quercus robur*. Only few anthers produced embryogenic callus containing small globular multicellular structures, which after transfer to media with reduced hormone concentration produced globular embryos.

Somatic embryogenesis in forest tree cultures offers the potential to produce large numbers of embryos. However, before synthetic seeds are used for forest tree clonal propagation, a number of problems need to be solved. Culture methods that promote initiation and development of vigorous somatic embryos need to be refined. Efficient and economical production of synthetic seeds must be achieved before the method can be used on a practical scale. The delivery system should satisfy a number of criteria including high genetic fidelity and good germination of synthetic seeds.

FOREST TREE IMPROVEMENT BY RECOMBINANT DNA TECHNOLOGY

The development of recombinant DNA technology and its application to woody plants opens new possibilities in forest tree improvement. The important advantage of application of genetic transformation methods to forestry is a time shortening of the breeding programs. Effective regeneration of trees from transformed cells still remains a serious obstacle to the utilization of genetic transformation methods in forestry. Another problem is an unusually large genome of forest tree species and our unadequate knowledge of nuclear DNA of forest tree species (Kriebel 1988, Ahuja 1988).

Existing technologies for transporting DNA into plant cells involve use of infectious agents and vectorless delivery systems which include electroporation, microinjection, laser microbeam and high-velocity microprojectils. The genetic engineering of some broadleaved forest trees with a recombinant Ti-plasmid can be achieved by cocultivation of leaf segments with a suspension of *Agrobacterium* cells. In our experiments with *Populus tremula* and *Betula pendula*, shoots were produced from leaf segments (Chalupa 1985b, 1989b) and experiments continue with the aim to regenerate plants from transformed cells. Using *Agrobacterium tumefaciens* as a vector, plant transformation and regeneration system was developed for *Populus* species (Fillatti et al. 1987).

Direct DNA delivery system does not rely on infection. Transformation of plant cells by electroporation requires, however, to remove enzymatically cell wall and incubate protoplasts with foreign DNA. Promising method is a DNA microinjection which can give high efficiencies. The method of microperforation of plant tissue with UV laser microbeam (Weber et al. 1988) does not require the digestion of cell wall and DNA can be introduced directly into selected plant cells. A novel mechanism for transporting biological substances into living cells has been developed recently (Klein et al. 1987). High -velocity microprojectils carrying biological substances can penetrate cell wall and enter cells in a nonlethal manner. Microprojectils can deliver nucleic acids in a biologically active form to a large number of cells simultaneously.

Genetic transformation of a number of forest tree species is achievable in a near future. Insertion of new important genes into forest tree genome may contribute to the production of more resistant and productive genotypes of forest trees. Important traits which could be incorporated into forest trees are herbicide tolerance, pest and disease resistance and environmental stress tolerance (drought, temperature extremes, salinity, waterlogging, and heavy metal tolerance).

CONCLUSIONS

Biotechnology in forest tree improvement will be used for production of more resistant and productive genotypes in the near future. Great progress was achieved in the field of *in vitro* propagation of forest tree species by organogenesis and somatic embryogenesis. Experiments with micropropagation of important European conifers and broadleaved tree species demonstrated that micropropagated forest trees can be successfully hardened and transferred to soil and grown in the field. The field growth of micropropagated trees was comparable with the growth of trees produced from seeds. Micropropagated trees exhibited normal growth and appearance. Biotechnology will contribute to the improvement of forests by multiplication of existing superior genotypes, by propagation of hybrids obtained by sexual crossing and by cytogenetic manipulations and by propagation of transgenic trees obtained by recombinant DNA technology. Transgenic trees with economically useful traits may be produced in the near future. Important traits which could be incorporated into forest trees are hebricide tolerance, pest and disease resistance and environmental stress tolerance.

REFERENCES

Ahuja, M.R., 1988, Gene transfer in woody plants: perspectives and limitations. In: Somatic Cell Genetics of Woody Plants, M.R. Ahuja, ed., Kluwer Academic Publishers, Dordrecht, pp. 83-101.
Chalupa, V., 1977, Organogenesis in Norway spruce and Douglas-fir tissue cultures. Commun. Inst. For. Cechosl., 10:79-87.
Chalupa, V., 1979, *In vitro* propagation of some broadleaved forest trees. Commun. Inst. For. Cechosl., 11:159-170.
Chalupa, V., 1981a, *In vitro* propagation of birch (*Betula verrucosa* Ehrh.). Biol. Plant., 23:472-474.
Chalupa, V., 1981b, Clonal propagation of broadleaved forest trees *in vitro*. Commun. Inst. For. Cechosl.,12:255-271.
Chalupa, V., 1983a, *In vitro* propagation of willows (*Salix* spp.), European mountain-ash (*Sorbus aucuparia* L.) and black locust (*Robinia pseudoacacia* L.). Biol. Plant., 25:305-307.
Chalupa, V., 1983b, Micropropagation of conifer and broadleaved forest trees. Commun. Inst. For. Cechosl., 13:7-39.
Chalupa, V., 1984, *In vitro* propagation of oak (*Quercus robur* L.) and linden (*Tilia cordata* Mill.).Biol.Plant., 26:374-377.
Chalupa, V., 1985a, Somatic embryogenesis and plantlet regeneration from cultured immature and mature embryos of *Picea abies* (L.) Karst. Commun. Inst. For. Cechosl., 14:57-63.
Chalupa, V., 1985b, *In vitro* propagation of *Larix*, *Picea*, *Pinus*, *Quercus*, *Fagus* and other species using adenine-type cytokinins and thidiazuron. Commun. Inst. For. Cechosl., 14:65-90.

Chalupa, V., 1987a, Effect of benzylaminopurine and thidiazuron on *in vitro* shoot proliferation of *Tilia cordata* Mill., *Sorbus aucuparia* L. and *Robinia pseudoacacia* L.. Biol. Plant., 29:425-429.

Chalupa, V., 1987b, European hardwoods. In: Cell and Tissue Culture in Forestry, Vol. 3,J.M. Bonga and D.J. Durzan, eds, Martinus Nijhoff, Dordrecht, pp. 224-246.

Chalupa, V., 1988, Large scale micropropagation of *Quercus robur* L. using adenine-type cytokinins and thidiazuron to stimulate shoot proliferation. Biol. Plant., 30:414-421.

Chalupa, V., 1989a, Micropropagation of *Larix decidua* Mill. and *Pinus sylvestris* L., Biol. Plant., 31 (in press).

Chalupa, V., 1989b,Micropropagation of mature trees of birch (*Betula pendula* Roth..) and aspen (*Populus tremula* L.). Lesnictví, 34 (in press).

Chalupa, V. and Durzan, D.J., 1973, Growth of Norway spruce (*Picea abies* (L.) Karst.) tissue and cell cultures. Commun. Inst. For. Cechosl., 8:111-125.

Durzan, D.J., 1982, Somatic embryogenesis and sphaeroblasts in conifer cell suspension. In: Plant Tissue Culture, A. Fujiwara, ed., Jap. Assoc. for Plant Tissue Culture, Tokyo, pp. 113-114.

Durzan, D.J., 1987, Improved somatic embryo recovery. Bio/ Technology, 5:636-637.

Durzan, D.J., 1988, Process control in somatic polyembryogenesis. In: Molecular Genetics of Forest Trees, J.E. Hällgren, ed., Swedish Univ. Agricult. Sciences, Umea, pp. 147-186.

Durzan, D.J. and Gupta, P.K., 1987, Somatic embryogenesis and polyembryogenesis in Douglas fir cell suspension cultures. Plant Sci, 52:229-235.

Fillatti, J.J., Selmer, J., McCown, B., Haissig, B. and Comai, L., 1987, *Agrobacterium* mediated transformation and regeneration of *Populus*. Mol. Gen. Genet., 206:192-199.

Gupta, P.K. and Durzan, D.J., 1986a, Plantlet regeneration via somatic embryogenesis from subcultured callus of mature embryos of *Picea abies* (Norway spruce).*In Vitro*, 22:685-688.

Gupta, P.K. and Durzan, D.J., 1986b, Somatic polyembryogenesis from callus of mature sugar pine embryos. Bio/ Technology, 4:643-645.

Hakman, I., Fowke, L.C., von Arnold, S. and Eriksson, T., 1985, The development of somatic embryos in tissue cultures initiated from immature embryos of *Picea abies* (Norway spruce). Plant Sci., 38:53-59.

Klein, T.M., Wolf, E.D., Wu, R. and Sanford, J.C., 1987, High-velocity microprojectiles for delivering nucleic acids into living cells. Nature, 327:70-73.

Kriebel, H.B.,1988, Molecular biology in forestry research : when is it relevant and how can we use it?In: Molecular Genetics of Forest Trees, J.E. Hällgren, ed., Swedish Univ. Agricult. Sciences, Umea, pp. 5-18.

Nagmani, R. and Bonga, J.M., 1985, Embryogenesis in subcultured callus of *Larix decidua*. Can. J. For. Res., 15-1088-1091.

Weber, G., Monajembashi, S., Greulich, K.O. and Wolfrum, J., 1988, Microperforation of plant tissue with a UV laser microbeam and injection of DNA into cells. Naturwissenschaften, 75:35-36.

BIOTECHNOLOGY IN FOREST TREE IMPROVEMENT: TREES OF THE FUTURE

Jochen Kleinschmit and Andreas Meier-Dinkel

Lower Saxony Forest Research Institute
Department of Forest Tree Breeding
D-3513 Staufenberg-Escherode, F.R. Germany

INTRODUCTION

Forest tree improvement has to handle long living organisms which start flowering late and which grow in a heterogeneous environment. In modern tree improvement programs, two main aims have to be fulfilled:

- Conservation of the genetic variation of the species under consideration for future needs and as a measure of protection of natural resources.
- Improvement of economic important characteristics of the trees to fit better to human needs. In face of the growing world population and the depletion of natural resources this is especially true for production characteristics.

It is obvious that conventional breeding and propagation methods can only give very slow progress in forest tree species. The turnover of generations needs 30 - 50 years in some species and the field testing in most temperate forest trees faces more than 20 years. Therefore, every effort has been made in the past to improve propagation methods which allow a rapid transfer of improved plant material to practical application. Since most of these methods are developed to allow high propagation rates, the problems connected are obvious:

- a rapid reduction of genetic variability is possible,
- a narrow genetic base increases production risks,
- insufficient testing increases the production risks on large areas.

Every advanced technology has its problems. This is true for biotechnology too, and has to be taken into account in tree improvement programs. Gene conservation becomes even more important under these conditions, a thorough testing is more crucial and a responsible planning of utilization programs, including risk considerations, is necessary.

BIOTECHNOLOGY IN FOREST TREE IMPROVEMENT

Biotechnology is defined as the knowledge of production using biological means. BAJAJ (1986) states that biotechnology has come to a stage where it can replace conventional breeding techniques, and produce novel and improved plants. He summarizes techniques of cellular and

Plant Aging: Basic and Applied Approaches
Edited by R. Rodríguez *et al.*
Plenum Press, New York, 1990

subcellular engineering, such as gene splicing and recombinant DNA, cloning, hybridomas and monoclonal antibodies, production of human insulin, protein engineering, industrial fermentation, artificial insemination, cryopreservation and ovum transfer, plant tissue culture and somatic hybridization, nitrogen fixation, phytomass production for biofuels etc. under this topic. Figure 1 gives some examples of biotechnology in tree improvement. Biotechnology can - due to definition - cover topics as cutting propagation and flower induction too, if plant material is used in technical processes to produce a special product (e.g. rooted cuttings, seed from otherwise not flowering trees). This concept extends the term 'biotechnology' compared to the classical meaning which summarizes processes that result in new products with the help of organisms (fungi, bacteria, plant cells) (DAMBROTH, 1985). Most recent publications use the term biotechnology in connection with in vitro culture (DURZAN, 1985; AHUJA, 1986; BONGA and DURZAN 1987; DUNSTAN, 1988 e.g.).

In forest tree improvement, obviously the most simple approaches for propagation - like cutting propagation and flower induction - are not used to a very large extent today. Conventional in vitro propagation methods are just entering forest plant production in a limited frame. In biotechnology, we are mainly discussing advanced methods including gene transfer which are far from application. If we have in mind a sufficient time for field testing which is more than 10 years in most species and which must be even longer with genetically manipulated plants, it becomes clear that we are discussing problems which will enter forestry to a considerable extent after the year 2000.

Subsequently, we will proceed from the conventional to the more advanced methods:

<u>Cutting propagation</u> has been developed for most temperate forest tree species, partly with a high technical standard of production, e.g. in Norway spruce propagation at the Hilleshög company in Sweden. Even hardly rootable hardwoods like oak and beech can be handled with modern equipment.

The propagation rates are comparatively low and high numbers of genotypes can easily be propagated. Therefore the danger of narrowing the genetic base is less severe. If planned thoroughly selection, testing and production can be closely integrated and maintain a balance between test information, genetic variability and production potential. A restriction in many species is still the maturation process which can be slowed down by hedging or serial propagation (LIBBY et al., 1972, ST. CLAIR et al., 1985).

Juvenility can be regained in some hardwood species and in sprouting conifers by in vitro propagation. Another possibility to restore juvenility is the production of somatic embryos from mature trees. However, here exist only a few examples.

<u>Flower induction</u> by environmental control and phytohormone treatment is another possibility of speeding up tree breeding. Outstanding juvenile plants can quickly be reintroduced into generative reproduction and early seed production is possible before the natural flowering starts.

Stress treatment is a classical approach. Root pruning, girdling, and strangulation have such an effect. Root soaking, reduction of oxygen and water supply or increase of temperature usually increase readiness to flower. Especially in some conifers, the application of gibberellic acid (GA 4/7) can induce flowering at an early age. The possibilities of hand-

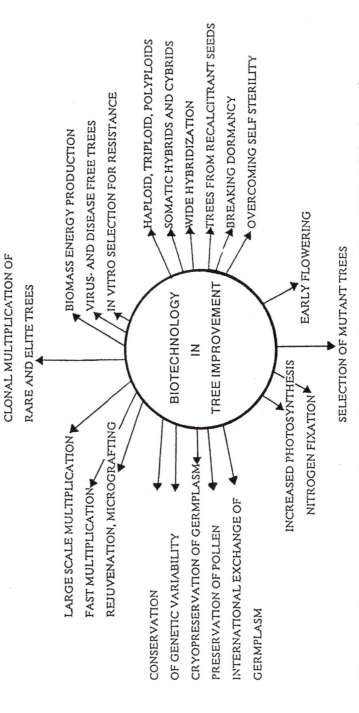

Fig. 1: Application of biotechnology in forest tree improvement (from BAJAJ, 1986, changed).

ling flowering has another important aspect. In some tree species, flowering consumes a considerable amount of assimilates and is reflected in reduction of vegetative growth. From a production point of view, selected trees will have an advantage if they do not flower at all, and if flowering can only be turned on when generative reproduction is desired. Since flowering is under genetic control too, this seems basically to be possible.

In forest tree breeding programs, the use of flower induction greenhouses becomes more and more common. This system has been developed in Finland and can be used not only for breeding populations but also for large scale seed production (e.g. in birch, KLEINSCHMIT et al., 1988). It is obvious that methods of generative reproduction with comparatively high technical input only pay if the resulting seed can be used for mass propagation. This is possible by a combination with vegetative bulk propagation.

In vitro techniques offer new possibilities for the manipulation of trees in breeding programs:

- Mass production of interesting genotypes
- Early in vitro selection for resistance
- Rejuvenation for classical vegetative propagation
- Production of virus and disease free plants
- Production of haploid plants for breeding purposes
- Induction and selection of mutants
- Conservation of genetic variation (cold storage and cryopreservation)
- Somatic embryogenesis and artificial seed production
- In vitro somatic hybridization via protoplast culture
- Gene transfers by recombinant DNA technology

Mass production of interesting genotypes can be carried out by means of in vitro culture which is an important support of forest tree breeding. A considerable number of tree species can be propagated in vitro today and the number is steadily increasing (BAJAJ, 1986; VIEITEZ et al., 1987; BONGA and DURZAN, 1987). Some outstanding hybrids cannot be reproduced by other means than vegetative propagation. Since propagation rates are high (e.g. MEIER-DINKEL, 1986), the genetic base can quickly be reduced drastically. This is a problem which must be controlled in the programs and balanced by conservation of genetic variability. The more testing experience is available, the more a reduction of genetic variation can be justified. Since costs for in vitro propagation are usually higher than for conventional propagation methods, a combination of both techniques in breeding programs makes sense.

Early in vitro selection for resistance can be an efficient and economic tool of large scale screening if close correlations exist between in vitro resistance and field resistance. Selection for tolerance to high levels of salt in the soil is one example.

Rejuvenation of adult material is a central problem in any vegetative reproduction. In vitro propagation has proven the potential for rejuvenation in some hardwood species. Advanced tested material can be reintroduced into cutting propagation. Somatic embryogenesis will allow rapid and comparatively less expensive mass propagation. So far most of the work in tree species started from zygotic embryos. However, there are first results with somatic embryogenesis starting from filament tissue (JÖRGENSEN, 1989). If this can be further developed for more tree species, it serves as an optimal tool for mass propagation as well as for any manipulation of adult genotypes.

Production of virus and disease free plants is another option which is employed for example in fruit trees including Citrus and grapevine. Some species - like Ulmus - may be conserved disease free in vitro until the pathogene has disappeared in certain regions. There are many other options in this field as described by BAJAJ (1986).

Production of haploid plants from anther culture was successful in some tree species (JÖRGENSEN, 1988; BAJAJ, 1986) . This offers interesting perspectives for tree breeding. Homozygous plants can be produced by diploidization which allow a quick selection against lethal alleles. Probably, this selection will be done in vitro too and the haploid tissue will be used for somatic hybridization by protoplast fusion in future. This would be a short cut way for an inbreeding - outcrossing program which would take in forest trees some hundred years by conventional methods and even some decades with flower induction methods. In vitro this could be performed within a few years.

Conservation of genetic variation becomes even more urgent with advanced propagation techniques which allow a rapid reduction of genetic variation. Conservation can be done in vitro with reduced temperature and light intensity or by cryopreservation of pollen, tissues, cells, embryos or seed. This is not direct tree improvement, but it can be a considerable support to tree improvement programs.

In vitro somatic hybridization is an interesting tool to get hybrids from combinations which do not occur in nature. Up to now, not much work has been done on protoplast fusion of trees. But tree protoplasts have been obtained from different species and it only seems to be a question of time until more progress is made.

Gene transfer in tree species is possible from a technical point of view too and has been done in Alnus and Betula by MACKAY et al. (1988) and SEGUIN et al. (1988). Quite intensive research was started at different places of the world. A main problem in this field remains the necessity for long testing of the products which makes it quite improbable that these techniques will play an important role in near future.

FUTURE TRENDS AND CONCLUSION

With the help of biotechnology, tree breeding will become much more efficient in future. However, there are risks which one has to consider when using these techniques:

- A rapid depletion of genetic variation is possible.
- Insufficient testing due to the pressure coming from propagation techniques can be seen already today. Long term testing is even more important with new somatic hybrids and genetically manipulated plants.

Therefore these techniques should only be practically applied within the frame of breeding or research programs. Only if a sufficient genetic variation is maintained and if the material is based on long term testing, a release to application is reliable. Even with these techniques, forest tree improvement remains a long term business.

The fascination of advanced techniques often surpasses necessary boundaries. Worldwide we observe an increasing consciousness concerning ecological questions. One should be very careful not to endanger possible progress in methods by unreflected application. If biotechnology is applied with care, the trees of the future will be adaptable (heterozygous), fast growing, of outstanding quality, growing in mixed stands with different genotypes, not early flowering (if we do not want), juvenile (if we want) and at the same time an important and charming part of our environment. Taken altogether, this is not at all an easy task.

SUMMARY

With improving propagation techniques, rapid mass propagation of outstanding genotypes becomes possible. This must be accompanied by the conservation of natural variability of forest tree species. The methods of flower induction and vegetative propagation including advanced in vitro techniques offer an excellent tool to speed up forest tree improvement considerably, to make it more flexible. However, the methods can not replace long term testing which is an important prerequisite for any tree improvement. Application of advanced techniques should be thoroughly controlled in a breeding program. Maintenance of genetic variation even in production populations for characters not under selection should be a major concern.

LITERATURE

Ahuja, M.R., 1986, Application of biotechnology to forest tree species and problems involved. Proceed. IUFRO Joint Meeting of WP S 2.04-05 and S 2.03-14. Mitteilungen der Bundesforschungsanstalt für Forst- und Holzwirtschaft, Hamburg, No. 154: 187-199.

Bajaj, Y.P.S., 1986, Biotechnology in Agriculture and Forestry. Springer Verl. Berlin/Heidelberg/New York/Tokyo (Vol. I): VII.

Bonga, I.M. and Durzan, D.I., 1987, Cell and Tissue Culture in Forestry. Martinus Nijhoff Publishers. Vol. 1.

Dambroth, M., 1985, Biotechnology in der Pflanzenzüchtung. HG.-Post, 3: 8-13

Dunstan, D.I., 1988, Prospects and progress in conifer biotechnology. Canad. Journ. For. Res., 18: 1497-1506.

Durzan, D.J., 1985, Biotechnology and the cell cultures of woody perennials. Forestry Chronicle: 439-447.

Jörgensen, J., 1988, Embryogenesis in Quercus petraea and Fagus sylvatica. J. Plant Physiology, 132: 638-640.

Jörgensen, J., 1989, Somatic embryogenesis in Aesculus hippocastanum L. by culture of filament callus, J. Plant Physiol. (submitted).

Kleinschmit, J., Hoffmann, D., Meier-Dinkel, A., Jörgensen, J., 1988, Biotechnologische Verfahren bei Generhaltung und Züchtung von Waldbaumarten. BioEngineering, 3: 236-239.

Libby, W.J., Brown, A.G. and Fielding, I.M., 1972, Effects of hedging Radiata pine on production, rooting and early growth of cuttings. New Zealand Journ. For. Sci., 2: 263-283.

Mackay, J., Séguin, A. and Lalonde, M., 1988, Genetic transformation of 9 in vitro clones of Alnus and Betula by Agrobacterium tumefaciens. Plant Cell Reports, 7: 229-232.

Meier-Dinkel, A., 1986, In vitro Vermehrung ausgewählter Genotypen der Vogelkirsche (Pruns avium L.). Allgem. Forst- und Jagdzeitung, 157: 139-144.

Séguin, A. and Lalonde, M., 1988, Gene transfer by electroporation in Betulaceae protoplasts: Alnus incana. Plant Cell Reports, 7: 367-370.

St. Clair, J.B., Kleinschmit, J. and Svolba, J., 1985, Juvenility and serial vegetative propagation of Norway spruce clones (Picea abies Karst.) Silvae Genetica, 34: 42-48.

Vieitez, A.M., Ballester, A., Vieitez, H.L., San Jose, M.C., Vieitez, F.J. and Vieitez, E., 1987, Propagacion de plantas leñosas por cultivo "in vitro". Diputacion Provincial Pontevedra, 97 p.

SELECTED POSTERS

MICROPROPAGATION OF <u>VITIS VINIFERA</u> FROM "VINHO VERDE" REGION OF PORTUGAL:

A METHOD FOR GRAPEVINE LEAFROLL VIRUS ELIMINATION

Margarida Casal   M. Salomé S. Pais

Lab. Biologia, Univ.   Dep. Biologia Vegetal, Fac.Ciências
Minho, 4700 Braga   de Lisboa, 1700 Lisboa, Portugal
Portugal

INTRODUCTION

One of the main conditionants of productivity and quality of "Vinho verde" is virus and virus-like diseases transmited by the traditional process of propagation, Grapevine Leafroll Virus (GLRV I) being one of the most widespread.

The economic importance of this severe disease is well known, due to the very significant effects, such as, decrease on yield production, late ripeness of grapes and decrease of sugar contents (Legin, 1972; Bass and Legin, 1981).

The obtention of virus-free plants via <u>in vitro</u> meristem culture can confer increase in the quality and competitiviness of this kind of wine in national and international markets. Attempts have been successfully made to eliminate some virus diseases of grapevines, using heat therapy followed by <u>in vitro</u> cultured shoot-tips (Gifford and Hewitt, 1961). Galzy (1964), submiting to heat therapy grapevine shoots during <u>in vitro</u> culture, has also obtained virus-free plants. With the same purpose, Barlass and Skene (1982), Barlass <u>et al.</u> (1982) and Barlass (1987) used shoot apex culture, for the adventitious bud formation from leaf primordia fragments (Barlass and Skene, 1978) instead of apical dome. Using this methodology, with or without heat therapy, these authors could obtain healthy clones from virus infected vines.

The aim of this study was the improvement of "Vinho Verde" productivity and quality by <u>in vitro</u> production of virus-free plants to be used for regeneration and rapid clonal propagation. Field growing plants, as well as <u>in vitro</u> produced plants were analyzed for virus infection, by means of serological tests, namely by ELISA. The virus checked were Arabis Mosaic Virus (ArMV), Grapevine Fanleaf Virus (GFV) and Grapevine Leafroll Virus I (GLRV I).

MATERIAL AND METHODS

1. Plant material

The biological material was collected during winter from "Vinho

Plant Aging: Basic and Applied Approaches
Edited by R. Rodríguez *et al.*
Plenum Press, New York, 1990

Verde" selected vineyards (cv. "Loureiro") and stored in plastic bags at 4°C.

Dormant cuttings were forced in water and maintained in the _in vitro_ incubation conditions. Two to three weeks later, the vegetative material was surface sterilized with a 10% (w/v) solution of calcium hypochlorite, for 15 minutes and rinsed 3 times in sterile distilled water.

2. Incubation conditions

Temperature: from 23°C in the dark to 25°C in the light.
Light source: flourescent tubes "Sylvania" Gro-Lux.
Day lenght: 16 h light, 8 h dark.

3. Micrografting and microcutting

Micrografting was performed according to Bass et al. (1976). The _Vitis_ seedlings were germinated at 38°C and 16 h of diffuse light. Shoot-tips were taken from surface sterilized scions and placed on decapitated seedling under aseptic conditions. If successful, the grafted plant was then tested for the presence of viruses.

Microcuttings of different ages were cultured in MS medium (Murashige and Skoog, 1962), solidified with agar 0.8% (w/v), pH 5.5, supplemented with sucrose 20 $g.l^{-1}$ and growth regulators according to table 1.

Table 1 - Concentrations of benzyladenine(BA) and
naphthaleneacetic acid (NAA) on MS medium.

Growth regulator $(mg.l^{-1})$	
BA	NAA
0.05	0.05
0.1	0.1
0.5	0.5
1.0	1.0

4. Meristem culture

A method previously described by Chee et al. (1984), for large scale propagation of _Vitis_, modified in the growth regulators composition and concentrations, was used in this step.

Apical and axillary meristems 0.1 mm long (containing the meristem and 2 leaf primordia), excised from sterilized shoots were established on MS medium, supplemented with sucrose 30 $g.l^{-1}$ and BA 2.0 $mg.l^{-1}$, pH 5.5, contained in Petri dishes (establishment phase). Shoot apices developed into rosettes of leaflike structures which, after 3 weeks, were cut in small pieces and transferred to 50 ml of fresh medium solidified with agar 0.6% (w/v), contained in 250 ml Erlenmeyer flasks (multiplication phase). Four weeks later, the shoots with 3 or

4 nodes could be transferred to rooting media. The remaining explant was subcultured in the conditions described above.

5. Rooting induction

The plantlets obtained in the multiplication phase were transferred to culture tubes containing 10 ml of MS basal medium, supplemented with sucrose 20 g.l^{-1}, pH 5.5. Different concentrations of indolebutyric acid (IBA), β-indole-3-acetic acid (IAA) and naphthaleneacetic acid (NAA) were used at the concentrations reported in table 2.

Table 2 - Auxins and respective concentrations tested on rooting medium.

CONCENTRATION (mg.l^{-1})	IBA	IAA	NAA
0.1	M1	M4	M7
0.5	M2	M5	M8
2.0	M3	M6	M9

6. Transfer to soil

The plants were transferred to sterile sand and during the first week they were kept under an high humidity atmosphere (plastic bags covering potts) and placed in the conditions described for in vitro culture. By opening the pastic bag, the humidity was slowly decreased during the following two weeks. The plants were watered regularly with a nutrient solution described by Huglin and Julliard (1964).

7. Serological tests for GFV, ArMV and GLRV detection

Serological tests for virus checking were performed on field plants, on forced cuttings and on in vitro propagated plants. The direct double antibody methodology (Clark and Adams, 1977) was used for the detection of GFV and ArMV. The indirect double antibody methodology (Van Regenmortel and Burckard, 1980) was used for the detection of GLRV I. Each analysis was repeated 9 times.

RESULTS AND DISCUSSION

1. Micrografting

According to several authors this technique, associated or not with heat treatment, enables the recovering of virus infected grapevines (Bass et al., 1976; Elgelbrecht and Schwerdtfeger, 1979; Walter, 1985). The assumption that the apical meristem may be virus-free, conditioned the utilization for micrografting of very small shoot-tip explants. However, according to Walter (1985) for the elimination of certain virus diseases, there is no general rule regarding the size of the explant. It depends on the disease and the duration of heat treatment.

In spite of the low efficiency of this technique, we succeed to obtain micrografted virus-free plants, which after 3 weeks could be transferred to the soil (Fig. 2).

2. Microcutting

The results showed that the behaviour of the explants depends on their age. The best reactions were obtained when the 3rd to 5th node of the forced cuttings were used (Fig. 2a). The presence of NAA in the medium slowed plantlets development when compared with the results obtained using the medium devoided of growth regulators. Twelve days after establishment the cuttings cultivated on media devoided of growth regulators developed 1 or 2 leaves. Those cultivated on media supplemented with NAA didn't develop. The explants growing on media supplemented with BA developed 2 or 3 leaves. After 24 days on medium supplemented with BA 0.5 mg.l^{-1}, it was possible to obtain plantlets ranging from 3.0 to 4.0 cm height, with 5 to 6 leaves (Fig.2a).

The presence of BA in the culture medium greatly increase plantlets development. The plantlets growth, estimated by the number of new leaves formed durind 24 days of culture increased with the increase of BA concentration until 0.5 mg.l^{-1}. The best results were achieved with 0.5 mg.l^{-1} of this growth regulator.

3. Meristem culture

This technique, once established, presents the highest possibilities for vegetative propagation comparatively with the two techniques described before. From one explant, after 3 subcultures on multiplication medium it is possible to obtain thousands of plantlets.

According to Barlass et al. (1982) the smaller the explant the better the chance of producing virus-free plants. However, the smaller the piece the greather the risk of failling to induce shoot regeneration. The procedure presented here involves shoot terminal and axillary meristems culture in a liquid nutrient medium, containing BA. By the meristem culture method, after 3 weeks for establishment phase, 4 weeks for multiplication phase and 3 weeks for rooting induction and shoot elongation, it is possible to produce continuously thousands of plants (Fig.2c). This methodology presents a great potential for industrial application if we take into account that, from a single vine, a nursery can obtain only 10 vines during an year using hardwood cuttings (Chee et al., 1984).

4. Rooting and aclimatization

In all media tested, root initiation starts within the second week after transfer to the rooting medium. The results were estimated in terms of shoot elongation, number of roots and maximum and minimum range values of root length. The results obtained for "Loureiro" are shown on Graph. 1 (a,b) and Fig.2d. This cultivar didn't require auxins for root initiation (medium MO). This result is in agreement with those described by Harris and Stevenson (1979) and Barlass and Skene (1981) for Vitis vinifera cultivars. However, the result was not uniform which conditioned the improvement of rooting conditions.

The addition of auxin 2 mg.l^{-1}, as well as NAA 0.5 mg.l^{-1}, stimulated calli formation. We verified that the IBA or IAA at 0.1 and 0.5 mgl^{-1}, induced similar results in terms of shoot length and number of roots (Fig. la). In terms of quality of radicular system, the medium supplemented with IBA, 0.5 mg.l^{-1} is the most favourable because it promotes the production of shorter roots (Fig. 1,a and b;Fig.2d) that are better for potting, avoiding the breakage and difficulties in replanting. Similar results have been reported by authors working with Vitis sp. (Chee et al., 1984).

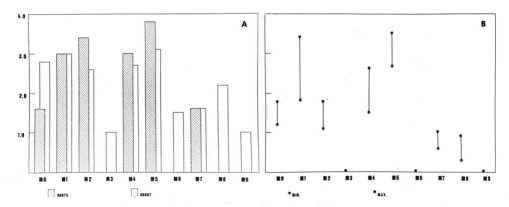

Figure 1

Rooting induction of plantlets derived from micropropagation.
Culture media composition: M0-M9 (see table 2)
A - Shoot length (cm)and number of roots.
B - Root length, maximum and minimum values (cm).

To avoid dessication during transfer to soil, the plantlets must be kept under 100% of relative humidity during the first week. During aclimatization period (3 weeks), the plants developed new leaves and exhibit a good shoot and root growth (Fig. 5).

5. Virus detection

Leaves of field selected plants were collected during September and stored at -20°C. These plants were free of ArMV and GFV, responsible for fanleaf disease of grapevines (Table 3). On the other hand it was possible to verify that 80% of the plants were infected by GLRV I.

One and three months after flushing, leaves from forced cuttings, collected from GLRV I infected plants, showed 100% of GLRV I elimination (table 3), suggesting that rejuvenation by flushing may be responsible for absence of virus replication in newly formed leaves.

Earlier studies provided experimental evidence that shoot apices culture of grapevine regenerate plants free from the graft-transmissible diseases, namely stem pitting, corky bark, yellow speckle, fanleaf, summer mottle or leafroll (Gifford and Hewitt, 1961; Galzy, 1964;Bass et al., 1976; Engelbrechet and Schwertfeger, 1979; Bass and Legin, 1981; Barlass et al., 1982; Barlass and Skene, 1982; Walter, 1985; Barlass 1987). By microcutting of field growing material, whithout heat treatment, it was possible to achieve 25% of recovered Vitis vinifera, inicially GLRV I infected (Bass and Legin, 1981). Our results show that, using forced material as primary explant, it is possible to obtain a higher value of recovered plants. As is shown in Table 3, in these conditions it was achieved 72% of GLRV I free plantlets. These results can be explained as a consequence of rejuvenation induced by flushing.

Using shoot-tip culture, Barlass and Skene (1982) achieved 100%

of GLRV I elimination. From our studies we could verify that plantlets derived from micropropagation, using either terminal and axillary meristems culture, revealed 100% of GLRV I elimination (table 3). Moreover, terminal and axillary meristems from flushed shoots appear to be a suitable material for virus elimination. Considering this result and the great amount of plantlets derived from one single meristem (thousands of plantlets after 3 months), we can consider that the use of rejuvenated material, as well as, the micropropagation of both apical and axillary meristems, are suitable for virus elimination in _Vitis vinifera_ (cv. Loureiro) for industrial purposes.

Table 3 - Results of serological tests concerning GLRVI, GFV and ArMV in leaves of field cultured plants (1), in leaves of flushed material (2) and micrografting (3), microcutting (4) and axillary and termind shoot meristems (5) derived plantlets.

ORIGIN OF MATERIAL	1	2	3	4	5
GLRV I DETECTION	80%	0%	0%	28%	0%
GFV DETECTION	0%	0%	0%	0%	0%
ArMV DETECTION	0%	0%	0%	0%	0%

———————————————————— Figure 2

a) 21 days old plant, obtained by micrografting of _Vitis vinifera_, cv. Loureiro shoot-tip meristem on a decapitated _Vitis_ seedling (arrow). It is possible to observe a well developed root and shoot system. The plantlet presents development suitable for transfer to aclimatization conditions.

b) Plantlets obtained by _in vitro_ culture of _Vitis vinifera_, cv. Loureiro, of forced derived cuttings, 24 days after establishment. The better growth, expressed in terms of number of leaves and shoot length, was observed on MS medium supplemented with sucrose 20 g.1^{-1} and BA 0.5 mg.1^{-1}. Nodes with different ages were established 2nd node (tube A), 3rd node (tube B and C), 4th node (tube D and E) and 5th node (tube F). The best development was observed for plantlets derived from 3rd node and older ones, until 5th node.

c) Shoots obtained by adventitious bud formation from leaflike meristems derived structures, after 28 days on the multiplication medium.

d) _In vitro_ rooting induction. 21 days after transfer to MS medium supplemented with sucrose 20 g.1^{-1} and: with no auxins (tube A), with IBA 0.1 mg.1^{-1} (tube B), 0.5 mg.1^{-1} (tube C) or 2.0 mg.1^{-1} (tube D).

e) "Loureiro" grapevine obtained through meristem culture, rooted on M2 media (table 2) and maintained during 12 days on aclimatization conditions.

REFERENCES

Barlass, M. and Skene, K.G.M. (1978) "In vitro propagation of grapevine (Vitis vinifera L.) from fragmented shoot apices" Vitis 17, 335--340.

Barlass, M. and Skene, K.G.M. (1982) "Virus-free vines from tissue culture". The Australian Grapegrower & Winemaker 224, 40-41.

Barlass, M., Skene, K.G.M., Woodman, R.C. and Krake, L.R. (1982) "Regeneration of virus-free grapevines using in vitro apical culture" Ann. Appl. Biol. 101, 291-295.

Barlass, M. (1987) "Elimination of stem pitting and corky bark diseases from grapevine by fragmented shoot apex culture" Ann. Appl. Biol., 110, 653-656.

Bass, P., Viuttinez, A. and Legin, R. (1976) "Improvement of grapevine thermotherapy by growing excised shoot tips on nutritive media or by grafting seedling aseptically cultivated in vitro" Abst. 6th ICVG Meeting, Cordoba, 1976, Monografias INIA, 1978, 325-332.

Bass, P. and Legin, R. (1981) "Thermotherapie et multiplication in vitro d'apex de vigne. Application à la separation ou a l'élimination de diverses maladies du type viral et a l'évaluation des dégats" C.R. Acad. Agric. 67, 922-933.

Clark, M.F. and Adams, A.N. (1977) "Characteristics of the microplate method of enzime linked immunosorbent assay for the detection of plant viruses" J. Gen. Virol. 34, 475-483.

Chee, R. Pool, R.M. and Bucher, D. (1984) "A method for large scale in vitro propagation of Vitis" New York Food and Life Sciences Bull. 109, 1-9.

Engelbrechet, D.J. and Schwertfeger (1979) "In vitro grafting shoot apices as an aid to the recovery of virus-free clones" Phytophylatica 11, 183-185.

Galzy, R. (1964) "Technique de thermothérapie des viroses de la Vigne" Annales des Epiphyties, 15, 245-256.

Gifford, E.M. and Hewitt, W.B. (1961) "The use of heat therapy and in vitro culture to eliminate fanleaf virus form grapevine" Am. J. of Enology and Viticulture, 12, 129-130.

Huglin, P. and Julliard, B. (1964) "Obtention de semis de vignes trés vigoureux à mise à fruits rapide et ses répercussion surl'amélioration génétique de la vigne" Ann. Amélior. Plantes 14 (3), 229-244.

Legin, R. (1972) "Experimentation pour étudier l'effect des principales viroses sur la végétation et la production de la vigne" Ann. Phytophatol. no hors série: 49-57.

Murashige, T. and Skoog, F. (1962) "A revised media for rapid growth and bioassays with Tobacco tissue culture" Physiol. Plant. 15, 473--497.

Van Regenmortel, M.H.V. and Burckard, J. (1980) "Detection of a wide sprectrum of tobacco mosaic virus strains by indirect enzyme immunosorbent assay (ELISA) Virology 106, 327-334.

IN VITRO CULTURE OF *PISTACIA VERA* L. EMBRYOS AND AGED TREES

EXPLANTS

A. González and D. Frutos

Departamento de Fruticultura
Centro Regional de Investigaciones Agrarias
30150-La Alberca (Murcia), Spain

INTRODUCTION

An increasing rate of pistachio (*Pistacia vera* L.) new
orchards has been detected in the mediterranean areas of the
World, where the interest of growers for this nut crop is currently crescent (Monastra et al., 1988). Altough the seedling
rootstock production doesn´t present any special problem (Casini y Conticini, 1979; Frutos y Barone, 1988), the grafting
or budding propagation doesn´t look so easy to do (Borisova-
Velkova, 1984; Avanzato et al., 1988; Romero et al., 1988); in
spite of it, topworking of *P. vera* cultivars on *P. terebintus*
and *P. kinjuk* adult trees (Kaska and Bilgen, 1988), and graft
nursery production on *P. atlantica* (Needs and Alexander, 1982)
have been noticed. Rooting *P. vera* softwood cuttings under
mist treated with very high auxins concentration during some
few seconds (Al Barazi and Schwabe, 1982), and *P. chinensis*
hardwood cuttings (Morgan and Maika, 1984) look possible, but
in both cases poor results were yielded.

In vitro micropropagation of pistachio species seems full
of suggestions because of its classic propagation troubles
above reviewed. At present some few micropropagation works
have been noticed (Barghchi and Alderson, 1983 and 1985; Martinelli, 1988), although its results, that have been useful to
initiate several experiments, don´t seem to bring into general
use because its genome and/or age of the micropropagated plant
materials look related to its different in vitro growth behaviour.

MATERIAL AND METHODS

P. vera seeds from adult seedlings grown in Torreblanca
(Murcia) where harvested, disinfected without endocarp, first
into a 70 p.100 ethanol solution for 45 seconds, then into a
30 p.100 commercial bleach solution of 40 g/l of active cholo-
rine for 10 minutes, and later three times rinsed with sterile
distilled water. After that the embryos were isolated and sown
on an aseptic Murashige and Skoog (1962) half strengthed medium (MS/2) without vitamins and without hormones, supplemen-

Plant Aging: Basic and Applied Approaches
Edited by R. Rodríguez *et al.*
Plenum Press, New York, 1990

ted with sucrose (30 g/l) and agar (8 g/l), and adjusted to pH 5.7. The in vitro embryos cultures were put into a climatic chamber at 24±1°C, 5.000 lux and 16:8 hours light: dark photoperiod. 5 weeks later, plantlets supplied with 4-5 leaflets where availables. Four micropropagation experiments to improve the medium were then established: 1) MS supplemented with 1.0 and 4.0 mg/l of 6-benzylaminopurine (BAP) in the above mentioned photoperiod and in the dark; 2) MS without vitamins, supplemented with BAP (2.0 and 4.0 mg/l) and with 0, 0.5 and 2.0 mg/l of gibberellic acid (GA3); 3) MS nitrate half strengthed, without vitamines and supplied with BAP (4.0 mg/l); and 4.0) Mc Cown and Lloyd's (1981) Woody Plant Medium (WPM) supplemented with BAP (4.0 mg/l) and GA3 (0 and 0.5 mg/l). In all the tested media, sucrose (30 g/l) and agar (8 g/l) were also added.

For every one of the prior treatments, two kinds of explant were used: a) shoot segments with 1-2 buds from proliferated embryos as above mentioned, and b) one bud shoot pieces out of adult trees. In this case, some dormant limbs in the field were chosen and disinfected with ethanol (97 p.100) and captafol (2 g/l), then were bagged with paper and several weeks later, the springshoots out of the chosen dormant limbs were excised, defoliated, disinfected into a 70 p.100 ethanol solution for 45 seconds followed by a 30 p.100 bleach solution (40 g/l of active chlorine) for 10 minutes, three times rinsed with sterile distilled water, segmented and used as showed in point b).

The shoots from the best proliferation treatments were used for rooting experiment. In this fase the next treatments were tried: 1) MS supplemented with 0, 1.0, 2.5, 5.0, 7.5 and 10.0 mg/l of indole-3-butyric acid (IBA); 2) MS supplemented with 0, 1.0, 2.5, 5.0, 7.5 and 10.0 mg/l of naphthyleneacetic acid (NAA); 3) MS without vitamins, supplemented with IBA (0, 0.5, 1.0, 2.5, 5.0 and 7.5 mg/l) and with BAP (0, 0.1, 0.25 and 0.5 mg/l); 4) MS without vitamins supplemented with NAA (0, 0.5, 1.0, 2.5, 5.0 and 7.5 mg/l) and BAP (0, 0.1, 0.25 and 0.5 mg/l); 5) WPM supplemented with IBA (0, 1.0, 2.5, 5.0, 7.5 and 10.0 mg/l); 6) WPM supplemented with NAA (0, 1.0, 2.5, 5.0, 7.5 and 10.0 mg/l); 7) Driver and Kuniyuki (1984) medium (DKW) supplemented with IBA (0, 1.0, 2.5, 5.0, 7.5 and 10.0 mg/l); and 8) DKW supplemented with NAA (0, 1.0, 2.5, 5.0, 7.5 and 10.0 mg/l). All the rooting media tested were supplemented with sucrose (30 g/l) and agar (8 g/l) and adjusted to pH 5.7.

The four precedent experiments were placed in 3.000 lux, 24±1°C and 16:8 hours light: dark environmental conditions. A last rooting experiment on MS medium without vitamins supplemented with IBA (2.5 mg/l) were placed in the dark for the first week, and then in the above refered climatic conditions. In all cases the rooting shoots were recultured every two weeks on fresh medium.

RESULTS AND DISCUSSION

Stages I and II: establishment and clonal propagation of P. vera in aseptic tissue culture.

a) Explants from in vitro cultured embryos plantlets.

When the embryos culture were started on MS medium supplemented with 4.0 mg/l of BAP a good propagation rate were observed, but 3-4 weeks later, independently of light or dark conditons, a vascular necrosis were observed and shoots died. To solve this trouble a vitaminless MS medium were tried, and then the vascular necrosis appeared any more. However, several weeks later, shoots vitrifiction came up. Vitrification problems were corrected when WPM supplemented with 4.0 mg/l of BAP and 0.5 mg/l of GA3 were used. For a start, the in vitro plant material for clonal propagation consisted on shoot segments removed from in vitro cultured embryos plantlets. The behaviour of these materials were very different in the last medium according to the mother plantlet. Therefore, a certain relation could exist between the genotipe and the growth rate of the micropropagated plant materials. So, for we to avoid this different behaviour on the same medium, all the next trials were established by using clonal material from the same mother plantlet.

b) Adult trees explants.

To start the clonal propagation of adult bud explants was only usefull the WPM supplemented with BAP (4.0 mg/l) and GA3 (0.5 mg/l) to break its dormant state. But when MS medium were used, the dormant buds died before sprouting and the translucent and clean color of the MS medium become darkish and blurred.

The results a) and b) seem differents of those got by Bargchi and Alderson (1983 and 1985) and by Martinelli (1988), when they make reference to MS as a good medium for pistachio species micropropagation. On the other hand, it has been observed that the leaflets from embryos plantlets are singles while these from adult trees are composed. A certain correspondence could be consider between the juvenile state for singles and adult state for composed leaflets, respectively.

Stage III: Rooting of micropropagated shoot cuttings.

The rooting trials were carried out only with clonal shoot cuttings from seed. At present the results are not definitive, but can be advanced that roots emerge when MS without vitamins supplemented with IBA (2.5 mg/l) was used. Whether during the first week the culture were kept in the dark, 50 p.100 of rooting shoots were noted, while when rooting were promoted since the begining in light: dark 16:8 hours, only 5 p.100 rooting were observed.

CONCLUSIONS

The best results for clonal propagation of *P. vera* were produced when a WPM supplemented with BAP (4 mg/l) and GA3 (0,5 mg/l) were used so much for seed plantlets as for adult explants.

Even considering a not definitive results for rooting, it looks possible to promote the root formation on juvenile plant lets from seeds when MS medium without vitamins is used and during the first week the rooting material is kept in the dark and then is growed in the climatic chamber.

REFERENCES

Al Barazi, Z.; Schwabe, W.W. 1982. Rooting softwood cuttings of adult *Pistacia vera*. Journ. Hort. Sci. 57 (2) 247-252.

Avanzato, D.; Monastra, F.; Corazza, L. 1988. Altivita di ricerca in corso sul pistacchio e primi risultati. Rapport EUR 11557. CEE. Colloque AGRIMED-GREMPA. Reus: 299-316.

Borisova-Velkova, D. 1984. Grafting pistachio trees. Fruit growing, 63 (11): 15-16.

Casini, E.; Conticini, L. 1979. Prove di germinabilita di semi delle specie *Pistacia vera* L. e *Pistacia terebinthus* L. Riv. Agric. Subtrop. e Trop. 73 (3/4): 223-240.

Driver, J.A.; Kuniyuki, A.H. 1984. In vitro propagation of *Paradox* walnut rootstock. HortScience 19 (4): 507-509.

Frutos, D.; Barone, E. 1988. Germinacion de *Pistacia vera* L. y primer crecimiento de las plantas de semilla tratadas con acido giberelico (GA3). Rapport EUR 11557. CEE. Colloque AGRIMED-GREMPA Reus: 289-298.

Kaska, N.; Bilgen, A.M. 1988. Top-working of wild pistachios in Turkey. Rapport EUR 11557. CEE. Colloque AGRIMED-GREMPA. Reus: 317-325.

McCown, B.H.; Lloyd, G. 1981. Woody plant medium (WPM). A mineral nutrient formulation for microculture of wood plant species. HortScience 16: 453 (Abstr).

Monastra, F.; Avanzato, D; Lodoli, E. 1988. Il pistacchio nel mondo. Confronto tra la pistacchicoltura delle aree tradizionali e quella emergente degli Stati Uniti. Rapport EUR 11557. CEE. Colloque Agrimed-Grempa. Reus: 271-288.

Morgan, D.L.; Maika, S. 1984. Propagation of a mature *Pistacia chinensis* BUNGE by stem cuttings. PR. Texas Agric. Experiment Stat. ISSN 0099-5142, NO, 4260.

Murashige, T.; Skoog, F. 1962. A revised medium for rapid growth and bioassays with tobacco tissue cultures. Physiol. Plant 15: 473-497.

Needs, R.A.; Alexander, D.M. 1982. Pistachio, a technique for chip budding. Australian Hort. 80 (10): 87-89.

Romero, M.A.; Vargas, F.J.; Aleta, N.; Batlle, I. 1988. Multiplicacion y manejo de plantas en pistachero. Rapport EUR 11557. CEE. Colloque AGRIMED-GREMPA. Reus: 327-335.

TISSUES CULTURE AND REGENERATION IN JOJOBA

Sylvain Unique

Laboratoire d'Histophysiologie Vegetale - U.A. 1180 C.N.R.S.
Universite Pierre et Marie Curie
12, rue Cuvier - 75005 PARIS - FRANCE

INTRODUCTION

Up to now very few studies were carried on tissue cultures of Jojoba
(Simmondsia chinensis Link). A few years ago, ROST and HINCHEE (1)
proposed a technique for node propagation and a method in order
to obtain undifferentiated callus tissue. WANG and JANICK (2) and
LEE and THOMAS (3) reported successful somatic embryogenesis from
very immature zygotic embryos. Never embryogenesis nor organogenesis
were obtained from any vegetative part of the plant. The present
report concerns these possibilities. The preliminary work consists
to compare the embryogenic and organogenic capacities of different
juvenile parts of the plant : immature and mature ovules and ovaries,
immature embryos.

MATERIAL AND METHODS

The flower buds were harvested from male and female plants growing
in an orchard located near Bastia (Corse). They were sterilized by
dipping for fifteen minutes in 0.4 W/V Benlate ((butylcarbamoyl-1
benzimidazolyl-2) carbamate de methyle) and for ten minutes in
calcium hypochloride (4%W/V). Then they were rinsed four times in
sterile water. Sepals, embryos, ovaries, ovules were excised and
transfered to petri dishes (60mm diameter). The basal medium for
all cultures contained the following substances : inorganic salts
according to MURASHIGE and SKOOG (modified 1962) (4), vitamins of
B5 medium (GAMBORG) (5), casein (0.1W/V), polyvinylpyrolidone
(0.05W/V), myo-inositol (1 mM), sucrose (0.08 M), aminoacid
supplementation according to CHU and HILL (6) and agar (0.65W/V).
PH was ajusted to 6.5 with KOH. The growing factors were used at
different concentrations and combinations : Naphthylacetic acid
(N.A.A.) (3 to 6 μM), Benzylaminopurine (B.A.P.) (1 to 20 μM),
2-4Dichlorophenoxyacetic acid (2-4D) (2 to 3 μM), Zeatin (1 to 7 μM).
The cultured material was maintained under continuous low red light,
the temperature was 24ºC. For histological studies the explants were fixed
in F.A.A., embedded in resin, sectionned and colored with pyronin.

RESULTS

A - Organogenesis from flower explants

We first examined callogenic capacities of different flower explants

Plant Aging: Basic and Applied Approaches
Edited by R. Rodríguez *et al.*
Plenum Press, New York, 1990

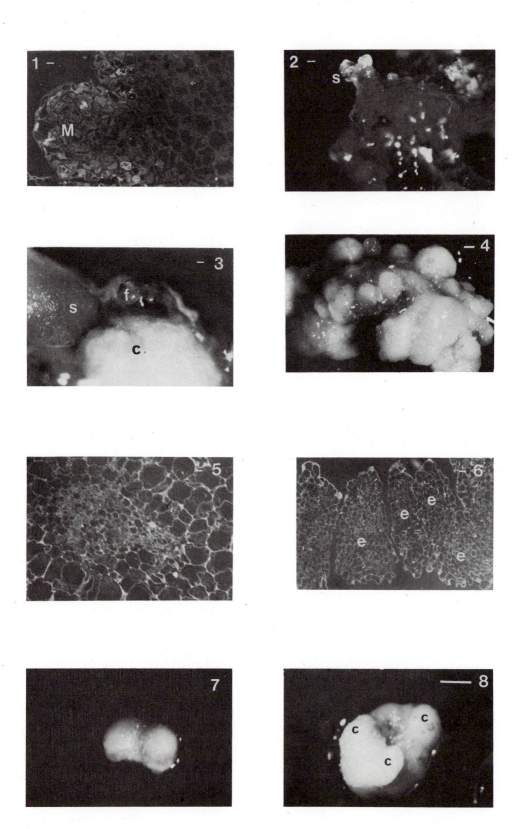

PLATE 1

Picture 1 : Neoformation of meristematic cells (M) on callus surface of
ovary (bar 3 μ)

Picture 2 : Shooting apices developed (s), in presence of Zeatin in
nutrient medium, on callus surface of ovary (bar 0.2mm)

Picture 3 : Development of leaf tissue (s : stigmate, c : ovary callus,
f : foliar structure - bar 0.1mm)

Picture 4 : Embryoid development from upper part of cotyledon (bar 0.5mm)

Picture 5 : Embryoid structure appeared inside cotyledon tissue
(bar 60 μ)

Picture 6 : Somatic embryos juxtaposition on the upper part of cotyledon.
Note irregular shape due to the high structures density.
(e : somatic embryo - bar 50 μ)

Picture 7 : Upper view from an heart shape embryo of 2mm wide

Picture 8 : Somatic embryo on basal medium with B.A.P. (1 μ M). Note the
3 developing cotyledons embryo (c : cotyledon - bar 1mm)

according to their developmental stage. The stigmates sometimes developed
some little calli on the basal medium supplemented with 2-4D (10 µ M).
The sepals cultured in vitro became thick, but did not produce any kind
of callus tissue. The immature or just fecundate ovules and the ovaries
(4-6mm wide) developed a green massive callus containing undifferentiated
cells. Small ovaries (0.5mm wide),harvested from flowers at the dormancy
stage did not develop any callus. The ovaries from 1.5 to 2mm wide produced
globular calli after one month of culture in the basal medium supplemented
with 2-4D (2 µ M), Zeatin (7µM) and N.A.A. (8µM) or 2-4D (10 µ M) alone.
Occasionally meristematic structures (Picture 1), shoot apices and foliar
primordia (Pictures 2-3) appeared. Such calli were transfered in a fresh
medium deprived of 2-4D in order to promote the meristematic structures
expression.

B - Mature tissues were tested for their organogenic capacities
    Mature ovaries and leaf tissue only gave undifferentiated calli with
white large and vacuolated cells. These calli were initiated from internal
cells. The epidermic cells did not express any reaction.

    Table 1   Somatic embryogenesis : Correlation between zygotic embryo
size and growth factors combinations added to the nutrient medium.(The
percentage represent the number of explants producing somatic embryos
after 40 days of incubation)

Zygotic embryos size(mm) Growth factors number		2-4D : 10 µ M	N.A.A. : 8 µ M Zeatin : 7 µ M 2-4D : 2 µ M	N.A.A. : 5 µ M B.A.P. : 6 µ M 2-4D : 2 µ M
0.3 to 0.7	Embryogenic Total Embryog/total	1 4 25%	12 20 60%	3 15 20%
0.8 to 2	Embryogenic Total Embryog/total	2 8 25%	4 11 36%	3 11 27%
3 to 10	Embryogenic Total Embryog/total	3 5 60%	5 7 71%	2 5 40%
Total number Total Embryogenic		17 35%	38 55%	31 25%

C - Somatic embryos from zygotic embryos

    The zygotic embryos were highly heterogenous, there were no relations
between the external size of the ovary and the size of the embryo inside.
For this reason, it was not possible to excise a similar number of embryos
for each class of embryo size (Table 1). Concerning the cytokinin added
we found that Zeatin was more efficient than B.A.P.. This is especially
important with very immature embryos (0.3-0.7mm). We also found that
embryos from 0.3 to 0.7mm developed much better if they were cultured in
their embryo sac. For each hormonal composition, a better result was
obtained with bigger embryos (3-10mm), but, on the contrary to WANG and
JANICK, we did not noticed a better embryogenesis with the intermediate
size embryo (2-4mm). The biggest embryos (7-10mm) gave a very good
embryogenesis rate. Even more, when these embryos were cutted transversally

every 2-3mm, it stimulates their reactivity and a higher embryogenesis level was obtained up to 80%. It was observed that this embryogenesis occured only at the surface of the cotyledon (Picture 4) of the embryo. By means of histological analysis, some meristematic zones were evidenced inside the cotyledon tissue. They seemed to be embryogenic but their development was stopped by high cotyledonary cell density (Picture 5). On the contrary, the upper part of the cotyledons showed a rapid cell proliferation which can explain the high level of embryogenic structures (Picture 6).

D - Development studies of somatic embryos

In order to grow somatic embryos, the basal medium was used with low cytokinin concentrations (B.A.P. or Zeatin 1 µ M). Somatic embryos have a different way to develop compared to zygotic embryos. They stay on globular or heart shape stage till a large size (Picture 7). Such a latency in embryo development could be a consequence of an inadequation of the nutrient medium (MONNIER) (7). However, when these embryos were transfered on the expression medium, they often produced more than two cotyledons (Picture 8).

CONCLUSION

The regeneration capacities are mainly correlated with the origin and stage of maturity of the cultured explants. Young ovaries formed organogenic calli and immature embryos produced somatic embryogenesis. These tissues are characterized by an high rate of growth.

REFERENCES

1 - ROST T.L. and M.A.W. HINCHEE, 1982. Preliminary report of the production of callus, organogenesis and regeneration of Jojoba in tissue culture. Journal of Horticultural Science, 55 (3) : 299-305
2 - WANG Y.C. and J. JANICK, 1986. Somatic embryogenesis in Jojoba. J. Amer. Hort. Sci., 111 (2) : 281-287.
3 - LEE C. W. and J. C. THOMAS, 1985. Jojoba embryo culture and oil production. Hort. Science, 20 (4) : 762-764.
4 - MURASHIGE T. and SKOOG, 1962. A revised medium for rapid growth and bioassays with Tobacco tissue cultures. Physiol. Plant, 15 : 473-497.
5 - GAMBORG O.L., R.A. MILLER and K. OJUMA, 1968. Nutrient requirements of suspension cultures of soybean root cells. Exp. Cell Res.,50 : 151-158.
6 - CHU C.C. and R.D. HILL, 1988. An improved anther culture method for obtaining higher frequency of pollen embryoids in triticum aestivum L. Plant Science, 55 : 175-181.
7 - MONNIER M., 1988. Embryogenese zygotique et somatique. In : Culture de cellules, tissus et organes vegetaux. J.P. ZRYD ed.,Presses polytechniques romandes, Lausanne.

EFFECT OF COLD TEMPERATURE ON SHOOT REGENERATION IN VITRO FROM

AGED CULTURES OF GF-677 (*PRUNUS PERSICA* x *PRUNUS AMYGDALUS*)

K. Dimasi-Theriou and A. Economou

Department of Horticulture
Aristotle University
GR-54006 Thessaloniki, Greece

ABSTRACT

In vitro cultures of GF-677, with arrested growth after a number of subcultures, were placed at 3°C for 4 weeks and then at 22°C. After 4 weeks an average number of 17.6 new shoots per culture were developed, whereas no shoots were formed in cultures which had not received the 3°C temperature treatment. Similar numbers of new shoots were achieved when the cold treatment of 3°C was applied for 3, 4 or 5 weeks, while 1 or 2 weeks in 3°C had no effect on shoot formation. When the cultures of GF-677 with the arrested growth received temperature treatment of 15, 11, 7 or 3°C, a linear response of shoot formation was found, with the minimum number of shoots at 15°C and the maximum one at 3°C.

INTRODUCTION

In vitro cultures of the fruit tree rootstock GF-677 (*Prunus persica* x *Prunus amygdalus*) after a number of sub-cultures show signs of aging, leading to a decline in new shoot formation. Such cultures maintained under constant long photoperiod and relatively high temperature cease to produce new shoots undergoing a period of quiescence. Decline in shoot formation in vitro was reported also in magnolia (De Proft et al., 1985) and azalea cultures (Economou and Read, 1986) and in species of Rosaceae (Norton and Norton, 1986) after continuous subculturing. Growth restoration in vitro is important for preserving species or building up stock plant material for propagation purposes. In this work application of low temperature to cultures of GF-677 with arrested growth is examined for its effect on restoration of new shoot formation.

MATERIALS AND METHODS

Culture establishment

Shoot tip explants of GF-677 after 4 weeks in culture

produce a cluster of new axillary shoots which are harvested and used for rooting. The remaining original explant tissues are subcultured undivided on fresh medium for new axillary shoot production. This procedure is repeated until the cultures of GF-677 cease to produce new axillary shoots. Such cultures were used for the experiments. They were grown for 16 h daily under cool-white inflorescent light of approximately 40 $\mu mol \cdot s^{-1} \cdot m^{-2}$ (400-700 nm) at a constant temperature of 22°C.

The nutrient medium used was the Woody Plant Medium (Lloyd and McCown, 1980) with the addition of 30 g/l sucrose and 6 g/l Oxoid agar (No.1) plus 5 µM benzyladenine (BA), 0.05 µM indoleacetic acid (IAA) and 0.03 µM gibberellic acid (GA_3). The medium pH was adjusted to 5.8 before adding the agar. Aliquots of 20 ml of medium were dispensed into 55x70 mm glass jars which were sealed with aluminum foil and sterilized by autoclaving at 121°C for 18 min.

Effect of cold temperature

The cultures with the arrested growth were kept for 4 weeks in the growth chamber at 3°C and then the temperature was switched to 22°C. After 4 weeks at this temperature the new shoots formed were evaluated. In the control the cultures were maintained at 22°C continuously for 8 weeks. Fifty cultures were used per treatment and the experiment was repeated twice.

Effect of the duration of cold temperature

The cultures of GF-677 received cold temperature treatment of 3°C for 1, 2, 3, 4 or 5 weeks and then the temperature was raised to 22°C. After 4 weeks at this regime the new shoots formed were counted. The control cultures, which did not receive the cold temperature treatment, were kept at 22°C for the equivalent period of time. Thirty cultures were ascribed per treatment in 2 replications.

Effect of various low temperatures

Cultures of GF-677 were placed at 3, 7, 11 or 15°C for 4 weeks and then were moved to 22°C where they remained for 4 weeks. At the end of the 4-week culture period at 22°C the new shoots formed were scored. There were 30 cultures per treatment in 2 replications.

RESULTS

Effect of cold temperature

An average number of 17.6 new shoots with a mean length of 1.9 cm were produced in cultures of GF-677 which had received the temperature treatment of 3°C. No shoots were formed in cultures which had not received the 3°C temperature treatment but they remained at 22°C (Fig. 1 and 2).

Effect of the duration of cold temperature

Similar numbers of new shoots were achieved when the

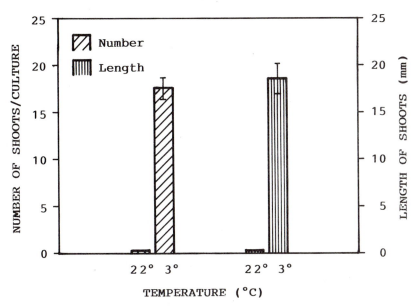

Fig. 1. After-effect of cold temperature on new shoot formation in vitro from GF-677 cultures

Fig. 2. New shoot formation in GF-677 cultures in vitro. Left: Culture not received the 3°C treatment; Right: Culture which received the 3°C treatment

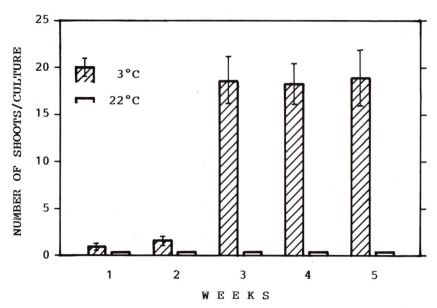

Fig. 3. After-effect of the cold temperature duration on new shoot formation in vitro
from GF-677 cultures

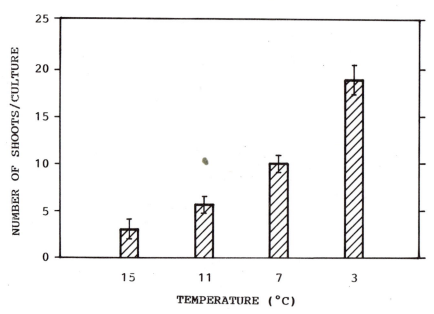

Fig. 4. After-effect of various low temperature
treatments on new shoot formation in
vitro from GF-677 cultures

cold treatment of 3°C was applied for 3, 4 or 5 weeks, while 1 or 2 weeks of 3°C had no effect on shoot formation (Fig. 3). Again cultures which had maintained at 22°C for 1, 2, 3, 4 or 5 weeks did not produce new shoots.

Effect of various low temperatures

When the GF-677 cultures received temperature treatment of 15, 11, 7 or 3°C a linear response of new shoot formation was found, with the minimum number of shoots at 15°C and the maximum one at 3°C (Fig. 4).

DISCUSSION

Constant long photoperiod and relatively high temperature for long periods of time cause a reduction in new shoot formation in the subsequent subcultures of the original explant of GF-677 cultures and finally a kind of dormancy. Such cultures regained their capacity for new shoot formation after receiving a 3°C temperature treatment for at least 3 weeks. This after-effect of the cold temperature treatment was revealed when GF-677 cultures were moved back to normal temperature of 22°C for 4 weeks. The cold temperature may break the dormancy of the existent axillary buds in GF-677 cultures and consequently new shoot formation occurs. The needs for a cold temperature of 3°C were saturated with 3 to 5 weeks of cold temperature treatment or otherwise with 500 h to 840 h which are within the limits of the cold temperature requirements for releasing almond and peach (parents of the hybrid GF-677) buds from dormancy in nature (Porlingis, 1965; Erez and Lavee, 1971). From the various low temperatures tested that one of 3°C enhanced new shoot formation effectively while those of 7, 11 and 15°C had a gradually reducing effect. Restoration of shoot proliferation in cultures of GF-677 is important for the micropropagation of this fruit tree rootstock in commercial tissue culture laboratories.

REFERENCES

De Proft, M.P., Maene, L.S., and Deberg, P.C., 1985, Carbon dioxide and ethylene evolution in the culture atmosphere of *Magnolia* cultured in vitro, Physiol. Plant., 65:375-379.
Economou, A.S., and Read, P.E., 1986, Microcutting production from sequential reculturing of hardy deciduous azalea shoot tips, HortScience, 21:137-139.
Erez, A., and Lavee, S., 1971, The effect of climatic conditions on dormancy development of peach buds. I. Temperature, J. Amer. Soc. Hort. Sci., 96:711-714.
Lloyd, G.B., and McCown, B.H., 1980, Commercially-feasible micropropagation of mountain laurel, *Kalmia latifolia*, by use of shoot-tip culture, Proc. Intern. Plant Prop. Soc., 30:421-437.
Norton, M.E., and Norton, C.R., 1986, Change in shoot proliferation with repeated in vitro subculture of shoots of woody species of Rosaceae, Plant Tissue Organ Cult., 5:187-197.
Porlingis, I., 1965, The dormancy of deciduous fruit trees, Geoponika, 130-131:130-140.

AGE AND MERISTEM "IN VITRO" CULTURE BEHAVIOUR IN FILBERT

Fernández, Mª.T.; Rey, M.; Díaz-Sala, C. and Rodríguez, R.

Cátedra de Fisiología Vegetal. Fac. de
Biología. Universidad de Oviedo. Spain

INTRODUCTION

One of the most important problems for vegetative propagation of
tree species, is associated with slow growth, also with changes between
the juvenile and adult state, and the gradual loss of morphogenetic
capacity (Libby, 1974).

Reversion of adult to juvenile material, resulting in an increase
in morphogenic capacity, is very important to scientific research and to
commercial applications, which have grown quickly in recent years.

In order to overcome survival problems during micrografting
experiments, optimal conditions for filbert meristem culture were
studied. Futhermore, the survival, growth and morphogenic responses of
meristems taken from adult, juvenile, "in vitro" plants and seedlings
were taken into account; if we accept maturation of meristem
(Schaffalitzky de Muckadell, 1959), propagules from meristems taken from
adult material are physiologically older than those taken from juvenile
plants (Bonga, 1981).

MATERIALS AND METHODS

Bud meristems from shoots of juvenile, adult, "in vitro" plants
and seedlings, were used in this work. Explants were prepared under cool
light in a stereomicroscope. They consisted of an apical dome and the
two first leaf primordia. Cultures were maintained in a growth chamber
at $25 \pm 2^{\circ}C$ under a photoperiod of 16 h. provided by a light of 2800
lux. Meristems were cultured on MS medium (Murashige and Skoog,
1962).The following growth-regulator combinations were tested (in mg/l):

```
A : 1 BAP; 0.05 IAA; 0.1 GA3
B : 0.5 BAP; 0.05 IAA; 0.1 GA3; 2 Ascorbic acid
C : 1 BAP; 0.1 IBA; 1 GA3
D : 1 BAP; 0.1 IBA; 0.5 GA3
E : 1 BAP; 0.05 IAA
```

We tested solid and liquid media, as well as modified MS medium
(Vieitez et al, 1985).Also (in mg/l): 100 Myo-inositol, 1 Nicotinic

acid, 1 pyridoxine-HCL and 2 Ascorbic acid, were added as organic supplements.

To minimize a high percentage of oxidation, very common in this material, after its sterilization, explants were rinsed with an anti-oxidant agent (DIECA, 2 g/l) which was sterilized by filtration (Martínez et al, 1979).

RESULTS AND DISCUSSION

We have observed a high percentage of survival and an optimum establishment of meristems, without any further development.

The pattern of development was usually the same in all materials, and growth of the second leaf primordium was only observed.

Explant survival seems to be related to BAP concentration and to the age of the material, most of all in meristems taken from embryonic material, where the age of seedlings determines the success of the culture, probably due to the coincidence with the active growth phase of this material.

Futhermore, high concentrations of gibberellic acid can inhibit the development of meristems. However, it is necessary for the development of meristems and its absence lowers the survival rate of meristems in liquid medium (Table nº1). When the solid medium was used, survival of explants decreased dramatically, dying a few days after establishment, this effect was probably due to the physical composition of the medium.

Development of filbert meristems taken from seedling and juvenile plants was improved in the presence of high BAP concentrations. Meristems from adult plants did not grow in these media, undergoing no further changes after culture establishment (Fig nº 1a).

Table nº 1. Effect of the growth-regulators on meristem survival. Meristems were taken from seedling, juvenile, in Vitro and adult plants. Results were quantified after 10 days of culture. (* = Died 10 days later).

PLANT AGE	GRW-RG	SURVIVAL (%)	MERIST.REACTIVITY (%)
seedling		70	56
juvenile	A	100	100
adult		0	0
seedling		62	54
juvenile		50	25
in vitro	B	90	90
adult		73	54
seedling		60	50
juvenile	C	--	--
adult		--	--
seedling		87	62
juvenile	D	71	71
adult		22	0
seedling		--	--
juvenile	E	75*	0
adult		10*	10

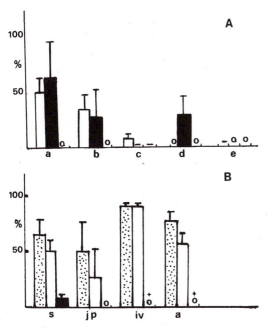

Fig. 1.

A) Effect of media composition on meristem development. Explants were taken from seedlings, ◯ juvenile ● and adult plants ⊙.
Results were quantified after 10 days of culture.

B) Relative percentage of survival, ⊙ reactivity ◯ and adventitious morphogenesis ●
of meristems taken from seedlings (s), juvenile (jp), in vitro (i.v.), and adult (a) plants. Results were quantified after a culture period of 20 days in medium B.
(*) only callus induction was observed.

Adventitious morphogenesis was observed in the culture of seedling meristems in the medium B (Fig nº 1b). A progressive growth was observed only in apical meristems; also a big leaf (Fig nº 2a) developed which underwent changes in its structure. Callus was observed in the apical and basal ends of this leaf. Later cultural stages revealed the induction and development of whitish embryo-like structures (Fig nº 2b) with only shoot growth.

These observations allowed us to establish the different morphogenetic potentialities of explants from filbert meristems, "in vitro" and adult plants in the same conditions.

Explants from adult plants underwent callus induction some months later, without any further adventitious responses (Fig nº3). Meristems taken from "in vitro" plants gave rise to multiple buds and basal callus formation (Fig nº 4). Cluster elongation did not occur in this medium. Long-term observation of the culture failed to show any further morphogenic manifestations. In juvenile explants, only "shoot" growth was observed. Adventitious responses or callus induction were not detected during the culture period.

The combined effect of ascorbic acid and DIECA, not only resolved the problems of oxidation, but also ascorbic acid could be responsible for organogenic responses (Joy, 1988; Murashige, 1961; Torphe, 1980; Torphe et al, 1986).

Ascorbic acid (AA) in plants, increases metabolic activity due to faster synthetic processes, amylase activation, rapid breakdown of reserves and accelerated release of sugars. These have all been shown in other systems in which ascorbate has been experimentally applied (Joy, 1988). AA, has been shown to eliminate the toxic effects of various hormones including GA3 in xylem formation in Pinus banksiana (Berlyn et al, 1980), and it also prevented browning and subsequent necrosis during secondary shoot formation in Picea mariana (Rumary et al, 1984), although the precise mechanism of AA is presently unknown.

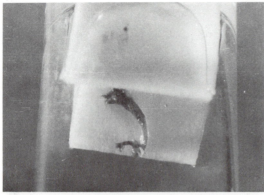

A

Fig. 2. Induction of adventitious morphogenesis in the culture of seedling meristems in the medium B.
A) Big leaf
B) Callogenic structures in the apical and basal ends of the leaf

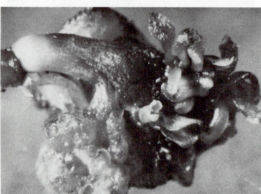

B

Fig. 3. Callus induction in explants taken from adult plants.

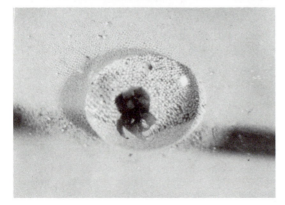

Fig. 4. Culture of juvenile meristems. Multiple buds and basal callus.

One possible mechanism could be through ascorbate protection of endogenous phytohormones produced by interactions of polyphenol oxidases, ascorbic acid oxidase and IAA-oxidase (Rao et al, 1982) .

We must not forget the genetic and epigenetic factors, which together with physiological and metabolic factors are responsible for the intrinsic cellular capacity to influence its own specialised activity.

The behaviour of meristems taken form adult plants is different to the behaviour of meristems taken from other materials, with a lower survival rate than juvenile material, in the same cultural conditions.

The age of the meristems could have influenced the organogenic responses, since it is easier to change genetic expression and to obtain newly-formed organs that did not exist in the plant before.

ACHNOWLEDGEMENTS

This work has been supported by CICYT (grant nº 860/84)

LITERATURE

Berlyn, G.P.; Beck, R.C., 1980, "Tissue culture as a technique for studing meristematics activity". In: "Control of shoot growth in trees".Proceedings of the joint workshop of IUFRO working parties in xylem and shoot growth physiology. Fredericton. New Brunswick. pp 305-324.

Bonga, J.M., 1981, "Organogenesis in vitro of tissues from mature conifers". In Vitro 17:511-518

Joy, R.W.; Patel, R.P. and Thorpe, T.A., 1988, "Ascorbic acid enhacement of organogenesis in tobacco callus".Plant Cell Tissue and Organ Culture 13:219-228

Libby, W.J., 1974, "The use of vegetative propagules in forest genetics and tree improvement".N. Z. J. For. Sci. 4:440-453

Martínez, J.; Hugard, J. and Jonard, R., 1979, "Sur les différentes combinasions de greffages des apex réalisés in vitro entre Pecher, Abricotier et Myrobolan". C. R.Acad. Sc. Paris 288:759-762

Murashige, T., 1961, "Supression of shoot formation in cultured tobacco cells by gibberellic acid". Science 134:280

Murashige, T.and Skoog, F., 1962, "A revised medium for ratid growth and bioassays with tobacco tissues culture". Physiol. Plant. 15:473-497

Rao, N.R.; Jasdanwala, R.T.; Shing, Y.D., 1982, "Changes in phenolic substances and ascorbic acid turnover during early stages of fibre differentiation". Beitr.Biol.Pflanzen. 57: 359-368.

Rumary, C.; Thorpe, T.A., 1984, "Plntlet formation in black and white spruce .I.In vitro techniques". Can.J.For.Res. 14: 10-16

Schaffalitzky de Muckadell, M., 1959, "Investigations on ageing of apical meristems in woody plants and its importance in silviculture". Forstl. Forspg. Dan. 25:310-455

Thorpe, T.A., 1980, "Organogenesis in vitro: Structural, physiological, and biochemical aspects".Int. Rev. Ctol. Suppl. 11A:71-111

Vieitez, A.M., San Jose, C. and Vieitez, E., 1985, "In vitro plantlet regeneration form juvenile and mature Quercus robur L." J.Hort.Sci. 60(1):99-106.

INFLUENCE OF EXPLANT SOURCE ON IN VITRO AXILLARY SHOOT

FORMATION IN OAK (Quercus robur L.) SEEDLINGS

H. Volkaert

Laboratory of Forestry
Catholic University of Leuven
Kardinaal Mercierlaan 92, B-3030 Heverlee, Belgium

INTRODUCTION

In this study, the possibility of propagating juvenile oak seedlings in vitro was investigated with special emphasis on the influence of shoot development and explant position on propagation efficiency.

MATERIALS AND METHODS

Culture initiation

Seedlings were raised in a hydroponic culture system under a controlled environment with a 12-h photoperiod, 5000 lux irradiance and 30/20°C (day/night) temperature cycle. Whole shoots were then trimmed into 15-mm-long apical and nodal segments. Axillary shoot formation was induced on a nutrient medium containing WPM macro-elements (Lloyd and McCown, 1980) and Murashige and Skoog (1962) micro-elements and vitamins with 100 mg/l inositol, 22.5 or 45 g/l sucrose, 2 g/l activated charcoal, and 5.6 g/l agar. The pH was adjusted to 5.8 $\pm$ 0.1 before autoclaving. The culture environment was maintained at 28 $\pm$ 2°C, and illumination of 4500 lux was provided continuously.

Explant source and explantation timing

Shoot growth in oaks is episodic and predetermined (Hanson et al., 1986). In order to study the influence of the development of the source shoot, five different stages were distinguished during flush growth:
  (1) the fifth leaf is 10 mm long (rapid shoot elongation);
  (2) the last leaf of the growth flush is 10 mm long (decrease of the shoot elongation rate, start of leaf expansion);
  (3) the last leaf of the growing flush is 40 mm long and is still light green (shoot elongation has stopped while the leaves are expanding);

(4) all leaves are fully expanded and dark green (rest period);

(5) the apical bud starts to elongate (3-5 mm), start of the next flush.

Shoots were collected during each of these developmental stages (DS).

Because every shoot has a different architecture, four reference positions were selected as explant source:

(1) the shoot tip with 3-4 leaves and axillary buds;

(2) a nodal segment with one normal leaf and axillary bud;

(3) the second nodal segment without a leaf, but only two stipules; and

(4) the basal segment which has 2-3 axillary buds but no leaves.

Design and analysis of the experiment

For both the first and second growth flush and for each of the five developmental stages, four shoots with four explants per shoot were cultured on four different nutrient media in a latin square configuration. The four media were a factorial combination of two macro-element levels (WPM at 75% strength + 22.5 g/1 sucrose or WPM at 150% strength + 45 g/1 sucrose) and two cytokinin concentrations (2.10^{-3} or 10^{-2} mM benzyladenine). There were five replications. The number of days an axillary bud needed to develop into a shoot large enough to be rooted, i.e. 20-25 mm length and at least one expanded leaf, was recorded as the dependent variable.

The analysis of variance was performed using a general linear model. The Ryan-Einot-Gabriel-Welsh multiple range test was used to separate differences among means at the 5% significance level. A chi-square test was used to search for significant differences among frequencies. The deviation from the expected frequency was assessed using adjusted residuals.

RESULTS AND DISCUSSION

In most explants a morphogenic response soon became evident. Some explants (20%) formed a 20-25 mm long shoot with 3-4 leaves within 12-25 days and could be rooted successfully at that time. During the same period, other explants (55%) formed a very short shoot (0.5-5 mm) with only one leaf. These explants were maintained on the nutrient medium until they formed a second growth flush in vitro. This occurred after 40-60 days or even later. Overall, 596 out of 800 explants formed a shoot that could be rooted. Eighty four explants were contaminated with fungi and had to be discarded. The remaining 15% of the explants did not yield a rootable shoot.

The developmental stage of the shoot, the position of the explant within the shoot and the nutrient level significantly influenced the time an axillary bud required to develop into a rootable shoot.

The age of the seedling (first or second growth flush) and the cytokinin concentration had no significant effect on axillary shoot formation.

Table 1 shows that the growth of axillary buds was slowest during the period of rapid elongation of the source shoot (DS 1 and 2) but increased with further development of the flush and was most rapid when elongation had stopped and the leaves were fully expanded (DS 4 and 5). A similar pattern of in vitro development has been found in Halesia carolina L. (Brand and Lineberger, 1986). In Halesia, the period of most rapid shoot elongation corresponded closely to the period of little or no proliferation potential of nodal shoot segments.

Table 1 also shows that axillary shoot formation was significantly slower for apical than for nodal segments. A basipetal increase of morphogenic response was observed, which is pronounced during early developmental stages, as shown in fig 1. Because elongation of the basal internodes slows down and stops before the elongation of the more apical internodes (Hanson et al., 1986), it may be better to correlate the development of axillary shoots with the rate of elongation of the corresponding internode. Vieitez et al. (1985) reported that mid-stem nodal explants from 3- to 4-month-old seedlings failed to develop on different media, while only the terminal buds could be established in vitro. Using in vitro subcultured shoots, San José (1986) showed that middle and basal segments elongated faster and multiplied at a higher rate than apical segments. This may suggest that the shoots were still in an early developmental stage.

A high level of nutrient salts and sucrose in the medium significantly enhanced the growth of axillary shoots. There was no clear interaction between nutrient level and developmental stage or position.

Table 1. Axillary shoot formation of explants on media containing activated charcoal. Different letters indicate significant differences.

factor and level		time to reach 25 mm length (1) days	frequency of outgrowth (2) %	rate of development (2),(3) days^{-1}
position	apical	69.91 a	67 a	1.430 a
	subapical	49.43 b	88 b	2.599 b
	leafless	45.86 b	90 b	2.923 bc
	basal	38.01 c	84 b	3.081 c
developmental stage	1	64.99 a	75 a	1.708 a
	2	58.61 ab	82 a	1.797 a
	3	48.24 bc	80 a	2.433 ab
	4	42.75 c	85 a	2.976 bc
	5	31.85 d	92 b	3.681 c
nutrient	high	44.07 a	84 a	2.802 a
	low	53.82 b	82 a	2.195 b
flush	first	49.49 a	88 a	2.734 a
	second	48.09 a	78 b	2.279 a

(1) only explants that produced a shoot
(2) only non-infected explants
(3) explants that did not produce a shoot were rated = 0

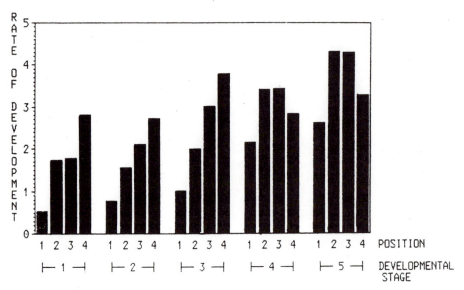

Figure 1. Effect of the developmental stage of the source plant on the rate of development of an axillary bud into a shoot. The influence of the position of the explant within the shoot is shown for each of the developmental stages

REFERENCES

Brand, M. H. and Lineberger, R. D., 1986, In vitro Propagation of Halesia carolina L. and the Influence of Explantation Timing on Initial Shoot Proliferation. Plant Cell Tissue Organ Cult., 7:103-113

Hanson, P. J., Dickson, R. E., Isebrands, J. G., Crow, T. R. and Dixon, R. K., 1986, A Morphological Index of Quercus Seedling Ontogeny for Use in Studies of Physiology and Growth. Tree Physiol., 2:273-281

Lloyd, G. and McCown B., 1980, Commercially Feasible Micropropagation of Mountain Laurel, Kalmia latifolia, by Use of Shoot-Tip Culture. Comb. Proc. Intl. Plant Prop. Soc., 30:421-427

Murashige, T. and Skoog, F., 1962, A Revised Medium for Rapid Growth and Assays with Tobacco Tissue Cultures. Physiol. Plant., 15:473-497

San José, M. C., 1986, Influencia de la situaciòn del explanto en la planta y del tamaño del tubo de cultivo en la multiplicaciòn in vitro de Quercus robur L. Phyton, 46:33-38

Vieitez, A. M., San José, M. C. and Vieitez, E., 1985, In vitro Plantlet Regeneration from Juvenile and Mature Quercus robur L. J. Hort. Sci., 60:99-105

SEGMENTATION EFFECT OF IMMATURE SPIKE ON TRITICALE CALLI

INDUCTION[*]

H. Guedes-Pinto, O. Pinto-Carnide and F. Leal

Genetics and Plant Breeding Division
University of Trás-os-Montes e Alto Douro
Apt. 202, 5001 Vila Real Codex, Portugal

INTRODUCTION

Calli induction is a stepwise process to obtain regenerated plants via somatic cells, cell suspension, genetic transformation, mutagenesis, etc. (see review of Maddock, 1985; Vasil, 1987 and Conger et al.,1988). Immature spikes proved to be a fine explant to the induction of calli in triticale (X *Triticosecale*) and other cereals (Ozias-Akins and Vasil, 1982; Rangan and Vasil, 1983; Eapen and Rao, 1985; Fedak, 1985; Sun and Zhu, 1987).

In previous studies besides differences on the number of calli per spike and calli diameter due to the triticale genotypes (Guedes-Pinto and Pinto-Carnide, 1987; Guedes-Pinto et al., 1989), it was observed in all the studied triticales that the calli induction was affected by the position on the spike, showing a higher value of calli number at the base of the immature spike which is associated to site where the spike was cut (Guedes-Pinto et al., 1989).

In the present study, the main purpose of the study was to search for the effect of the immature spike segmentation on calli induction.

MATERIALS AND METHODS

Immature spikes, with leaves which surrounded them, were surface sterilized with ethanol (70%).

Eighteen segmented and twenty-seven non-segmented (whole) immature spikes of 6x-triticale, cv. "Clercal", were cultured in M+S medium with 1.0 mg/l of 2.4 D at 25°C and 16 hours of light/day. The segmented spikes were cut in three equal size fragments (basal, middle and extreme end). In each fragment, the calli position was also considered (base, middle and upper site).

Spike lengths ranged from 10 to 25 mm, with averages of 18.94 mm for segmented and of 16.48 mm for non-segmented spikes, however the average spike lengths were not statistically different.

The number and diameter of calli according to their position on the spike, the position on the spike fragment, percentage of spikes with calli and number of calli with roots and shoots were registered during the 8 weeks period. All the calli observations were made with stereoscopic microscope.

―――――――――
[*]Granted by Junta Nacional de Investigação Científica e Tecnológica Proj. No. 87 394.

RESULTS

At the 2nd week, 92.86% of segmented spikes showed <u>calli</u> while non--segmented spikes at that very time only reached 41.66%. At the 4th week, all of the segmented spikes presented a <u>calli</u> induction. At the non-segmented spikes this condition was only observed at the 6th week (Fig. 1).

Considering the total <u>calli</u> number per spike at the 8th week on <u>in vitro</u> culture a negative correlation, statistically significant, between length of the immature spike and the total <u>calli</u> number per spike was found both for segmented (P<0.01) and non-segmented spikes (P<0.01) (Fig. 2).

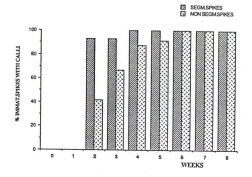

Fig.1- Percentage of segmented and non-segmented immature spikes with <u>calli</u> according to time.

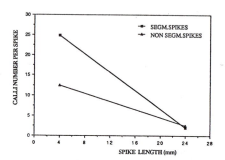

Fig.2- Regression between immature spike length and total <u>calli</u> number per segmented and non-segmented spikes.

It was also observed a higher <u>calli</u> induction in the segmented spikes during the first weeks. In the 2nd week an average of 3.11 <u>calli</u> per segmented spike and 1.00 <u>calli</u> per non-segmented spike were registered. However, as time goes by those differences become smaller and at the 8th week the average of <u>calli</u> number was 6.17 for the segmented and 5.74 for non--segmented spikes (Fig. 3). In many cases it was observed fusion of <u>calli</u> originating a larger one. The effect of time (weeks) and type of spikes was statistically significant (P<0.001).

Also, the average of the <u>calli</u> diameter was always higher and statistically significant (P<0.001) in the segmented spikes than in the whole spikes. But oppositely to what happen with <u>calli</u> number, the differences between <u>calli</u> diameter average of segmented and non-segmented spikes tend to be greater with time (Fig. 4).

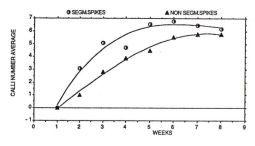

Fig.3- <u>Calli</u> number average per segmented and non-segmented immature spikes according to time.

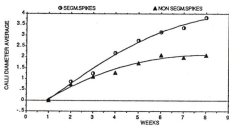

Fig.4- <u>Calli</u> diameter average per segmented and non-segmented immature spikes according to time.

Regarding calli position in the spike it was observed higher average value of calli number at the base and the extreme end of the segmented spikes than in the non-segmented ones. In the middle segment of the spikes, non-segmented spikes showed higher value than segmented ones and after the 6th week the calli number in the middle of the non-segmented spike was the highest value observed in this type of spike (Fig.5a and 5b). Effect of the spike type, calli position on the spike and time were statistically significant as well as the interaction type of spike x position (P<0.0001).

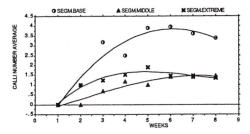 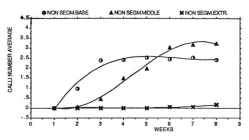

Fig.5a- Calli number average in the segmented immature spikes according to the position in the spike and time.

Fig.5b- Calli number average in the non-segmented immature spikes according to the position in the spike and time.

The average of calli diameter at the base of the spike was bigger than that at the middle or extreme end of the spike, either in segmented or non-segmented spikes.

In segmented spikes, calli were registered according to their position in the fragment (base, middle and upper of the fragment). In all of the 3 fragments the highest calli number were observed at the base with values 3 to 6 times higher than those which occurred at the middle or upper site of the fragment. These differences were statistically significant (P<0.01) (Fig. 6).

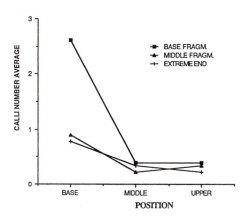

Fig.6- Calli number in the spike fragments (base, middle and extreme end) according to their position in the fragment (base, middle and upper).

Calli diameter showed in fragmented spikes the highest average values at the base fragment and the lowest values in the middle fragment.

Inside the three fragments the higer calli diameter were seen at the base of the base and middle fragments, but not at the extreme end segment on which the highest value occurred in upper site. However, this average value concerns only the diameter of three calli.

Concerning shoots, segmented spikes showed an average of 0.29 shoots per spike and non-segmented spikes an average of 0.16 at the 8th week of in vitro culture (N.S.).

Oppositely, the number of calli with roots registered higher average value in non-segmented spikes, with 0.72, than in segmented spikes with 0.20 (P<0.001).

DISCUSSION AND CONCLUSIONS

Length of the immature spike, which is generally associated with its age within the same genotype cultivated under the same condition, plays a major role in the response of in vitro culture. The younger and smaller the spikes are the higher their calli induction is. The effect is much more stronger in segmented than in non-segmented spikes.

Segmentation of the immature spikes seems to induce higher calli number in the earlier weeks of in vitro culture. This is probably due to a better access of the segmented explants to the nutrients and growth regulators of the medium culture, namely the procambial/vascular tissues near which the first cell divisions very often start (Vasil, 1987). On the other hand, wounding cells by segmentation may play itself a major role on auxin transport (see van der Linde, in this Proceedings). Other methods involving wounding cells like micro-cross sections of the leaf midvein in hybrid poplars (Ostry et al., in this Proceedings) and thin cell layer (TCL) (Tran Than Van et al., in this Proceedings) give also good in vitro culture response. In association with the higher calli number, bigger calli in segmented spikes were observed due to an earlier calli growth and/or fusion.

Although after the 6th week of in vitro culture of the non-segmented spike the highest value of calli number was registered at the middle of the spike, which is not in accordance with our previous data (Guedes--Pinto et al., 1989), considering the two types of spikes it seems that position on the spike has a major role in calli formation, showing a gradient of calli induction from the base to the extreme end of the spike. So, the highest values were seen at the base and middle segments, showing a polarity effect. The same position effect occurred in the fragment of the segmented spikes.

Position effect on calli induction was also observed in leaf fragment of other gramineae species, showing an induction gradient from the base to the extreme of the leaf (Saalbach and Koblitz, 1978; Conger et al., 1983; Vasil and Vasil, 1984).

According to Vasil (1987) it is likely that the endogenous pools of plant growth regulators, such as IAA and ABA, are casually related to this gradient observed in young leaves and inflorescences of grass species.

In conclusion, youngest immature spikes segmentation is an appropriate way to increase earlier and bigger calli induction on triticale and probably in other cereals inflorescences.

REFERENCES

Conger, B.W., Trigiano, R.N. and Gray, D.J., 1988. Cell Culture of the Poaceae (Gramineae). In: Plant Cell Biotechnology. Ed. M. Pais, F. Mavituna & J.M. Novais. NATO ASI Series, Cell Biology (18): 49-61.

Conger, B.W., Hanning, G.E., Gray, D.J. and McDaniel, K., 1983. Direct embryogenesis from mesophyll cells of orchardgrass. Science, 221 : 850-851

Eapen, S. and Rao, P.S., 1985. Plant regeneration from immature inflorescence callus culture of wheat, rye and triticale. Euphytica, 34 : 153-159.

Fedak, G.,1985. Propagation of intergeneric hybrids of Triticeae through callus culture of immature inflorescences. Z. Pflanzenzuchtung, 94 (1) : 1-7.

Guedes-Pinto, H. and Pinto-Carnide, O., 1987. Plantlets regeneration from in vitro culture of 6x-triticale immature spikes. Ciência Biológica (Supl.), 12(5A) : 208.

Guedes-Pinto, H., Pinto-Carnide, O. and Alpoim, F., 1989. Calli induction from Triticale Immature Spikes. Vorträge für Pflanzenzüchtung, Heft 15-1 : 7-14.

Maddock, S.E., 1985. Cell culture, somatic embryogenesis and plant regeneration in wheat, barley, oats, rye and triticale. In Cereal Tissue and Cell Culture, Ed. S. W.J. Bright and M. G.K. Jones, Martinus Nighoff / Dr. W. Junk Publ., Dordrecht: 131-174.

Ozias-Akins, P. and Vasil, J.K., 1982. Plant regeneration from cultured immature embryos and inflorescences of Triticum aestivum L. (Wheat), evidence for somatic embryogenesis. Protoplasma, 110: 95-105.

Rangan, T.S. and Vasil, I.K., 1983. Somatic Embryogenesis and Plant Regeneration in Tissue Culture of Panicum miliaceum L. and Panicum miliare Lamk.. Z.Pflanzenphysiol, 109 : 49-53.

Saalbach, G. and Koblitz, H.,1978. Attempts to initiate callus formation from barley leaves. Plant Science Letters, 13:165-169.

Sun, J.S. and Zhu, Z.Q.,1987. Culture of explants from immature inflorescence of octoploid triticale (Abs.). Wheat, Barley and Triticale Abstracts, Feb., 4(1) : 89.

Vasil, I.K., 1988. Developing Cell and Tissue Culture Systems for the Improvement of Cereal and Grass Crops. J. Plant Physiol.,128: 193-218.

Vasil, V. and Vasil, J.K., 1984. Induction and maintenance of embryogenetic callus cultures of Gramineae. In Cell Culture and Somatic Cell Genetics of Plants, Vol. 1, Ed. J.K.Vasil, Academic Press, Inc., Orlando : 36-42.

CUTICLE DEVELOPMENT IN Dianthus caryophyllus PLANTLETS

M.A. Fal, P. Bernad (*), R. Obeso (*) and
R. Sánchez Tamés

Lab. Fisiología Vegetal, Dpt. B.O.S.,
Univ. Oviedo. 33005, Oviedo, Spain.
(*) Dpt. Química Organometálica,
Univ. Oviedo. Spain.

ABSTRACT

Cuticle development on "in vitro" developed leaves is important in order to prevent excessive water loss during acclimatization of "in vitro" micropropagated plants. Therefore quantitative and qualitative content of epicuticular wax on leaves of Dianthus caryophyllus has been studied from the "in vitro" juvenile stage to adult plants acclimatized and grown in greenhouse conditions.

The results show there are different cuticles all through their development. Thus "in vitro" developed leaves have less epicuticular wax than adult, greenhouse adapted plants, and the chemical composition was also different after acclimatization.

INTRODUCTION

Successful transfer of regenerated plants from in vitro culture to the greenhouse is essential for micropropagation. This transfer stage is often problematic, because plants are susceptible to excessive desiccation immediately after removal from culture, as a consequence irreversible tissue damage or even death may result (Sutter, 1988).

The most important function of the cuticle is probably to supplement the action of the stomata in regulating the passage of water from within the plant to its environment. Cuticle thickness and chemical characteristics of cuticle wax are related to plant age and environment (Martin and Juniper, 1970).

In vitro subcultures may favour juvenile characteristics in plants, and high relative humidity is found in "in vitro" micropropagation vessels. Both

Plant Aging: Basic and Applied Approaches
Edited by R. Rodríguez *et al.*
Plenum Press, New York, 1990

circumstances cause a poorly developed cuticle on the leaves of these plantlets (George y Sherrington, 1984; Yeoman, 1986 ; Ziv, 1986), so that excessive water loss makes it neccesary to regulate transpiration during the process of acclimatization to external conditions, in order to avoid the loss of plants at this stage.

The present paper reports quantitative and qualitative wax cuticle changes from "in vitro" juvenile to adult stage in leaves of Dianthus caryophyllus L.

MATERIAL AND METHODS

Plant material

Dianthus caryophyllus L. cv. Improved White plantlets were micropropagated "in vitro" from meristem as has been reported by Casares et al (1987), on MS medium (Murashige and Skoog, 1962), whithout hormones, these plantlets grew and developed roots. At this stage the first sample of leaves for wax extraction was taken.

Rooted plantlets were transferred to soil in a greenhouse and excessive transpiration was prevented with a plastic top which was removed after total acclimatization. Once acclimatized the plants developed new leaves with anadult aspect. These leaves were the second sample for wax extraction.

Plants were grown in the greenhouse up to flowering. At this stage the third sample of leaves was taken.

Amount and composition of epicuticular waxes

Most studies of cuticular waxes were done on the easily extractable epicuticular lipids from leaf surfaces. The common way to extract these waxes is by dipping leaves and other plant parts in a non-polar solvent for a few seconds (Martin and Juniper, 1970). Epicuticular leaf waxes from "in vitro" plantlets (A), just after acclimatization (B), and adult greenhouse plants (C) were extracted by dipping the leaves of each sample for 10 s in chloroform. After evaporation of the solvent the lipid extract was weighed. Leaf areas were determined by image analysis and the relationship wax weight/leaf area was calculated.

The extracts were then re-dissolved in chloroform and a small volume of each sample solution (about 250 μg wax) was applied to thin-layer chromatography plates (silica gel 60A, with preadsorbent area interface, 250 μm thickness, Whatman LK6D). The plates were developed up to 8 cm in chloroform:methanol (98:2 v:v) and, after drying, re-run up to 16 cm in petroleum ether:toluene (80:20 v:v) according to Larsson and Svenningsson (1986). In order to detect the compound classes separated the TLC plates were exposed for 5 min to iodine vapor.

TLC detection of oxo-compounds (for free aldehyde and keto groups) was readily accomplished spraying the plates with a solution of 2,4-dinitrophenylhydrazine (Krebs et al, 1969; Zweig and Sherma, 1972). The 2,4-dinitrophenylhydrazones formed were visible as yellow spots on developed plates.

Compounds separated on the TLC plates were recovered by washing with chloroform exhaustively and infrared spectroscopy was run for identification purposes.

RESULTS AND DISCUSSION

The amount of epicuticular waxes in "in vitro" leaves was lower than in the older leaves sampled. Acclimatized (B) and adult (C) leaves show no differences in the amount of epicuticular waxes (Table 1).

The thin wax surface in "in vitro" leaves is consistent with the high transpiration rate and the severe desiccation problems found at acclimatization of the "in vitro" plantlets, and suport the importance of epicuticular waxes in preventing excessive water loss at this stage as has been reported (George and Sherrington, 1984; Yeoman, 1986; Ziv, 1986).

In acclimatized carnation the plants that survive develop new leaves with an adult aspect and the amount of epicuticular waxes in these leaves was similar to the adult leaves. Thus, these leaves are perfectly adapted to external conditions, because of this, they develop cuticle thicker than "in vitro" leaves.

The epicuticular wax surface at each stage was also different in chemical composition (Fig. 1).

In "in vitro" plantlets (A) epicuticular waxes are more polar than adapted (B) and adult (C) plants. There are some different compounds for each stage sampled, and the main compound class in adult and adapted cuticle is not present in "in vitro" cuticular waxes.

Table 1. Amount of epicuticular waxes in relation to leaf area (5 replications-mean ± s.d.) at different stages in micropropagated carnation.

Plants stage	Amount of wax surface ($\mu g/mm^2$)
A: "in vitro"	0.13 ± 0.03
B: just adapted	1.32 ± 0.24
C: adult	1.13 ± 0.19

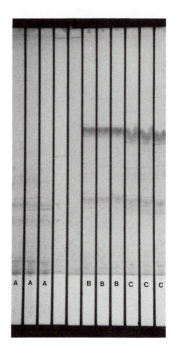

Fig. 1. Compound class separated by TLC. Starting line is at the bottom of the photograph and the front of the solvent is at the top. The A, B and C indicate the origin of wax extracts, "in vitro", just acclimatized and adult material respectively.

This main compound in adult cuticles is probably a ß-diketone (reaction with 2,4-dinitrophenylhydrazine and IR spectrum data). However, identification of all compounds requires further analysis by combined GC-MS.

The chemical composition of the wax might be important for two reasons: first, it has been suggested to be essential for wax conformation, second, different wax components are probably not equally efficient in reducing evaporation (Bengtson et al, 1978). Therefore, differences observed in chemical composition of cuticular waxes during development of carnation plantlets from "in vitro" to adult stage could produce differences in wax conformation as well as in the effectivity of its components in reducing evaporation and so, justifying the importance of (quantitative and qualitative) changes in epicuticular wax during acclimatization of "in vitro" plantlets.

REFERENCES

Bengtson, C., Larsson, S., and Liljenberg, C., 1978, Effects of water stress on cuticular transpiration rate and amount and composition of epicuticular wax in seedlings of six oat varieties,Physiol. Plant., 44: 319-324.
Casares, A., Estrada, O., Astorga, R., and Rodríguez, R., 1987, Aplicaciones del cultivo de tejidos I. Saneamiento de clavel mediante cultivo de meristemos,Rev. Soc. Esp. Hort., 6.882: 8-10.
George, E. F., and Sherrington, P. D., 1984, "Plant Propagation by Tissue Culture. Handbook and

Directory of Commercial Laboratories", Exegetics Ltd., Eversley, Basingtoke.

Krebs, K. G., Heusser, D., and Wimmer, H., 1969, Spray reagents, in: "Thin-Layer Chromatography. A Laboratory Handbook", E. Stahl ed., Springer Verlag, Berlin.

Larsson, S., and Svenningsson, M., 1986, Cuticular transpiration and epicuticular lipids of primary leaves of barley (Hordeum vulgare), Physiol. Plant., 68: 13-19.

Martin, J. T., and Juniper, B. E., 1970, "The Cuticles of Plants", Edward Arnold Ltd., Edinburgh.

Murashige, T., and Skoog, F., 1962, A revised medium for rapid growth and bioassays with tobacco tissue culture, Physiol. Plant., 15: 473-497.

Sutter, E., 1988, Stomatal and cuticular water loss from apple, cherry, and sweetgum plants after removal from in vitro culture, J. Amer. Soc. Hort. Sci. 113(2): 234-238.

Yeoman, M. M., 1986, "Plant Cell Culture Technology. Botanical Monographs. Vol. 23", J. H. Burnett, H. G. Baker, H. Beevers and F. R. Whatley ed. Blackwell Scientific Publications, Edinburgh.

Ziv, M., 1986, In vitro hardening and acclimatization of tissue culture plants, in: "Plant Tissue Culture and its Agricultural Applications", L. A. Withers and P. G. Alderson ed., Butterworths, Cambridge.

Zweig, G., and Sherma, J., 1972, "Handbook of Chromatography. Vol II", CRC Press, Cleveland.

ACKNOWLEDGEMENT

The authors gratefully acknowledge the sopport by grants from the CICYT and Ministerio de Educación y Ciencia of Spain.

SITKA SPRUCE BY LOWER CROWN PRUNING

C. Selby, R. Lee, and B. M. R. Harvey

Agricultural Botany Research Division, Department of
Agriculture for Northern Ireland, Newforge Lane, Belfast,
BT9 5PX, UK.

INTRODUCTION

Once trees have attained an age at which elite individuals can
reliably be selected they are usually too old to be vegetatively
propagated successfully, because of problems related to plant aging
(Bonga, 1987). Close proximity to a root system or to the tree base has
frequently been associated with the maintenance of juvenility. For
example, hedging (Libby and Hood, 1976) or serial propagation of cuttings
(St. Clair, Kleinschmit and Svolba, 1985) has been used to retain
juvenile rooting potential, whilst pruning (Copes, 1983) or production of
stump sprouts (Boulay, 1979) has been used to improve the rooting
potential or cultural properties of tissues.

In the present study the lower crowns of mature Sitka spruce [Picea
sitchensis (Bong.) Carr.] were pruned to force outgrowth of new shoots.
Rooting of stem cuttings, growth habit and morphology of rooted cuttings
and in vitro organogenesis on bud explants taken from the rooted cuttings
were used to assess the maturation state of shoots formed on the pruned
zones.

MATERIALS AND METHODS

Two lower crown pruning treatments were imposed on the basal
branches (2 m from the ground) of 20-year-old Sitka spruce trees in
Ballypatrick Forest, Co. Antrim in late March 1984. These were mild and
severe treatments in which the previous one or four years growth
respectively was cut away. A group of control trees was left unpruned.
Eight replicate trees were used per treatment (24 trees in all). At each
sampling date 160 cuttings were taken from each tree and rooted in four
randomised blocks each with 40 cuttings per plot. Cuttings were
collected in March before the treatments were imposed and from the
re-growth occurring on the pruned zones in March of 1986 and 1988.
Following the 1986 collection of cuttings all the new growth on the
pruned zones was cut away.

Cuttings were prepared and rooted under mist as described by Kennedy
and Selby (1984). Rooting was assessed after four months and the
resultant plants grown in a peat based compost (Selby, 1988) under

Plant Aging: Basic and Applied Approaches
Edited by R. Rodríguez *et al.*
Plenum Press, New York, 1990

glasshouse conditions without supplementary heat or light. Growth habit
assessments were made one complete growing season after rooting.

In March 1988 bud explants from plants rooted in 1984 and 1986 (ten
buds per cutting; three cuttings per tree per sampling time) and directly
from each of the mother trees in the study (30 buds per tree) were
assessed for their ability to form adventitious buds <u>in</u> <u>vitro</u>.   Culture
methods used were those of Selby and Harvey (1985) except that 5×10^{-6} M
6-benzylaminopurine was used in the bud induction medium and there was no
<u>in</u> <u>vitro</u> flushing passage.

RESULTS

In the first growing season after pruning small enscaled dormant
buds, concentrated mainly around branching nodes, were formed on the
pruned zones. These buds gave rise to shoots suitable for rooting as
cuttings the following year (1986). Similarly, a second batch of shoots
arose from the pruned zones two years later, after the treatments had
been re-imposed in 1986.

Neither of the pruning treatments significantly improved   the
percentage rooting of cuttings two or four years after pruning (Table 1).
A few of the cuttings from the severely pruned trees developed a more
extensive fibrous root system than the poor root systems normally formed
by mature cuttings but this was not statistically significant.

Table 1. Percentage rooting of cuttings prepared from shoot
re-growth on trees pruned in 1984 and growth habit of
the resultant plants after one growing season.

Pruning treatment	Year rooted			SEM
	1984	1986	1988	
Rooting (%)				
Unpruned (control)	58.1	58.2	64.8	
Mild pruning	60.0	58.1	72.4	4.72
Severe pruning	73.3	77.4	77.4	
Plagiotropism – angle from the vertical (degrees)				
Unpruned (control)	62.3	42.0	ठ	
Mild pruning	54.0	28.6	ठ	4.31
Severe pruning	54.1	23.8	ठ	
Percentage of lateral buds breaking in the terminal flush				
Unpruned (control)	0.00	1.07	ठ	
Mild pruning	0.00	1.26	ठ	0.645
Severe pruning	0.00	3.96	ठ	

ठ To be determined at the end of the 1989 growing season.

The growth habit of plants produced from the severely pruned trees in
particular showed several changes compared to control plants.   Plants
from five out of the eight trees in this treatment were virtually
orthotropic (< 20° from the vertical; Table 1). Similarly, plants from
several of the severely pruned trees showed active free growth in excess
of that predetermined in their resting buds. Furthermore, lateral buds

in the terminal flush grew out in the year in which they were laid down
(Table 1). These plants also developed a more radial distribution of
lateral branches and needles than did plants from the control trees,
which continued to show typically mature bilateral symmetry in both these
characters.

Explants from rooted cuttings failed to show an improved organogenic
capacity in vitro over explants taken directly from the mother trees
regardless of whether or not the trees had been pruned (Table 2). All of
the adventitious shoots produced in the study showed poor elongation and
remained stunted.

Table 2. In vitro organogenesis on bud explants excised directly
          from mother trees and from rooted cuttings from the
          same individual trees pruned in 1984.

Pruning Treatment	Source of explants			SEM
	Direct from mother trees	Cuttings rooted in		
		1984	1986	
Percentage explants forming adventitious buds				
Unpruned (control)	11.8	24.2	22.5	
Mild pruning	7.8	7.5	5.8	6.52
Severe pruning	9.8	16.7	15.0	
Adventitious buds per cultured bud				
Unpruned (control)	1.23	2.57	2.18	
Mild pruning	1.04	0.61	0.14	0.80
Severe pruning	0.96	1.08	1.54	

DISCUSSION

Changes in the growth habit of rooted cuttings prepared from shoot
re-growth on pruned lower branches indicate at least partial
rejuvenation. These changes were not only physiological, caused by the
improvement in nutrition, since cuttings from unpruned trees did not
exhibit comparable morphological modifications.

The lack of improvement in the rooting properties of the cuttings
reflects the incompleteness of the rejuvenation achieved by pruning.
Correspondingly, the stunted nature of adventitious buds produced in
vitro is typical of tissues from mature gymnosperms and is in marked
contrast to the vigorous in vitro elongation of adventitious buds induced
on juvenile tissues. A greater degree of rejuvenation may have been
achieved by more severe pruning to force out shoots from even closer to
the tree base. More probably a sequence of several dissimilar treatments
may be required to achieve total reversion of all aspects of development
to the juvenile condition. After such treatments tissues from mature
trees should be fully amenable to vegetative propagation.

A better understanding of the molecular mechanisms regulating the
processes of plant maturation will assist in the interpretation of
partial rejuvenation phenomena such as those reported here. Undoubtedly,
use of recombinant DNA and gene cloning technology will help to assess

the level of rejuvenation attained by tissues and to determine the involvement of differential gene expression in plant maturation and rejuvenation.

REFERENCES

Bonga, J.M., 1987, Clonal propagation of mature trees: Problems and possible solutions, in: "Cell and Tissue Culture in Forestry, Volume 1, General Principles and Biotechnology," J.M. Bonga and D.J. Durzan eds., Martinus Nijhoff, Dordrecht. pp 249-271.

Boulay, M., 1979, Multiplication et clonage rapide du Sequoia sempervirens par la culture in vitro. Ann. AFOCEL, 12: 49-55.

Copes, D.L., 1983, Effects of annual crown pruning and serial propagation on rooting of stem cuttings from Douglas-fir. Can. J. For. Res., 13: 419-424.

Kennedy, S.J. and Selby, C., 1984, Propagation of Sitka spruce by stem cuttings. Rec. Agric. Res. (Dept. Agric. N. Ireland), 32: 61-70.

Libby, W.J. and Hood, J.V., 1976, Juvenility in hedged radiata pine. Acta Hortic., 56: 91-98.

Selby, C., 1988, Micropropagation of Sitka spruce [Picea sitchensis (Bong.) Carr.]. Ph.D. Thesis, The Queen's University of Belfast, UK.

Selby, C. and Harvey, B.M.R., 1985, The influence of natural and in vitro bud flushing on adventitious bud production in Sitka spruce [Picea sitchensis (Bong.) Carr.] bud and needle cultures. New Phytol., 100: 549-562.

St. Clair, J.B. Kleinschmit, J. and Svolba, J., 1985, Juvenility and serial vegetative propagation of Norway spruce clones (Picea abies Karst.). Silvae Genet., 34: 42-48.

PHASE CHANGE IN *SEQUOIADENDRON GIGANTEUM*

MONTEUUIS O., BON M.C.

ASSOCIATION FORET-CELLULOSE (AFOCEL)
Domaine de l'Etançon
77370 Nangis FRANCE

1 - INTRODUCTION

The phase change phenomenon is classically defined as the successive
different changes which occur in plants during the transition from the
juvenile to the mature condition (ROHMEDER, 1957 ; ROBINSON and WAREING,
1969 ; HACKETT, 1983, 1985). It includes all the characteristics
affected by the maturation process associated with ontogenetic
development during time course, such as changes in morphology, in
physiological condition, and other differences which could be subtler to
discern. Although phase change concerns all vegetal organisms, the
expression of the phenomenon appears most obviously in woody perennials
plants and particularly in arborescent species (HACKETT, 1983).

This theme was investigated on *Sequoiadendron giganteum* Buchholz,
commonly named giant sequoia, species which gave rise to the world
biggest and oldest plant specimen. By contrast with *Sequoia sempervirens*,
the redwood, the giant sequoia, like most arborescent species, does not
sprout from the base of its trunk. For our investigations, the mature
material consisted in 100 year-old trees which were compared to 2-3 year-
old seedlings used as juvenile reference.

2 - PHASE CHANGE RELATED TO ONTOGENETIC DEVELOPMENT

The ontogenetic development in giant sequoia is associated with the
exhibition of number of different changes that seems worth reviewing.

21. Morphological and cytomorphological aspects

A salient contrast in foliar morphology exist between young and
mature giant sequoias, in the same way as for other species (ROHMEDER,
1957 ; SCHAFFALITZKY de MUCKADELL, 1959). The juvenile material exhibits
big and thin leaves, whereas adult trees bear small, dark-green colour,
hard and thick scale-like foliar structures (MONTEUUIS, 1985).

The possibility to investigate cytophotometrically cell walls
revealed that the mesophyll of leaves belonging to the mature forms
contained more polysaccharides than the juvenile material ones. Further
investigations indicated that this marked difference could be mainly due

to higher contents in hemicellulosic and cellulosic compounds (MONTEUUIS and GENESTIER, in press).

Such results incited us to pursue our analyses at the meristem level owing to the fact that leaves, with their particular characteristics, originated from the shoot apex. Relevant cytological observations demonstrated that the juvenile material meristems were significantly wider, exhibiting a larger basal surface than those belonging to the adult trees (MONTEUUIS, 1987a). Concurrently, referring to concomitant infrastructural analyses, it seemed reasonable to assume that ageing process was associated with a significant decrease of the meristem diameter/height ratio. Moreover, the meristematic domes of young trees were found to be bigger and to contain more cells with higher nucleoplasmic ratio than those of the mature material (MONTEUUIS, in press).

22. Physiological aspects

Physiological demonstrations of phase change in *Sequoiadendron giganteum* found expression in the following features :

* Decrease of capability for organogenesis.

As it could be presumed (BIONDI and THORPE, 1984) ability for adventitious budding passed off very early since up to now only the vegetative organs that characterize the foremost stages of ontogeny - hypocotyl, cotyledons, epicotyl- responded positively in *in vitro* conditions. Considering axillary budding, we observed that the material exhibiting mature morphological characters (adult foliage) was almost totally recalcitrant to exogenous stimulations like hedging or pruning manipulations, by contrast with the juvenile forms (MONTEUUIS,1985).

Conjointly, attemps to clone young and mature giant sequoias through in vitro meristem cultures showed that the juvenile material was much more reactive, and in the same time less dependent on macromineral solution than the mature material (MONTEUUIS, 1987b).

In *in vitro* conditions, the juvenile material exhibited somewhat stronger potentialities for growth (BON, 1988a ; MONTEUUIS, 1988), connected with significant differences in polyphenol contents between the bottom and the top of the microcuttings. This contrast did not exist in the mature material (BON et al., 1988).

Such differences in capacity for vegetative regeneration remained valid when considering ability for adventitious rooting. Ageing process resulted in a progressive but noticeable deterioration of the quality of the neoformed root system, while concurrently enhancement of topophysis and of intraclonal variability could be observed as a preliminarly to the total loss of reactivity. At this occasion, it was demonstrated that the ability for adventitious rooting of cuttings was satisfactorily correlated with the foliar morphology of the ramet (MONTEUUIS, 1985). By another way, physiological compounds, theoretically known to be involved in the rooting process such as peroxidases and phenolic substances, were proved to fluctuate too greatly to be considered as reliable indicators of phase change (MONTEUUIS and BON, 1986 ;MONTEUUIS et al., 1987).

* Attainment of the flowering stage.

Giant sequoia have to reach generally an age of about 25 to 30 year-old before bearing first flowers and cones (MONTEUUIS, 1985), although differences may exist according to the genotypes and the environmental parameters, and keeping in mind, furthermore, that some cases of neoteny were observed on one to two year-old seedlings (SKOK, 1961 ; MONTEUUIS, unpublished).

23. Biochemical aspects

Phase change in *Sequoiadendron giganteum* was shown to be associated with a general decrease of ATP/NTP as well as RNA/DNA ratio, excepted during the period of budbreak (MONTEUUIS and GENDRAUD, 1987). This result was confirmed by BON (1988b) who found in addition higher levels of GDP, GTP and greater potentialities for aminoacylation of tRNA in apices belonging to the juvenile trees compared to the mature ones. Moreover, according to this author, capacity for translation should be as efficient in the two different aged materials only during a short duration, beyond which differences appeared in favor of the young trees.

As a complementary remark relative to foliar dimorphism, we observed consecutively to micrografting manipulation that the apical meristems of the mature form contained more abundant protein population, especially from the acid protein viewpoint, than those belonging to the morphologically rejuvenated shoots (BON and MONTEUUIS, 1987). These original results restricted up to now to micrografted scions need nevertheless to be further verified in the case of shoots removed directly from the initial ortets.

But so far, the prevailing result remains the recent discovery of a 16 kD membran associated polypeptide that characterizes undoubtedly the juvenile status (BON, 1988c). The so-called "J.16" protein was proved indeed to be a reliable marker of maturation process, since it was totally independent of a possible disturbing effect due to genotypic differences nor significantly influenced by fluctuating physiological state.

3 - POSSIBLE REVERSION TO THE JUVENILE PHASE : REJUVENATION

From the foregoing it appears that phase change in giant sequoia is characterized by number of various demonstrative criteria which state the reality of the phenomenon in relation to the ontogenetic development.

Attentive observations mainly based on judicious and fine investigation means pointed out nevertheless that during the short period corresponding to budbreak most of phase change characteristics tended to the juvenile traits, to reappear as soon as shoot expansion occurred again (MONTEUUIS, 1988). Relevant results were sufficiently demonstrative to support the concept of a reiterative maturation process related to each shoot flush included in the general pattern of phase change applied to the plant in its whole (FRANCLET, 1983 ; MONTEUUIS, 1988). As a matter of fact, the sequential occurrence of juvenile potentialities according to shoot ontogeny should become more and more space-time restricted as tree develops during time course an increasing architectural complex. Nevertheless, the phenomenon remains most of the time morphologically unperceived most likely because of hypothetic repressive effects due to presumed correlative systems (ROBINSON and WAREING, 1969) which are known to be able to act at very short distances.

In this situation, the removal of the shoot apical meristems from the mature ortet at the suitable physiological period to place it in favorable in vitro culture conditions should be helpful to counteract the negative influence of an inhibitory physiological context *in situ*.

Table 1. RECAPITULATIVE OF THE MAIN PHASE CHANGE
CHARACTERISTICS IN *SEQUOIADENDRON GIGANTEUM*

	Juvenile	Mature
Morphological and cytomorphological criteria		
*Leaves		
form	long and thin	small, hard and thick
mesophyll cell wall composition		higher levels of poly-saccharidic substances hemicelluloses and celluloses
*Meristem		
conformation	wide	sharper
volume		smaller
number of cells		less numerous
nucleoplasmic ratio		lower
Physiological criteria		
Ability for budding	high	low
in vitro reactivity of excised meristems	high	low with great influence of the macromineral solution
growth capacity of microcuttings	better on average, with significant differences in inner repartition of endogenous phenolic compounds	
adventitious rooting capability	high	recalcitrant
flowering ability	none	fully
Biochemical criteria		
ATP/NTP ratio		lower
RNA/DNA ratio		lower
GDP and GTP contents		lower
potentialities for tRNA aminoacylation		lower
"J16" concentration	high	too low to be detected

Additional remark : During the short period corresponding to budbreak, number of phase change characteristics were observed to tend to the juvenile traits.

As a concrete illustration, a recent study based on the possibilities for true-to-type cloning of mature giant sequoias established that *in vitro* culture of meristems removed during budbreak time was the only mean to achieve this goal through the obtention of a spectacular rejuvenation case (MONTEUUIS, 1988).

4 - CONCLUSION

Throughout the different traits exposed, *Sequoiadendron giganteum* could be objectively considered as a good model of phase change phenomenon of arborescent species, taking account of the fact that most of the presented results and especially those involving new investigation techniques, were assumed to be transposable to other tree species. Currently we found indeed many hopes in the use of powerful biochemical investigation means. Markers such as the "J 16" protein appear of primordial interest for instance to quantify by immunoassay the juvenility degree. Another attractive aspect is the possibility to detect by immunolocalization eventual juvenile remaining territories within mature tissues.

In addition, the micropropagation resorts, with special reference to *in vitro* meristem culture or micrografting, remains an undeniable advisability to progress in the way of a better understanding of phase change through significant features. For instance, the striking rejuvenation obtained on giant sequoia stands for a highlight demonstration of phase change reversion, referring to the classically recognized definition of this term.

REFERENCES

BON M.C., 1988 a. - Aspects biochimiques du clonage de séquoias géants jeunes et âgés. PhD Thesis Univ. Blaise Pascal, Clermont-Ferrand, France,150 p.

BON M.C., 1988 b. - Nucleotide status and protein synthesis *in vivo* in the apices of juvenile and mature *Sequoiadendron giganteum* during budbreak. Physiol. Plant., 72, 796-800

BON M.C., 1988 c. - "J 16" : an apex protein associated with juvenility of *Sequoiadendron giganteum*. Tree Physiol, in press

BON M.C., GENDRAUD M., FRANCLET A., 1988. - Roles of phenolic compounds on micropropagation of juvenile and mature clones of *Sequoiadendron giganteum* ;influence of activated charcoal. Sci. Hort., 34, 283-291

BON M.C., MONTEUUIS O., 1987. - Application de la technique micro 2 D PAGE au microgreffage de *Sequoiadendron giganteum* Buchholz. C.R.Acad. Sc.Paris, 224, (3), 667-370

FRANCLET A., 1983. - Rejuvenation : Theory and practical experiences in clonal silviculture. In : Clonal Forestry : its impact on tree improvement and our future forests. XIXth Meeting of the Canadian Tree Improvement Association, 22-26/8/1983, Toronto, 96-134

HACKETT W.P., 1983. - Phase change and intra-clonal variability. Hort. Science, 18, (6), 12-16

HACKETT W.P., 1985. - Juvenility, maturation and rejuvenation in woody plants. Hort. Rev., 7, 109-155

MONTEUUIS O., 1985. - La multiplication végétative du séquoia géant en vue du clonage. Annales AFOCEL 1984, 139-171

MONTEUUIS O., 1987 a. - Profils méristématiques de séquoias géants (*Sequoiadendron giganteum* Buchholz) jeunes et âgés durant les stades de repos végétatif et de débourrement. C.R. Acad. Sc. Paris,305, (III), 715-720

MONTEUUIS O., 1987 b. - *In vitro* meristem culture of juvenile and mature *Sequoiadendron giganteum*. Tree Physiol., 3, 265-272

MONTEUUIS O., 1988. - Aspects du clonage de séquoias géants jeunes et âgés. PhD Thesis Univ. Blaise Pascal, Clermont-Ferrand, France, 190 p.

MONTEUUIS O., BON M.C., 1986. - Microbouturage du séquoia géant. Annales AFOCEL 1985, 49-87

MONTEUUIS O., BON M.C., BERTHON J.Y., 1987. - Micropropagation aspects of *Sequoiadendron giganteum* juvenile and mature clones. Acta.Hort., 212, 489-497

MONTEUUIS O., GENDRAUD M., 1987. - Nucleotide and nucleic acid status in shoot tips from juvenile and mature clones of Sequoiadendron giganteum during rest and growth phases. Tree Physiol., 3, 257-263

ROBINSON L.W., WAREING P.F., 1969. - Experiments on the juvenile-adult phase change in some woody species. New Phytol., 68, 67-78

ROHMEDER E. von, 1957. - Alterphasenentwicklung der Waldbäume und Forstpflanzenzüchtung. Silv. Genet., 6, 136-142

SCHAFFALITZKY de MUCKADELL M., 1959. - Investigations on aging of apical meristems in woody plants and its importance in silviculture. Kandrup and Wunsch's Bojtrykkeri, Copenhagen, 307-455

SKOK J., 1961. - Phtoperiodic responses of *Sequoia gigantea* seedlings. Botanical Gazette, 1234, 63-70

THORPE T.A., BIONDI S., 1984. - Conifers. Handbook of plant cell culture. Sharp W.R., Evans D.A., Ammirato P.V. et Yamada Y. eds, Macmillan Publishing Company, New York, 435-470

REJUVENATION OF ADULT SPECIMENS OF <u>CASTANEA SATIVA</u> MILL:
THROUGH <u>IN VITRO</u> MICROPROPAGATION

José A. Feijó and M.Salomé S. Pais

Secção de Biologia Celular, Departamento de
Biologia Vegetal, Faculdade de Ciências de
Lisboa R,.Ernesto de Vasconcelos, Ed. C2,
1700 LISBOA   PORTUGAL

INTRODUCTION

Castanea sativa is a woody plant difficult to propagate by
conventional methods. Cuttings are difficult to root, and, due
to the almost complete autoincompatibility, the levels of
heterozygoty are very high, thus compromising programs of
propagation by seeds. Until recently some success was obtained
on the micropropagation of adult plants of C.sativa x C.crenata
hybrids (reviewed by Vieitez et al.,1986). However, as stressed
by several authors trees and shrubs can only be selected for
cloning when they are adult (cf. Bonga and Durzan,eds.,1982,
and Pierik,1987). As a consequence, efficient methods of
micropropagation of adult specimens are required if cloning of
a specific tree is needed. In C.sativa different attempts of
micropropagate adult material have failed systematically in the
rooting step (Chevre, 1985; Mullins, 1987). The presence of
rooting inhibitors in mature material may be one of the
reasons of this failure, since this kind of compounds seems to
be absent from juvenile and etiolated material (Gesto et
al.,1977; Vieitez et al.,1987). Since C.sativa lacks other
precise juvenility markers, the ability of rooting could
therefore be considered as a rough rejuvenation marker.
Recently a complete process of micropropagation of adult
material of C.sativa has been described (Feijó and Pais,1988;
Feijó and Pais,1989). This paper describes some results on this
process related to a possible rejuvenation occurring in vitro.

MATERIAL AND METHODS

Detailed conditions of all the assays performed are
described elsewhere (Feijó and Pais, 1989). Material used was
exclusively apical and axillary buds of mature (20-100 years
old) elite trees. Final conditions of the micropropagation
include the use of the full Gresshof and Doy (1972) medium in
all the steps of the cycle, use of 1,0 mg/l of benzyladenine
(BA) in the establishment step, use of 5,0 mg/l BA plus 1,0
mg/l of indolacetic acid (IBA) on the multiplication step, and

of 0,2 mg/l of BA in the elongation step. Rooting was achieved both by incorporation of the hormone on the basal medium (best results with IBA 1,0 mg/l) or by dipping in concentrated hormone and transfer to hormone-free medium (best results with 1,0 mg/ml of IBA during at least 15 min.). Final acclimatization was performed in highly aerated soils containing peat as the basic substrate, in a ultrasonic mist chamber with white light.

RESULTS AND DISCUSSION

Some results show a clear difference between juvenile and mature material in the first steps of in vitro establishment. As reported before (Vieitez et al.,1986; Mullins,1987; Marques, 1988) juvenile material has a very good response in nearly all media tested, ranging from high salt media like the Murashige and Skoog (1962) medium, to very low salt medium like White's (1934) medium. With adult material the effect of the initiation medium was striking, as expressed in fig.1.

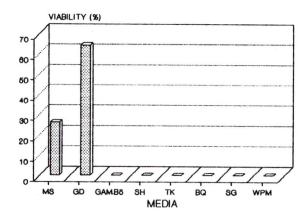

Figure 1.Influence of the different media in the viability rates during the in vitro establishment. Viability was defined as the number of living and actively growing buds 4 months after culture initiation, over all the buds that survived one month after culture initiation. In all media 1,0 mg / l of benzyladenine were added.

The Gresshof and Doy medium was clearly the best medium, this fact having been confirmed in all the subsequent steps of the process. However after some months of subculture the sensitivity to the medium was not so intense, and, for instance, the use of a full MS medium gave also good results.
The evolution of the rooting rates during the different number of subcultures revealed an initial rise in the in vitro capacity of rooting (fig.2).
With the continued subcultures, and optimization of technical details rooting rates in the range of 90% were obtained, which corresponds to some of the best results obtained for juvenile material (Vieitez et al.,1986; Mullins, 1987). Whether or not this graph shows a real rejuvenation can be arguable since, admitting that the lack of rooting capacity

is due to the presence of inhibitors in adult material (Gesto et al.,1977; Vieitez et al.,1987) there is always the possibility of a progressive dilution of these factors during the transfers.

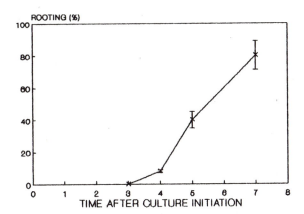

Figure 2.Evolution of the rooting capacity during the course of the sequential subcultures. Values expressed are the percentage of rooted shoots over all the shoots transferred to rooting media ± S.E.. All the obtained from similar rooting conditions, with 1,0 mg/l IBA incorporated in the basal GD medium.

Thus a progressive regaining of the rooting capacity would occur based on the removal of such inhibitory compounds. However in vitro rejuvenation is a very difficult process to characterize and in the absence of other characteristics (namely leaf changes) increase in rooting potential can be considered, by itself, as a characteristic of rejuvenation, particularly in woody plants (Moncousin, 1982; Pierik, 1987).

A particular problem of the in vitro culture of woody plants is the browning of the tip, generally reported as apical necrosis. This characteristic has been sometimes associated with a possible aging, and is particularly important during the rooting stage. Following some recent studies on the role of boron ion in the apical growth in plants (Lovatt,1985; Ali and Jarvis,1988), the influence of the concentration of this micronutrient on the apical necrosis during rooting was studied (fig.3).

These results show clearly the importance of this ion in the appearance of apical necrosis, which decreases with the increase of concentration of borate. Unfortunately this response is also accompanied, from a certain point, by a decrease in the growth capacity, here shown by a decrease in the rooting capacity, but also observed in the normal conditions of growth.

Other characteristics of the advantage of the in vitro micropropagation method for Castanea sativa Mill., that can account for a possible rejuvenation hypothesis, are the general aspect and the growth rates of the shoots when transferred to soil (fig.4). Has shown in this graph after an initial lag period, plants start to grow at very good rates, better than most seed material in equivalent culture conditions.

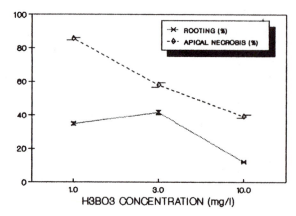

Figure 3.Influence of the concentration of boric acid in tip browning during rooting. Values are expressed as mean ± S.E.

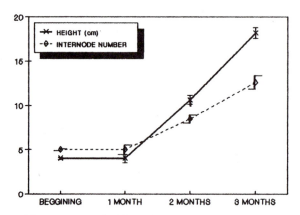

Figure 4.Growth rates of the _in vitro_ regenerated plants in soil. Values represent mean ± S.E..

Again, if this characteristics reflects, or not, a true rejuvenation can be argued since a general revigoration of the material, due to the culture in aseptic conditions, could be responsible for such changes. However, as stressed by several authors, the process of rejuvenation is not still completely understood or adequately defined (cf.Bonga and Durzan,eds.,1982,1987; Pierik,1987). So, the doubt remains that, having in mind the difficulties always present when micropropagating woody mature material, namely in the case of

<u>C.sativa</u>, successes over traditional methods difficulties (namely rooting difficulties) could always be explained by a rejuvenation of the material in the course of the sequential <u>in vitro</u> subcultures. Similar opinion has been stressed by other authors, namely Pierik (1987).

REFERENCES

Bonga, JM, Durzan, DJ, eds., 1982, <u>Tissue Culture in Forestry</u>, Martinus Nijhoff & W.Junk Publ.,The Hague, 1,420

Bonga, JM, Durzan, DJ, eds., 1987, <u>Cell and Tissue Culture in Forestry</u>, vols. 1-3, Martinus Nijhoff Pub., Dordrecht

Chevre, AM, 1985, <u>Recherches sur la multiplication vegetative in vitro chez le chaitaigner</u>, These Doct., Univ. Bordeaux, Bordeaux

Gesto, MDV, Vasquez, A, Vieitez, E, 1977, "Rooting substances in water extracts of <u>Castanea sativa</u> and <u>Salix viminalis</u>", <u>Physiol. Plant.</u>, 40: 265, 268

Gresshof, PM, Doy, CH, 1972, "Development and Differentiation of haploid <u>Lycopersicum esculentum</u> (tomato)" <u>Planta</u>, 17: 161, 170

Feijó, JF, Pais, MS, 1988, "Micropropagaçāo de <u>Castanea sativa</u> Mill. a partir de gemas axilares de espécimes adultos", <u>Proc.Reuniao Soc.Port.Biotecnol.</u>, SPB, Coimbra

Feijó, JF, Pais, MS, 1989, "<u>In vitro</u> propagation of adult trees of <u>Castanea sativa</u> Mill, through axillary bud culture", <u>Submitted</u>

Marques, CM, 1988, "Estabelecimento de <u>Castanea sativa</u> in vitro", <u>Relatório de projecto</u>, Fac. Ciências de Lisboa, Lisboa.

Moncousin, J, 1982, <u>Proc.5th Int. Cong.Plant Tissue Cell Culture</u>, 5: 147,148

Mullins, KV, 1987, "Micropropagation of chestnut (<u>Castanea sativa</u> Mill.)", <u>Acta Horticulturae</u>, 212:525,530

Pierik, 1987, <u>In Vitro culture of higher plants</u>, Martinus Nijhoff Pub., Dordrecht

Vieitez, AM, Vieitez, ML, Vieitez, E, 1986, "Chestnut (<u>Castanea</u> spp.)" in <u>Biotechnology in Agriculture and Forestry</u>, YPS Bajaj, ed., Springer-Verlag, Berlin, Vol.1 (Trees), 393, 414

Vieitez, J, Kingston, DGI, Ballester, A, Vieitez, E, 1987, "Identification of two compounds correlated with lack of rooting capacity of chestnut cuttings", <u>Tree Physiol.</u>, 3: 247,255

EMBRYOGENIC CULTURES OF TOBACCO POLLEN AS A MODEL SYSTEM TO STUDY PLANT

REJUVENATION

Oscar Vicente, Dolores Garrido, Norbert Eller, Rosa M. Benito
Moreno, Anna Alwen, and Erwin Heberle-Bors

Institut für Mikrobiologie und Genetik, Universität Wien
Althanstraße 14, A-1090 Wien, Austria

INTRODUCTION

Formation of pollen in angiosperms takes place in the anthers, the male
sexual organs. During normal male gametophytic development in vivo, each diploid microspore mother cell generates, through meiosis, four haploid microspores; after one or two mitosis and maturation, they will eventually give
rise to mature pollen grains. However, in in vitro cultures of isolated anthers or pollen, pollen grains are able to follow a different, sporophytic
developmental pathway, leading to the formation of embryos and haploid plants
directly and asexually. Recent advances in the technique of isolated pollen
culture in the model plant tobacco, allow the strict control of pollen development in both the sporophytic and the gametophytic direction.

Here we describe a highly efficient in vitro system for pollen maturation **and** embryogenesis, in which the nutritional status of the immature
pollen grain, at a particular developmental stage, provides the trigger for
its development towards one or the other pathway.

These in vitro cultures of tobacco pollen will enable us to investigate the molecular regulatory mechanisms underlying the induction of pollen
embryogenesis, that is, the processes of dedifferentiation of the male ga-
metophyte, derepression of cell division activity, and recovery of embryogenic capacities, closely related to the phenomenon of plant rejuvenation.

IN VITRO MATURATION OF POLLEN

Isolated microspores or immature pollen grains of tobacco can continue
in vitro their maturation process, if cultured in appropiate conditions.
Pollen isolated at the mid-binucleate stage can be cultured to maturity
within two days in a relatively simple "rich" medium, containing Murashige
and Skoog macro and micro minerals, sucrose (0.25 M), and glutamine (3 mM),
at pH 7.0 (Kyo and Harada, 1986; Benito Moreno et al., 1988a). This process
is very efficient, as shown by in vitro germination tests: about 80% of the
in vitro matured pollen grains produce pollen tubes in a medium containing
sucrose and boric acid.

Complete maturation in vitro from earlier developmental stages (late uninucleate, e.g. microspore, and early binucleate pollen) requires more complex conditions, and the germination frequency is reduced to about 20% (Benito Moreno et al., 1988a, 1988b). The morphology of the in vitro matured pollen (size, triangulat shape, cells fully packed with starch grains) and the time required for in vitro maturation, are similar to those for pollen that matures in vivo.

In situ pollination with the in vitro matured microspores or pollen, indicates that those pollen grains have acquired full gametophytic function, i.e. in vivo fertilization of embryo sacs (female gametophytes) and seed set. Genetic tests with marker genes revealed that it was indeed the in vitro matured pollen that performed fertilization (Benito Moreno et al., 1988a).

These experiments show that the influence of the anther wall on pollen development, at least from the late uninucleate stage on, can be completely replaced by an artificial medium. It might be possible to reproduce in vitro the complete development of pollen, from meiosis to maturity, simulating the action of the anther wall by providing the appropiate in vitro conditions at the specific stages of development, but further work will be required to establish those conditions for the earlier stages, meiocytes, tetrads and early microspores.

EMBRYOGENIC CULTURES OF ISOLATED POLLEN

As mentioned above, mid-binucleate pollen undergoes in vitro normal male gametophytic development very efficiently, when cultured in a "rich" medium. However, the same pollen grains can be directed towards cell division and embryogenesis, also in a nearly quantitative way. A hunger signal is responsible for this reprogramming of pollen development. Mid-binucleate pollen is cultured in a medium without nutrients (MS macro minerals, 0.4 M manitol, pH 7.0) (Kyo and Harada, 1986; Benito Moreno et al., 1988b). After 6-8 days in this starvation medium a characteristic type of cells have been formed, the so-called embryogenic pollen grains or P-grains. If left in this medium, P-grains die. Transfer to a sugar-containing medium induces cell division, after a lag period, and high numbers of globular embryos are formed. In earlier experiments (Kyo and Harada, 1986), these embryos did not develop further. Using a different medium, containing macro and micro salts, sucrose (0.25 M) and Fe-EDTA (0.1 mM), pH 7.0 (Benito Moreno et al., 1988b), heart-shaped and torpedo embryos are formed, and haploid plantlets can be directly and very efficiently regenerated.

Formation of P-grains, induced by the starvation treatment of mid-binucleate pollen, is a prerequisite for cell division; if they are not formed, for example, when pollen at different developmental stages is used, pollen grains mature or die, but embryos never develop. Embryogenic pollen has characteristic features that distinguish it cytologically from normal gametophytic pollen of any stage, such as large vacuole-like, structural empty spaces formed by fusion of lysosomes, nuclei located in the centre of the cell, cytoplasmic strands conecting them with a thin layer of cytoplasm that can be observed along the pollen wall, and absence of starch grains (amyloplasts).

These results also provide the basis to understand the induction of embryogenesis in anther cultures, and the phenomenon of pollen dimorphism displayed by some species, including tobacco: the presence in mature anthers of embryogenic pollen grains formed in situ, together with normal gametophytic pollen. It seems now clear that starvation is a necessary requirement for the induction of P-grains, and that it can also occur in excised anthers or flowers and, under specific growth conditions, during pollen development in vivo (reviewed by Heberle-Bors, 1989). In these cases, only a fraction of pollen grains develop into embryogenic pollen, whereas in isolated pollen cultured in a sugar-free medium, P-grain formation is an almost quantitative process.

PROPERTIES OF THESE IN VITRO CULTURES OF TOBACCO POLLEN

Several properties make these in vitro cultures an attractive model system for the study of pollen development at the molecular level. Tobacco contains large numbers of pollen grains (about 200,000 per flower), relatively easy to isolate. In optimal conditions, pollen development is very synchronous, in vivo (within one anther and between anthers of the same flower), as well as in vitro; plants growing under non-optimal conditions, however, tend to produce poor-quality pollen showing asynchronous development and reduced viability. The stage of pollen development can be precisely established cytologically. Maturation and embryogenesis from mid-binucleate pollen are both very efficient, only limited by pollen viability. Embryos develop in vitro from single, isolated, cytologically well-defined cell (embryogenic pollen) and without the exogenous addition of hormons; this is in contrast to somatic embryogenesis (e.g. in carrot or alfalfa), in which the material competent for embryogenesis are not isolated cells, but cell clamps, induction of cell division requires a hormonal treatment and is quite unefficient, in the sense that only a few cells in those clamps actually divide. Finally, plants are regenerated without a callus phase, that can be a source of genetic and epigenetic variability.

Therefore, these cultures not only enable us to strictly control pollen development in vitro, but it is also possible to obtain highly homogeneous populations of cells, at specific developmental stages, in the amounts required for biochemical work.

EXPERIMENTAL CONTROL OF THE ALTERNATION OF GENERATIONS

Perhaps the most interesting aspect of these in vitro cultures of tobacco pollen is the efficient control of the switch from gametophytic to sporophytic development. Starting out from pollen grains at the mid-binucleate stage, pollen can be directed either towards maturation (in a rich medium) or towards embryogenesis (starvation treatment followed by embryogenesis medium). The most important developmental decision during plant development, the alternation of generations, is thus controlled in vitro by the nutritional status of the cells involved.

Similat switch mechanisms, triggered by a hunger signal, also occur in yeast, where starvation induces the vegetatively growing cells to undergo meiosis and produce spores, and in other lower eucaryotes. It remains to be seen whether starvation is a general trigger for phase change in plants, for example, during meiotic prophase or before cell division in the zygote, or if it could even have a more general effect, inducing cell division in other systems.

EXPERIMENTAL APPROACHES FOR THE STUDY OF POLLEN EMBRYOGENESIS AT THE MOLECULAR LEVEL

We have undertaken the study of the molecular regulatory mechanisms underlying pollen embryogenesis and its induction from immature pollen. Our final goal is the isolation and characterization of the genes that govern the switch from gametophytic to sporophytic development, and of the factors that regulate their expression. A drastic change in gene expression, including activation of cell cycle and embryo-specific genes, is likely to be involved in these processes.

As a first step in the study of differential gene expression in pollen, we are trying to select molecular markers for pollen development, at the protein and RNA levels, by i) analysis of patterns of proteins synthesized at different stages during in vitro pollen maturation (as a control), starvation and early embryogenesis, and ii) isolation of starvation-induced and embryogenesis-specific genes by differential screening of stage-specific cDNA libraries. The cDNA libraries can also be screened with heterologous probes, looking for specific genes that are likely to be expressed during pollen embryogenesis, e.g. cell cycle genes from yeast or embryogenesis-specific clones isolated from somatic embryos in carrot or alfalfa. Parallel studies can be done at the protein level, investigating changes during pollen development in particular enzymatic activities, such as protein kinases, known to be involved in general, conserved regulatory mechanisms in other systems. Finally, the search for factors or conditions (hormons, changes in temperature, etc.) that could substitute for, inhibit or improve the efficiency of the starvation treatment as trigger of pollen embryogenesis, may give some clues about the mechanisms mediating its effects.

PERSPECTIVES

These studies in the model plant tobacco should provide valuable insights into the molecular basis of pollen embryogenesis; if some general regulatory mechanisms can be established, they could eventually be extended to other species. It would also be very interesting to develop this type of isolated pollen cultures in species of more economical importance, and in those still recalcitrant to regeneration from protoplasts or other explants (cereals, legumes, woody species).

In vitro cultures of isolated pollen should be very useful for the study of different, basic and applied aspects of plant reproductive development. For commercial haploid production, the simpler technique of anther culture appears to be the method of choice, but in some species that are still recalcitrant, pollen culture may turn out to be the only technique for obtaining haploids. On the other hand, in vitro pollen maturation systems will facilitate the study of the influence of the sporophytic tissues in the anther on pollen development. This interaction is the basis of phenomena of considerable importance in plant evolution and plant breeding, as well as in human pathology, such as cytoplasmic male sterility (cms), self-incompatibility and pollen wall formation (including the incorporated pollen allergens). Seed production after pollination with in vitro matured pollen may overcome fertility barriers, both in male steril lines and in self-incompatible plants. In those types of cms where post-meiotic development is affected, pollen rescue and maturation in vitro could be a means to self-fertilize cms plants. In self-incompatible plants, in vitro matured pollen may be able to self-fertilize, since the self-incompatibility antigens would be absent.

Finally, the immature pollen grain may be used as a target for gene transfer. The transformed pollen could then either be induced to form embryos, or matured in vitro and used as a "super vector" in its normal biological function, i.e. fertilization of embryo sacs. Both pathways would lead to the formation of transgenic plants. These alternative routes would be of interest in those species in which it is not possible to regenerate whole plants from protoplasts.

REFERENCES

Benito Moreno, R.M., Macke, F., Alwen, A., and Heberle-Bors, E., 1988a, In situ seed production after pollination with in vitro matured, isolated pollen, Planta, 176:145-148.

Benito Moreno, R.M., Macke, F., Hauser, M.-T., Alwen, A., and Heberle-Bors, E., 1988b, Sporophytes and male gametophytes from in vitro cultured, immature tobacco pollen, pp 137-142, in "Sexual Reproduction in Higher Plants", M. Cresti, P. Gori, and E. Pacini, eds., Springer, New York Berlin Heidelberg.

Heberle-Bors, E., 1989, Isolated pollen culture in tobacco: plant reproductive development in a nutshell. Sex. Plant Reprod., 2:1-10.

Kyo, M., and Harada, H., 1986, Control of the developmental pathway of tobacco pollen in vitro, Planta, 168:427-432.

THE MERISTEMATIC CALLI OF MAIZE: A MAINTENANCE SYSTEM OF TISSUE JUVENILITY

J.M. Torné and M.A. Santos

Dpto. Biología Molecular y Agrobiología, C.I.D. (C.S.I.C.)

Jordi Girona nº 18-26, 08034 -Barcelona (Spain)

Definition and Morphology of Meristematic Calli

Meristematic callus is an organized tissue obtained from cauline meristem after culture in medium with the auxin 2,4-dichlorophenoxyacetic acid, (2,4-D). It is derived from the hypertrophic development of a meristem caused by the action of 2,4-D in a specific phase of its development (Torné et al., 1984; Santos et al., 1984).

It is globular or multilobular in shape with a typical green colour and a relatively high chlorophyll concentration. The size varies according to age. The number of globular structures is related to its capacity of proliferation (Torné and Santos, 1987).

Initially, the globular surface is smooth, but during the culture period, small wart-like eruptions which are clearly visible to the naked eye develop. The new globular forms or plantlets emerge from these eruptions. Growth and differentiation take place in the green zone and the absorbent system is located in the lower zone (Torné, 1981).

Properties

Meristematic calli have the following properties:
(a) Ready production of active growth points in the surface of the calli and as a result new meristematic calli with the same structure and function. Young calli derive from old calli and from their active growth points.
(b) This peculiarity allows young calli to be used in subculture and maintains the juvenility of the system for a long time.
(c) Plant regeneration, via organogenesis, starts when the callus is subcultured in a medium containing a low or zero 2,4-D concentration.

Regeneration is improved when naphtaleneacetic acid, (NAA), and 6-(y,y-dimethylallylamino)purine, (2iP), hormones are added to the medium. During the differentiation process every active growth point can produce one plantlet and consequently one callus can regenerate many plants. It is advantageous to isolate the different shoots or plantlets from the callus in order to increase its efficiency (Torné, 1984).

(d) In certain cases, it is possible to obtain embryogenic calli from

Fig. 1. A typical meristematic callus (three years old) in the maintenance medium with 2 mg/l of 2,4–D. Bar = 1 cm.

the meristematic calli after long periods of subculture.

(e) Meristematic calli with opportune manipulation and in MS modified medium with 2 mg/l of 2,4–D, (Green and Phillips, 1975), are capable of maintaining themselves for a long time without change in their morphogenic capacity.

There is apparently a relationship between growth speed and active growth points on the surface of the calli. When the external surface remains smooth and the green colour of calli becomes dark the capacity to increase the number of calli during subcultures is reduced. Variations in the callus colouring facilitate a suitable selection of samples, which maintains the culture juvenility.

Manipulation of calli

The peculiar growth pattern of these calli calls for an appropiate manipulation. In general, it is desirable not to cut the calli because injury reduces growth capacity and produces necrosis; it is advisable to detach the new calli when possibility of injury is reduced. The maintenance of large calli is preferable to damaging them through clumsy manipulation. In general, large calli can divide spontaneously after one subculture period. It is recommendable not to change the initial position of the calli in the medium because of their histological pattern. Finally, the use of culture tubes is more effective than Petri dishes; it seems that humidity and the volume of the container are necessary for good callus growth.

In other words, the maintenance process of calli is potentiated by: (i) a spontaneous division into fractions (ii) a suitable size of the calli and (iii) the avoidance of culture with necrosed tissues.

Histological and ultrastructural pattern

The histological observations revealed:

(a) A peripheral zone formed by several layers of clustered, small cells with large nuclei and swollen apical tips with initial leaf primordia

396

in areas of the surface.

(b) A middle or internal zone formed by large irregular cells with small nuclei and in some areas clusters of small cells with large nuclei which make up the vascular bundles.

(c) A lower zone formed by very vacuolated cells and anomalous roots in certain cases, (Castellví et al., 1984; Torné, 1985).

The ultrastructural observations of young meristematic calli revealed:

(a) A very active peripheral layer with small cells, larger nuclei, prominent nucleolus and thin walls. In certain cases it is possible to observe cell division. The number of these layers of peripheral cells is related to age. The younger the tissue, the greater the number of layers. The cytoplasm in these cells is very dense.

(b) The possibility to detect vascular bundles between the cells of the calli. These vascular bundles form the phloem surrounded by parenchyma cells with an almost empty cytoplasm and thicker walls. There are typical chloroplasts in the middle zone with thylacoid, grana and sometimes starch.

Mitochondria, endoplasmic reticulum, rough endoplasmic reticulum, vacuoles, etc. are also visible as in other vegetal cells.

Methods to obtain meristematic calli

The standard method used in our laboratory to obtain meristematic calli is the "atrophic tissue culture", (Torné et al. 1984). The sequence is as follows:

Immature embryos between 15 and 25 days of postpollination are sterilized, extracted and placed in the modified medium MS with 1 mg/l of 2,4-D with the scutellum upwards and in the light. After the first period of culture, the immature embryos germinate and in some cases anomalous germinations are produced (inflexions,curvatures, twisted stems, and in the nodal zone, green tissues similar to calli) constituting the "atrophic tissues" which are used for the culture. Finally, the meristematic calli are formed from them.

There are also other possibilities of obtaining meristematic calli:

(a) From seedling segments of germinated seeds.
(b) Using mesocothyl of immature embryos.
(c) Using nodal stem sections from buds or young plantlets obtained "in vitro". With this system it is possible to obtain, in optimal conditions almost 50% of meristematic calli neoformations.
(d) Obtaining the calli from the somatic embryos during the subculture of embryogenic calli in a medium with 2 mg/l of 2,4-D.

Other considerations

We observed that some conditions relating to the embryo (size or age) may have an influence on the appearance of "atropic tissues". In general, the young or small embryos produce anomalous germinations only at certain ages, (Santos, 1986).

The genotype or cultivar also exerts an influence.

Moreover, other influential factors include the environmental conditions of the field where the cultivars are grown, (Santos and Torné, 1986).

Similarly, variations in the total number of calli among different calli-clones or subclones may appear even within the same genotype and after various subcultures. The capacity to maintain the calli may also be influenced by the initial callus-clone chosen from the cultures, (Torné and Santos, 1987).

Applications

Owing to their particular characteristics and ease of manipulation, meristematic calli have been successfully used, to date, in subsequent studies eg. selections of methomyl resistance, (Torné and Bervillé, 1986); agroinfection with the MSV virus, (unpublished); polyamines action Fdez.-Tiburcio et al., in this book) and other physiological and molecular studies, (Vilardell, et al., 1989; Ludevid et al., 1989).

Note

These experiments were carried out with the financial support of C.S.I.C. through several projects.

References

Castellví, M., Santos, M.A., and Torné, J.M., 1984, Histologie du cal méristématique du maïs , 4th Congress of the FESPP, Strasbourg,193.

Fernandez-Tiburcio, A., Figueras, X., Claparols, I., Santos, M.A., and Torné, J.M., 1989, in this book.

Green, C.E., and Phillips, R.L., 1975, Plant regeneration from tissue cultures of maize, Crop Sci., 15:417-421.

Ludevid, M.D., Ruiz-Avila, L., Stiefel, M.P., Torrent, M., Torné, J.M., and Puigdomenech, P., 1989, (submitted).

Santos, M.A., Torné, J.M., and Blanco, J.L., 1984, Methods of obtaining maize totipotent tissues I. Seedling segment culture, Plant Sci. Lett., 33:309-315.

Santos, M.A., and Torné, J.M., 1986, A comparative analysis between totipotency and growth environment conditions of the donor plants in tissue culture of Zea mays L., J.Plant Physiol., 123:299-305.

Santos, M.A., 1986, PhD Thesis, Univ. of Barcelona (Spain).

Torné, J.M., 1981, PhD Thesis, Univ. of Barcelona (Spain).

Torné, J.M., 1984, Autre modalité de regénération de plantes a partir de fragments de Zea mays L., 109 Congrés National des Sociétés savantes, Dijon (France), 187-195.

Torné, J.M., Santos, M.A., and Blanco, J.L., 1984, Methods of obtaining maize totipotent tissues. II Atrophic tissue culture, Plant Sci. Lett., 33:317-325.

Torné, J.M., 1985, Ultraestructura comparada de dos tipos de callos de maíz: verde y albino, VI Reunión SEFV, Valencia (Spain), 207.

Torné, J.M., and Bervillé, A., 1986, Cytoplasm T-maize tissue culture to obtain methomyl resistance, in: Nuclear Techniques and in vitro Culture for Plant Improvement, IAEA, Vienna, 1986, 299-303.

Torné, J.M. and Santos, M.A., 1987, Aspectos de la totipotencia en cultivo de callos de Zea mays L.: Mantenimiento y variabilidad de las estructuras totipotentes, Inv. Agrar., 2:111-120.

Vilardell, J., Goday, A., Freire, M.A., Torrent, M., Martinez, C., Torné, J.M., and Pages, M., 1989, (submitted).

GROWTH AND DEVELOPMENT IN IN VITRO LONG TERM CULTURES

Richard W. Joy IV and Trevor A. Thorpe

Plant Physiology Research Group, Department of Biological
Sciences, University of Calgary, Calgary, Alberta, Canada
T2N 1N4

INTRODUCTION

Long term storage and preservation of experimental material is useful
for in vitro research. However, callus cultures of most plant species lose
their regeneration capacity with age in culture (Gould, 1986; Halperin,
1986). In tobacco callus, that capacity is lost slowly, beginning during
the second year in culture (Murashige and Nakano, 1967) and being manifested
by a slower formation of shoots, followed by a sequential reduction in the
number of shoots formed per callus and finally by little or no shoot formation (Thorpe, unpublished). It is desirable to use the same tissue line
from year to year when investigating physiological parameters of morphogenesis. This would eliminate introduction of different genotypes and allow
comparisons of previous data without fear that differences are arising from
different callus lines.

When considering woody species, preservation of superior producing
genotypes, such as those found in Pinus radiata, may also be beneficial to
preserve superior qualities (Thorpe, 1988). Although a deterioration of
morphogenic capacity has not been investigated in white spruce, it may be
anticipated that over time in culture, maturation capabilities of somatic
embryogenic cultures will decline. Notwithstanding this conclusion,
superior producing callus lines of white spruce are known in-as-far as
embryo maturation (Joy and Thorpe, unpublished) and should be stored to
reduce the possibility of declining morphogenic capacity.

The investigation of the physiology and biochemistry in regards to long
term storage of in vitro material should allow us to understand the basic
controls of loss of morphogenic competence and the maintenance or induction
of the juvenile state in plants.

TOBACCO

The tobacco shoot-forming system has been established since the late
1950's. It was the discovery by Skoog and Miller (1957) that auxin-
cytokinin ratios were important in the control of morphogenesis in tobacco
which led to tobacco being used as a model system for organogenesis. They
found that high auxin to cytokinin ratios led to root formation, high
cytokinin to auxin ratios led to shoot formation and intermediate levels of
the two phytohormones produced callus (wound parenchyma).

The tobacco shoot-forming system as it is utilized today is reproducible within a defined temporal framework (Thorpe and Murashige, 1970). When maintenance callus is placed on shoot-forming (SF) medium (high cytokinin to auxin) we find preferential zones of division beginning to appear by day seven on the basal portion of the tissue in contact with the medium. By day nine, meristemoids begin forming and by day twelve, primordia are forming. Recognizable buds begin to emerge from the bottom of the callus, and these further develop to leafy vegetative shoots thereafter. The addition of gibberellic acid (GA_3) to the shoot-forming medium (SFG) is utilized as a control for physiological studies, since its addition inhibits shoot formation in the time frame we are making observations (ca. one month).

The objective of this study was to investigate the effect of ascorbic acid on morphogenesis using young (4-12 subcultures of ca. 1 month each) and old (>30 subcultures) tissue. The protocol used was to include ascorbate (AA) in SF and SFG medium and compare morphogenic responses of the young and old tissues after 35 days.

Preliminary studies indicated that ascorbate exhibited maximum effects in the range of 10^{-7}-10^{-3}M with the optimal results around 10^{-3}-10^{-5}M (Joy et al. 1988). We therefore focussed on the latter range for further studies. In general, both fresh weight and dry weights increased with 5 x 10^{-5} AA and then decreased with higher concentrations to 5 x 10^{-3}M. This trend was found for both young and old tissue on shoot-forming (SF) and non-shoot-forming (SFG) media.

Analysis of starch and free soluble sugars indicated that in the old callus without AA added, sugars remained high over a sixteen day analysis period and there was little change in starch content. This trend, however, was different when optimal levels of ascorbate were included in the medium. The starch and sugar profiles in the tissue mimiced those of young tissues (Thorpe and Meier, 1975). At ascorbate concentrations of 4-8 x 10^{-4}M, free soluble sugars increased to day four and then declined continuously thereafter. The starch profiles started increasing in concentration from days four to ten whereupon concentrations decreased, as is the norm with young tissues during the shoot formation (Thorpe and Meier, 1975). The shoots which grew from these treatments (old tissue) were morphologically indistinguishable from young, untreated tissues which were induced to produce shoots.

The most pronounced effect of ascorbate was that of shoot production (Fig. 1). Young tissues without ascorbate typically formed 19 shoots per callus. However, ascorbate levels (8 x 10^{-4}M) increased shoot numbers by 45% over controls. An increase of 450% in shoot number was found when old tissue was cultured with 8 x 10^{-4}M. This large increase in shoot numbers was also found in the GA_3 treated (SFG) tissues. Shoot formation increased from none without the addition of ascorbate to approximately 19 shoots per callus at an ascorbate concentration of 4 x 10^{-4}M (Fig. 1). Numbers of shoots on SFG medium declined with higher concentrations of ascorbic acid, but were still substantially higher than the control. Ascorbate increased shoot production with old tissue on SFG medium to the maximum concentration of ascorbate used (2 x 10^{-3}M). However, numbers of shoots remained quite low (3-4 per callus at 2 x 10^{-3}M ascorbate).

WHITE SPRUCE

Until recently gymnosperms were considered recalcitrant in-as-far as somatic embryogenesis was concerned. However, in the mid 1980's somatic embryos of _Picea glauca_ (Hakman and von Arnold, 1985) and _Larix decidua_

(Nagmani and Bonga, 1985) were produced using immature tissues. White spruce (<u>Picea glauca</u>) among other conifers, has also been successfully induced to produce somatic embryogenic callus as well as the production of plantlets (Hakman and Fowke, 1987; Lu and Thorpe, 1987). Somatic embryogenic (SE) white spruce callus was induced from immature embryos between the months of June and August. The induction medium was a modified von Arnold and Eriksson (1981) formulation with 10 µM 2,4-D and 10 µM picloram as added phytohormones (Lu and Thorpe, 1987). SE callus from initial explants was subcultured every four to five weeks thereafter. Callus cultures proliferated as proembryonal masses which comprised of a densely cytoplasmic head which was subtended by elongate, vacuolated suspensor cells.

The induction of development to mature embryos, comprising of an enlarged, expanded hypocotyl region and fully developed, elongated cotyledons was accomplished by a phytohormone change. The induction medium was von Arnold and Eriksson medium with 10 µM Abscisic acid (ABA). Development of proembryos to mature cotyledonary stage embryos took place over four to six weeks.

White spruce callus cultures were placed on a maintenance medium (Lu and Thorpe, 1987) in flasks and a gastight seal was formed by placing a serum cap on the top of the flask. Growth of the sealed cultures was minimal in a gastight environment. Controls, which were capped with foam bungs or aluminum foil increased in fresh and dry weights for at least six weeks and subsequently senesced.

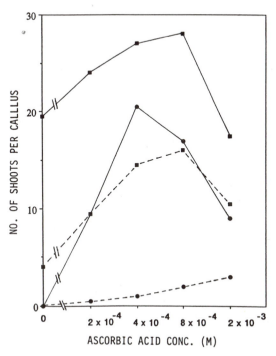

Fig. 1. Effect of ascorbate on number of shoots produced per callus in young (———) and old (————) tobacco callus. Shoots counted after 35 days in culture on SF (■) and SFG medium (●).

401

After one year in sealed flasks, 50-75% of cultures were still viable and could be subcultured. Comparisons between the morphology of calli subcultured every four to five weeks (CY) and that regrown from cultures stored for one year (RC) presented no evident differences. Growth data from callus cultures indicated that the RC callus grew slightly faster than the CY callus within the first three weeks after subculture, however, differences disappeared by four weeks in culture. The RC tissue easily produced cell suspensions which also had very similar growth characteristics as that of the CY line. Fresh weights were slightly higher for RC over a ten day culture period, whereas CY tissues were higher in dry weight. Packed cell volume and number of embryos per ml of medium over the ten day culture period for RC and CY tissue lines were similar. Development of embryos was accomplished by culturing maintenance callus to ABA-containing medium. The production of mature embryos took place normally and the efficiency of the RC line was similar to that of the original CY callus line.

CONCLUSIONS

Long term storage of in vitro material is important for the preservation of valuable germplasm, retention of desirable culture variation and to maintain morphogenic competence. Current methods employed to facilitate this storage are cryopreservation techniques. This type of storage may not be needed in cases such as tobacco and white spruce as has been illustrated above.

Recently, cryopreservation techniques have been employed for the long-term preservation of somatic embryogenic white spruce (Kartha et al., 1988). This technique utilizes highly specialized and expensive equipment, not common to most tissue culture facilities. In addition, cell cultures must be pretreated with cryoprotectants before storage in liquid nitrogen. The aim of cryopreservation is for the long-term storage of valuable germplasm which is prone to changes due to selective pressures from the culture environment as well as continuous subculturing.

In tobacco it was found that ascorbic acid enhanced not only morphogenic capacity in young but also in old tissue. Many endogenous factors may be considered to contribute to loss of regenerative capacity or possibly even recalcitrance. Inhibitors such as phenolics and gibberellins would be likely candidates for this type of epigenetic control. Gibberellic acid and phenolics are known to be natural endogenous products of plant material. Changes in these compounds either qualitatively or quantitatively may be responsible for loss of morphogenic competence. Since the one time addition of ascorbate to tobacco callus cultures initiated recovery of shoot-forming capacity in old tissue and increased shoot numbers in young tissues, the question of genetic aberrations causing loss of morphogenic capacity may be mostly discounted. Epigenetic or physiological factors are most likely the causative agents of morphogenic disfunction in this case. The buildup of these agents over time would explain the decrease in shoot formation in old tissue. In the young tissue, initial elimination of basal levels would further increase the numbers of shoots which were observed.

With white spruce, a gaseous environment which didn't allow exchange with the external environment seemed to play an important role in the storage of the tissues. From the data thus far acquired, it appears that lack of growth was detrimental to the sealed cultures, however, minimal growth (increases of 2-3 times initial subculture weight) was beneficial over one year in the sealed flasks. The tissues stored in such a way were also morphogenically competent, when induced to produce mature embryos on ABA-containing medium. It therefore appears that culture vessel gases (carbon dioxide, ethylene and oxygen) may play an important role in the storage of white spruce callus, which may subsequently eliminate possible epigenetic barriers in its growth and development. Whether such simple controls are operative in other cell culture systems remains to be determined.

REFERENCES

Gould, A. F., 1986, Factors controlling generation of viability in vitro, in: "Cell Culture and Somatic Cell Genetics of Plants,", Vol. 3, I. K. Vasil, ed., Academic Press, New York, pp. 549-567.

Hakman, I., and von Arnold, S., 1985, Plantlet regeneration through somatic embryogenesis in Picea abies (Norway spruce), J. Plant Physiol., 121:149-158.

Hakman, I., and Fowke, L. C., 1987, Somatic embryogenesis in Picea glauca (White spruce) and Picea mariana (Black spruce), Can. J. Bot., 65:656-659.

Halperin, W., 1986, Attainment and retention of morphogenic capacity in vitro, in: "Cell Culture and Somatic Cell Genetics of Plants,", Vol. 3, I. K. Vasil, ed., Academic Press, New York, pp. 3-47.

Joy IV, R. W., Patel, K. R., and Thorpe, T. A., 1988, Ascorbic acid enhancement of organogenesis in tobacco callus, Plant Cell Tiss. Org. Cult., 13:219-228.

Kartha, K. K., Fowke, L. C., Leung, N. L., Coswell, K. L., and Hakman, I., 1988, Induction of somatic embryos and plantlets from cryopreserved cell cultures of white spruce (Picea glauca), J. Plant Physiol., 132:529-539.

Lu, C.-Y., and Thorpe, T. A., 1987, Somatic embryogenesis and plantlet regeneration in cultured immature embryos of Picea glauca, J. Plant Physiol., 128:297-302.

Murashige, T., and Nakano, R., 1967, Chromosome complement as a determinant of the morphogenic potential of tobacco cells, Amer. J. Bot., 54:963-970.

Skoog, F., and Miller, C. O., 1957, Chemical regulation of growth and organ formation in plant tissues cultured in vitro, Symp. Soc. Exp. Biol., 11:118-131.

Thorpe, T. A., and Meier, D. D., 1975, Effect of gibberellic acid on starch metabolism in tobacco callus cultures under shoot-forming conditions, Phytomorphology, 25:238-245.

Thorpe, T. A., and Murashige, T., 1970, Some histochemical changes underlying shoot initiation in tobacco callus cultures, Can. J. Bot., 48:277-285.

Thorpe, T. A., 1988, Physiology of bud induction in conifers in vitro, in: "Genetic Manipulation of Woody Plants," J. W. Hanover and D. E. Keathley, eds., Plenum Publishing, pp. 167-184.

von Arnold, S., and Eriksson, T., 1981, In vitro studies of adventitious shoot formation in Pinus contorta, Can. J. Bot., 59:870-874.

PROTEIN PATTERNS ON <u>Corylus avellana</u> L ROOTING CAPACITY

A. González, R. S. Tamés and R. Rodríguez

Lab. Fisiología vegetal, Dpto B.O.S., Fac. Biología
Univ, Oviedo, 33005 Oviedo, Spain

INTRODUCTION

The final culture stage prior to acclimation of plantlets requires the induction of a root system. Optimum survival of plantlets depends on a good root to shoot ratio (Sommer and Caldas, 1981). Whilst root induction is usually controlled by hormonal treatments there is considerable evidence that other factor, including carbohydrate supply (e.g. Thorpe and Patel, 1984) auxin (Hartmann and Kester, 1975) and ethylene, are important in determining ultimate success. Since adventitious root formation involves dedifferentiation, cell division and cell enlargement, as well as differentiation, there is a basic need to investigate the temporal control of such aspect of metabolism.

Ethylene has repeatedly been assigned a major role in rooting, but the experimental results have been contradictory. Roy et al., (1972) did not find any effect of ethylene on rooting, whereas Mullins (1972) reported inhibition and Robbins et al., (1985) reported stimulation.

Control of protein metabolism during initiation and development of adventitious root and the effect of ethylene synthesis inhibitor and precursor on rooting have been the subject of relatively little work to date in woody species.

Molnar and Lacroix (1972) have reported that an increase in cellular protein precedes the first divisions of preformed initials in Hydrangea. In French bean, which has no preformed initials, IBA, at concentration which readily initiated the rooting, rapidly stimulated synthesis of protein (Kantharaj et al. 1979). However, other work has shown inhibitors of RNA and protein synthesis to enhance the rooting response of some cuttings (e.g. Anzai et al., 1971).

This paper presents information on the protein patterns of adventitious roots induced in cotyledons and shoot-segments of seedlings of <u>Corylus avellana</u> L. treated with inhibitors and precursors of ethylene synthesis .

Plant Aging: Basic and Applied Approaches
Edited by R. Rodríguez *et al.*
Plenum Press, New York, 1990

MATERIAL AND METHODS

Seed of <u>Corylus avellana</u> L., previously sterilized, were cultured in a half-strength K(h) (1/2 K(h) basal medium (Cheng, 1975)) in continuous dark. After 20 days of culture in basal medium, the cotyledons and shoot-segments were transplanted to rhizogenic medium (RM) consisting of 1/2 K(h) basal medium plus indole-3-butiric acid (IBA) 50 uM and kinetine (Kn) 5 uM (Pérez et al., 1984).

Aminoethoxyvinyl glycine (AVG) 30 μM, inhibitor of ethylene synthesis, 1-aminocyclopropane 1-carboxilic acid (ACC) 5 μM, and 2-chloroethylphosphonic acid (CEPA) 0.5 μM precursor and ethylene-releasing compound respectively were added to the rhizogenic medium at the start of the culture.

The number of roots per explant of each treatment were counted every five days after a rooting period of 15 days. Each treatment comprising 30 cotyledons or 15 shoot-segments was repeated three times.

Proteins were extracted by homogenizing, with a blender homogenizer, 0.5 gr. of cotyledons or shoot-segments with 5 ml of anhydre acetone (-20ºC) as described by (Ibraim and Cavia, 1975). Previously the lipid content was eliminated with chloroform-methanol (3:1 v/v) mixture (Rodriguez-Bujan, 1975).

The homogenate was centrifuged at 15000 g for 10 min. The pellet was dried with liquid nitrogen and the powder obtained was suspended in 1% (w/v) SDS at 100ºC for 2 min., centrifuged again at 14000 g for 10 min. Protein was determined by the method of Lowry et al. (1951) using bovine serum albumin disolved in 0.5 N NaOH as standard.

SDS-polyacrylamide gel electrophoresis (SDS-PAGE) was used to separate the proteins and the procedure carried out as described by Laemmli (1970). The resolving gel contained a gradient of 6 to 18 % (w/v) acrylamide and with 0.3 % (w/v) bisacrylamide. Aproximately 150 - 200 ugr of protein per track was used and run for 15 h at 90 v. per gel. Molecular weight markers (Sigma) were run simultaneously with 0.25 % (w/v) Coomassie Brilliant Blue in 45 % (w/v) methanol and 10 % (v/v) acetic acid in a Bio-Rad destainer.

The dry gels covered with a cellophane membrane were introduced on the platine for the lecture and quantified by transmission with a Joyce Lobel densitometer at 560 nm (A_{560}).

RESULTS AND DISCUSSION

Cotyledons and shoot-segments on rhizogenic medium (control) showed a high number of roots after 30 days of culture (12 roots/cotyledons, approximately). Rooting was higher for cotyledons for than shoots (Fig.1 and 2).

Adventitious root induction (ARI) was significantly reduced by AVG in both tissues but shoot-segments showed basal and apical necrosis and after 30 days they did not show any root (Fig.1 and 2). AVG presumably reduced the levels of endogenous ethylene and thereby reduced the amount of ethylene available for the induction of roots. High AVG concentration caused a non-specific inhibition of ARI and growth through a general suppression of protein and RNA synthesis (Miller and Roberts,1984).

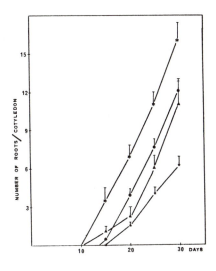

Fig. 1. Rhizogenic response in cotyledons after 30 days of culture in:

- ●——● rhizogenic medium
- ▲——▲ RM plus ACC 5.0 μM
- *——* RM plus CEPA 0.5 μM
- ★——★ RM plus AVG 30 μM

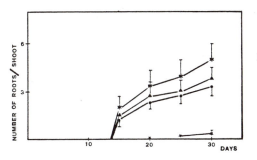

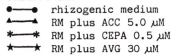

Fig. 2. Rhizogenic response in shoots after 30 days of culture in:

- ●——● rhizogenic medium
- ▲——▲ RM plus ACC 5.0 μM
- *——* RM plus CEPA 0.5 μM
- ★——★ RM plus AVG 30 μM

The addition of ACC (5.0 μM) to the RM did not seem to have any influence in rooting (Fig.1 and 2). More direct evidence that ethylene plays a positive role in ARI in hazel comes from the observation that the ethylene-releasing compound (CEPA, 0.5 μM) also stimulated rooting compared to IBA-Kinetin-induced control explant (Fig.1 and 2). Presumably, ethylene was released from the culture medium thus raising the "in vivo" level of ethylene (Miller and Roberts, 1984). Apparently, cotyledons and shoot explants cultured on RM produced ethylene necessary for adventitious root induction and when an increase occurs (as a result of CEPA), maximum rooting is achieved.

Analysis of protein changes in cotyledons or shoots was carried out simultaneously in all the samples by the single dimensional SDS-PAGE. Only rooting region of tissues (cotyledons or shoots) was taken for electrophoretic study .

The patterns produced from extracts of different samples were compared visually on the basis of differences in relative intensity or presence/absence of specific bands and were compared with a densitometer. The patterns of the bands obtained were different for each tissue (cotyledons and shoots). Higher protein content was obtained in cotyledons than in shoots.

During root induction, in cotyledons, there did not seem to be a synthesis of new proteins, only a degradation of stored proteins (Fig.3 a and b). Shibaoka et al., (1967) suggested that root regeneration may necessitate the inhibition of some synthesis of RNA, presumably mRNA, which is normally involved in suppressing root formation. Therefore Jarvis et al., (1985) showed a inhibitory effect of IBA on protein synthesis in the stem cutting of Phaseolus aureus. This compound enhances the development of adventitious roots.

When cotyledons were treated with AVG 30 μM, the main differences were amongst lower molecular weight proteins (Fig.3 c).

ACC treatment cotyledons showed strong differences in relation to the control. These differences were most marked amongst the proteins of molecular weight between 40 and 20 kd (Fig. 3 d). The bands of 35 and 23 kd were similar to those obtained at the beginning of root induction.

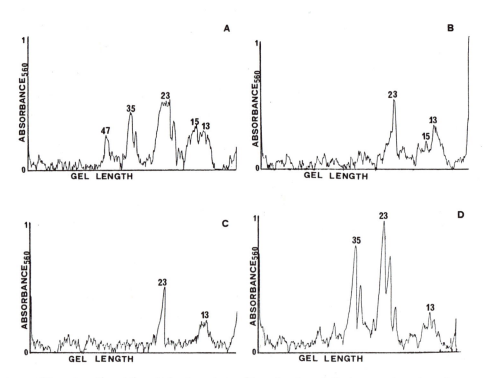

Fig. 3. Densitometric tracings (A_{560}) of cotyledons extracts of:
A.- At the beginning of root induction.
B.- After 30 days in rhizogenic medium
C.- After 30 days in rhizogenic medium plus AVG 30 μM
D.- After 30 days in rhizogenic medium plus ACC 5.0 μM

The protein patterns of the shoots were different to those obtained in cotyledons. During roots formation the number of bands were similar but the protein content were lower in shoots extracts after 30 days in rhizogenic medium (Fig. 4 a and b). Only bands of 35 kd and 23, 21 and 13 kd were similar to those ebserved in cotyledons.

AVG treated shoots showed a reduction of total protein content, mainly amongst proteins with molecular weight lower than 40 kd (Fig. 4 c).

Densitometric tracing obtained with shoots cultured for 30 days in rhizogenic medium suplemented with ACC showed higher protein content than AVG treatment. The Bands were similar to those obtained at the first rooting.

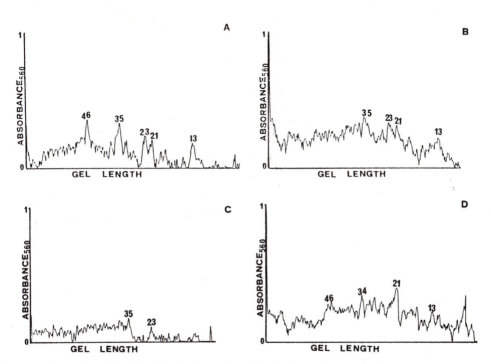

Fig. 4.- Densitometric tracing of shoots extracts of:
A.- At the beginning of root induction
B.- After 30 days in rhizogenic medium
C.- After 30 days in rhizogenic medium plus AVG 30 μM
D.- After 30 days in rhizogenic medium plus ACC 5.0 μM

REFERENCES

Anzai, T., Shibaoka, H. and Simokoriyama, M.1971. Increases in the number of adventitious roots caused by 2-thiouracil and 5-bromodeoxyuridine in Phaseolus mungo cuttings. Plant Cell Physiol. 12: 695-700.
Cheng, T. Y. 1975. Adventitious bud formation in culture of Douglas fir (Pseudotsuga menziensii Mirb. Franco). Plant Sci. Lett. 5: 97-102.
Hartmann, H. T. and Kester, D. E. 1975. Plant Propagation. Principles

and Practices.- Third edition, Prentice-hall Englewood Cliffs, N. J.

Ibraim, R. K and Cavia, E. 1975 Acrylamide gel electrophoresis of proteins from intact and cultured plant tissues.Can. J. Bot. 53 (5): 517-519.

Jarvis, B.C., Yasmin, S. and Coleman M.T. 1985. RNA and protein metabolism during adventitious root formation in stem cutting of Phaseolus aureus . Physiol. Plant. 64. 53-59.

Kantharaj, G. R., Mahadevan, S. and Padmanaban, G. 1979. Early biochemical events during adventitious roots initiation in the hypocotyl of Phaseolus vulgaris Biochemistry 18: 383-387.

Laemmli, V. K. 1970. Cleavage of the structural proteins during the assembly of the head of bacteriophage T_4.Nature 227: 680-685.

Lowry, O. H., Rosebrough, N.J., Farr, A:L. and Randall, R.J. 1951. Protein measurement with the Folin phenol reagent. J. Biol. Chem. 193: 265-275.

Miller, A. R. and Roberts, L. W. 1984. Ethylene biosynthesis and xylogenesis in Lactuca pith explants cultured "in vitro" in the presence of auxin and cytokinin: The effect of ethylene precursors and inhibitors. J: Exp. Bot. 35 (1544): 691-698.

Molnar, J. and Lacroix, L. 1972. Studies of the rooting of cutting of Hydrangea macrophylla: enzyme changes. Can. J. Bot. 50: 315-322.

Mullins, M. G. 1972. Auxin and ethylene in adventitious root formation in Phaseolus aureus (Roxb.). In: Plant Growth Substances. 1970 (D. J. Carr, ed.), pp. 526-533. Springer-verlag, Berlyn.

Perez, C., Fernandez, B., Rodriguez, R. and S. Tamés, R. 1984. Starch content and rooting capacity of filbert cotyledon "in vitro". Phyton 44 (2): 101-106.

Robbins, J.A., Reid, M.S., Paul, J.L. and Rost, T. L. 1985. The effect of ethylene on adventitious root formation in Mung bean (Vigna radiata) cuttings. J. Plant Growth Reg. 4: 147-157.

Rodriguez- Bujan, M. C. 1975. Aspectos fisiológicos de la germinación de semillas de Cicer arietinum L. Tesis doctoral. Universidad de Santiago.

Roy, N. N., Baser, R. N. and Bose, T. K. 1972. Interaction of auxins with growth-retarding, inhibiting and ethylene-producing chemicals in rooting of cutting. Plant Cell Physiol. 13: 1123-1127.

Shibaoka, A., Anzai, T. Mitsuhashi, M. and Shimokoriyama. M. 1967. Interaction between heliangine and pyrimidine in adventitious root formation of Phaseolus cutting. Plant Cell Physiol. 8: 647-656.

Sommer, H. E. and Caldas, L. S. 1981. "in vitro" methods applied to forest trees. In: T. A. Thorpe, ed. Plant Tissue Culture: Methods and Aplications in Agriculture. Academic Press, New York. pp. 349-358.

Thorpe, T. A. and Patel, K. R. 1984. Clonal propagations: adventitious buds. In: I.K. Vasil, ed. Cell Culture and Somatic Cell Genetics of Plants. Vol. 1. Academic Press. New York. pp 49-60.

INDUCING ARTIFICIAL TETRAPLOIDS FROM

DIPLOID MEADOW FESCUE (**Festuca pratensis** Huds.) VARIETIES

AND THE INVESTIGATION OF SOME MEIOTIC CHARACTERISTICS OF THEM*

Bilal Deniz

Div.Cytogenetics and Plant Breeding
Agricultural Faculty
Erzurum, Turkey

INTRODUCTION

Culture varieties of diploid meadow fescue are extensively used as forage plant in Europe and North America (Caputa, 1967). After the discovery of colchicine as a polyploidy effect producer, researches on polyploidy in many plant species were started. Many studies on meadow fescue by using colchicine have also been carried out (Simonsen, 1975; Easton, 1977; Sugiyama et al., 1979). Tetraploid plant rate, obtained by colchicine treatment, showed differences according to plant species and varieties. Ahloowalia (1967) and Sağsöz (1982) applied colchicine treatment of 0.2 % to different varieties of perennial ryegrass at different tempreratures and obtained different results according to different treatments and varieties.

In two different studies on the same autotetraploid perennial ryegrass (Crowley and Rees, 1968; De Roo, 1968), different results of bivalent and quadrivalent numbers in metaphase I (MI) were found. Simonsen (1975) reported mostly 7 bivalents in diploid plants and univalents, quadrivalents and trivalents in some others. It was understood that in C_2 tetraploids generally there were bivalents followed by quadrivalents. Small numbers of univalents and trivalents were also observed. Simonsen reported a general regularity in the distribution of anaphase I (AI) in diploid and tetraploid plants. In studies on different plant species, AI differences were taken as a criterion in determining

---

*This paper is a part of the doctoral thesis entitled: "Inducing Artificial Tetraploids from Diploid Meadow Fescue (**Festuca pratensis** Huds.) Varieties and the Comparison of Some Cytological and Morphological Characteristics of Them."

Plant Aging: Basic and Applied Approaches
Edited by R. Rodríguez *et al.*
Plenum Press, New York, 1990

411

regular meiosis. It was also found that tetrads, which lend themselves to easy examination, would be a criterion in finding regular meiosis and there was a relation between micronuclei in tetrads and MI and AI irregularities (Myers, 1945 ; Müntzing, 1951; Webster and Buckner, 1971 ; Sağsöz, 1983).

MATERIAL AND METHODS

In order to obtain polyploid plants, S.215 of English origin and Senu and Salfat of Denmark origin varieties were used. Of each variety 1500 seedlings with 3-5 mm root length were selected and these were divided into three equal groups. Each group were treated with 0.2 % colchicine at 24, 27 and 30°C for a period of three hours. After the selection of seedlings according to their morphological characteristics, their chromosomes were counted by using root tip-squash and feulgen staining method (Heneen, 1962).

Diploid and autotetraploid plants, grown in clons under greenhouse conditions, were examined comparatively in terms of seed set* and meiotic characteristics. For meiotic studies, panicles were fixed in Carnoy's solution (Simonsen, 1975) and then 2 % of acato-carmin was used for the preparation of slides (Hazarika and Rees, 1967).

RESULTS AND DISCUSSION

Despite tetraploid plants grew slowly at the beginning, they showed some stronger and more vigorous growth at the later stages. Varieties responded differently to colchicine treatment at different temperatures and the average tetraploid rates in Senu, Salfat and S. 215 varieties were found to be 3.9 %, 2.8 and 1.6 respectively. These differences may be attributed to the different relative division rates of the cells in the growth points of the seedlings. Similar results were obtained by Ahloowalia (1967) and Sağsöz (1982) who worked on perennial ryegrass plants. They proved that each variety had a specific response to colchicine treatment and this might change according to temperatures during the treatment.

All the MI cells examined in diploid Senu variety had only bivalent pairing whereas in the other two varieties univalents and quadrivalents were seldom found besides bivalents (Table 1). Senu and Salfat varieties had only ring (O) and rod (--) bivalents while in S.215 variety cross (X) shaped bivalents were also seen. General occurence of bivalent pairing and the abundance of ring bivalents indicate high rate of homology and the regularity of MI. On the other hand, besides other bivalent shaping, uni-valents and rare quadrivalents are considered to be the cause of irregular meiosis in diploid plants. Simonsen (1975), in a similar study on meadow fescue reported similar pairings in MI. In the MI cells of autotetraploid C_o plants, there occured mostly quadrivalent and bivalent and smaller

*Seed set characteristics are not given in this paper.

412

Table 1. The Shapes of MI Chromosome Pairing in Pollen Mother Cells of Three Diploid Meadow Fescue (**F. pratensis** Huds.) Varieties and Their Artificial Tetraploids

Varieties	Plant Number	Cell Number	Univalent Min.	Max.	Avr.	Bivalent Min.	Max.	Avr.	Trivalent Min.	Max.	Avr.	Quadrivalent Min.	Max.	Avr.	Hexavalent Min.	Max.	Avr.
S.215	20	131	0.0	2.0	0.046	6.0	7.0	6.947				0.0	1.0	0.015			
S.215 C°	20	97	0.0	4.0	0.61	0.0	14.0	2.99	0.0	2.0	0.19	4.0	7.0	4.58	0.0	2.0	0.05
Senu	23	117				7.0	7.0	7.0									
Senu C°	43	105	0.0	5.0	0.79	0.0	14.0	3.68	0.0	3.0	0.16	3.0	7.0	4.83			
Salfat°	21	151	0.0	2.0	0.331	5.0	7.0	6.781				0.0	1.0	0.026			
Salfat C°	43	119	0.0	7.0	1.19	0.0	14.0	3.43	0.0	1.0	0.11	3.0	7.0	4.84	0.0	1.0	0.03

Table 2. AI Separations and Tetrads Micronuclei in Pollen Mother Cells of Three Diploid Meadow Fescue (**F. pratensis** Huds.) Varieties and Their Artificial Tetraploids

Varieties	Plant Number	Cell Number	7/7	6/8	14/4	13/15	12/16	Lag.and Div.Uni.	Frag.	Bridge	Tetrads Cell Number	M/Q	%
S.215	19	90	76	2				5	2	5	137	0.235	79.7
S.215 C°	20	76			51	13	–	6	3	3	206	0.416	68.6
Senu	20	81	72	–				5	–	4	154	0.111	88.6
Senu C°	45	75			52	8	2	12	–	1	293	0.326	75.4
Salfat°	21	79	67	2				5	–	5	202	0.287	78.8
Salfat C°	44	97			45	11	2	37	–	2	196	0.839	53.6

number of univalent and trivalent pairings. In addition, small number of hexavalent pairing was seen in S.215 and Salfat varieties (Table 1). Most of the quadrivalent pairings occured in the shape of chain (oo), ring (O) and zig-zag (Z) . Ring-shaped and rod-shaped bivalent pairings were seen in equal frequency in bivalent pairings. The main reasons for irregular meiotic division, often observed in artificial tetraploids, are considered to be the univalent and trivalent pairings as well as other pairing abnormalities. Simonsen (1975) observed the majority of bivalents and quadrivalents and smaller number of univalent and tirivalent pairings in C_2 generation. Taking the results of many researches into consideration, it can be suggested that there is no significant difference between bivalent and quadrivalent pairings in terms of regular meiosis. However, in two separate studies, in which the same autotetraploid material was used, the findings were reported to be different in terms of the average number of quadrivalent and bivalent (Crowley and Rees, 1968; De Roo, 1968).

AI separations in all three diploid varieties were generally regular (7/7) (Table 2). In tetraploid plants, however, AI separations were relatively regular (14/14) in Senu and S.215 varieties while, in Salfat variety there occured high rate of irregularities (Table 2). In diploid and tetraploid plants, besides lagging and divided univalents, dicentric choromatid bridges and fragments caused irregular anaphase separations.In the light of the evidence available, it may be concluded that there was a close relationship between regular meiosis and AI separations. It was particularly considered that univalents of MI were correlated with the lagging and divided univalents seen in AI. Simonsen (1975), in his study on meadow fescue, reported similar AI irregularities in diploids and tetraploid C_2 plants.

In this study, tetrads of diploid and tetraploid were examined and different numbers of micronuclei were found in different varieties. It was observed that, in Senu variety, the percentage of tetrads without micronuclei (%) was higher while the average number of micronuclei in a tetrad (M/Q) was lower (Table 2). In the light of the evidence provided by experiments on the three meiotic periods of diploid and tetraploid plants, it could be said that the irregularities seen in meiotic divisions were interrelated. Based on this, in addition to AI separations, constituting a good criterion in determining the meiotic regularity, very easily examined tetrads would also be of much use. In studies on rye (Müntzing, 1951) and perennial ryegrass (Sağsöz, 1983) it was proved that to decide about the meiotic regularity could be easier with the average micronuclei number in tetrads as well as with AI separations. Similar relations were found by many other researchers who studied on various plant species (Myers, 1945; Webster and Buckner, 1971; Sağsöz, 1983).

REFERENCES

Ahloowalia, B. S., 1967 Colchicine induced polyploids in ryegrass, Euphytica,16: 46-60

Caputa, J., 1967, "Les plantes fourregéres", la Maison Rustique, 26 Rue Jacop, Paris 6e.

Crowley, J. G., and Rees, H., 1968, Fertility and selection in tetraploid lolium, Chromosoma, 24: 300–308

De Roo, R., 1968, Meiosis and fertility in tetraploid perennial ryegrass (**Lolium perenne** L.), Landbouwetenschap. Gent., 33: 247–258.

Easton, H.S., 1977, A comparative study of genetic effects in isogenic diploid and tetraploid plants of **F. pratensis** Huds. I. Introduction, methodology and mean character differences, Ann. Amelior Plantes, 27: 563–576.

Hazarika, M. H., and Rees H., 1967, Genotypic control of chromosome behaviour in rye. X. Chromosome pairing and fertility in autotetraploids, Heredity, 22: 317–332.

Heneen, W. K., 1962, Karyotype studies in **Agropyron junceum, A. repens** and their spontaneous hybrids, Hereditas, 48: 471–500.

Müntzing, A., 1951, Cyto-genetic properties and practical value of tetraploid rye. Hereditas, 37: 17–84.

Myers, W. M., 1945, Meiosis in autotetraploid **L. perenne** in relation to chromosomal behaviour in autotetraploids, Bot. Gaz., 106–:304–316.

Sağsöz, S., 1982, Farklı ingiliz çimi(**L.perenne** L.) çeşitlerinde poliploid bitki elde etme olanakları üzerinde bir araştırma, Atatürk Üni., Yay. No: 596, Erzurum, Turkey.

Sağsöz. S., 1983, Tetraploid ingiliz çiminde (**L. perenne** L.) anafaz I ayrılışlarının ve tetradlardaki çekirdekcik sayılarının fertilitede seleksiyon ölçüsü olarak kullanılma olanakları, İn:" VII. Bilim Kongresi TOAG Tebliğleri" TUBİTAK Yayınları No:552 Ankara,Turkey.

Simonsen, Q., 1975, Cytogenetic investigations in diploid and autotetraploid populations of **F. pratensis** Huds., Hereditas, 79:73–108.

Sugiyama, S., Takahashi, N.,and Goto, K., 1979, Studies on potential variability in Festuca. II. Polycross progeny test in an induced autoetraploid of **Festuca pratensis**, Plant Breed Abst., 1981, 51:455(5192).

Webster, G.T., and Buckner, R. C., 1971, Cytology agronomic performance of lolium-festuca hybrid derivatives, Crop Science, 11:109–112.

INACTIVATION OF PEROXIDASE: ITS ROLE IN PLANT SENESCENCE

M. Acosta, M.B. Arnao, J.A. del Río, J.L. Casas, J. Sánchez-Bravo and F. García-Cánovas[*]

Department of Plant Biology and [*]Department of Biochemistry. University of Murcia
Sto. Cristo, 1. 30001-MURCIA. SPAIN

INTRODUCTION

The activated metabolites of molecular oxygen (H_2O_2, O_2^-, $HO^·$) play an important role in cellular activity, because their accummulation can induce degradative processes in cellular components. This oxygen toxicity is caused by the imbalance of a sensitive equilibrium: molecular oxygen is needed to generate energy and maintain other biological functions, but at the same time it is the source of the activated forms of oxygen, which damage the cells (Fridovich, 1976).

Systems exist to protect the cells against these activated metabolites, such as enzymes (superoxide dismutases, catalases and peroxidases), and antioxidant metabolites (ascorbate, vitamin E, hydroquinones, etc.) (Rabinowitch and Fridovich, 1983).

Both peroxidase and peroxide levels in the tissues play an important role in fruit senescence and ripening. This hemoenzyme combines the destruction of H_2O_2 and other hydroperoxides with its capacity of oxidize a great variety of substrates.

Peroxidase undergoes a process of inactivation during its catalytic cycle (Ator et al., 1987; Acosta et al., 1988). This process might play an important role in senescence because it is a way of controlling the enzyme level and the capacity of the tissue to respond to the hydroperoxide levels.

In the present work, we study the process of inactivation which peroxidase undergoes during the destruction of either H_2O_2 or another hydroperoxide, at physiological pH, using 2,2'-azino-bis-(3-ethyl-benzthiazoline-6-sulphonic acid) (ABTS) as a reductant substrate. The "in vivo" role of this inactivation process and the importance of the reductant substrate/peroxide ratio in the enzyme mechanism are discussed.

MATERIAL AND METHODS

Purified horseradish peroxidase type VI, RZ (A_{403}/A_{275})= 3.1 (Sigma). ABTS (Boehringer-Mannheim). Hydrogen peroxide (Merck). m-chloroperoxybenzoic acid (m-CPBA) (Aldrich).

Rate of Peroxidase Inactivation

The inactivation of HRP was carried out at 25 °C in 1 ml-incubations of 50 mM Na-phosphate buffer (pH 6.3) containing the enzyme (1 μM). Two types of incubation were made: a) incubations in the presence of H_2O_2; b) incubations in the presence of m-CPBA. At specified time intervals, 10 μl aliquots of the incubation mixtures were transferred to cuvettes containing 2 ml of an assay mixture composed of 0.5 mM ABTS and 0.2 mM H_2O_2 in 50 mM Gly-HCl buffer (pH 4.3). The peroxidase activity was measured by the increase in the absorbance at 414 nm, which is characteristic for the ABTS oxidation product (Childs and Bardsley, 1975).

Determination of the Product Accumulated at the End of the Reaction (P_∞)

The reactions were made in 50 mM Na-phosphate buffer, pH 6.3 at 25 °C, followed as ΔA_{440}, using an estimated molar extinction coefficient for the ABTS oxidation product at 440 nm of 8,943 $\pm$ 100 M^{-1} cm^{-1}. HRP, ABTS and peroxide (H_2O_2 or m-CPBA) were added in variable concentrations, two at a time being maintained constant and the other changing. In these conditions it was possible to estimate maximum values for the product accumulated at the end of each reaction (P_∞), because the final values of A_{440} show a good stability.

RESULTS AND DISCUSSION

The inactivation of peroxidase by peroxide, in the absence of reductant substrates, was investigated by checking the residual enzymatic activity in a)

H_2O_2 –incubations; b) m–CPBA –incubations, both containing the enzyme (see Materials and Methods). Figure 1 shows a time- and peroxide concentration–dependent inactivation of the enzyme. The incubations with m–CPBA show, at 1 min, a higher loss of the enzymatic activity, m–CPBA showing itself to be a much more powerful inactivating substrate than H_2O_2, even at lesser concentrations.

Figure 2 shows a plot of $[P_\infty]$ versus [ABTS] for both peroxides, keeping the enzyme concentration constant. A high H_2O_2 concentration produces lower P_∞ values for each ABTS concentration. m–CPBA, at a concentration much less than that of H_2O_2 and with a higher concentration of enzyme, produces lower P_∞ values, m–CPBA again proving itself to be a stronger inactivator of the enzyme as in Fig. 1.

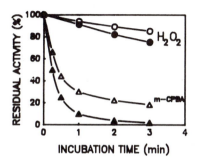

Fig. 1. Rate of peroxidase inactivation. The peroxide concentrations were: H_2O_2 : 1 mM (O) and 5 mM (●); m–CPBA: 0.01 mM (Δ) and 0.1 mM (▲). Details of the incubations are given under Materials and Methods.

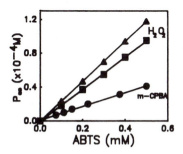

Fig. 2. Dependence of the P_∞ on the ABTS concentrations. The concentrations were: H_2O_2 10 mM (▲) and 20 mM (■), both with peroxidase 2.5 nM; m–CPBA 0.25 mM (●) with peroxidase 40 nM. The experimental details are given under Materials and Methods.

Figure 2 also shows that for a fixed peroxide concentration, the P_∞ values increase as the ABTS concentration increases. This protective role of ABTS towards enzymatic activity occurs with both peroxides, the values of P_∞ depending on the type of peroxide and on the ABTS/peroxide ratio.

The main function of peroxidases in plant cells is to eliminate the H_2O_2 generated by other enzymatic systems, such as superoxide dismutases, glycolate oxidase, xanthine oxidase. This protective role of the enzyme in oxygen metabolism is only possible if reductant compounds susceptible to oxidation by peroxidase are present.

The absence of reductant substrates provokes a rapid inactivation in the enzyme which thus loses its capacity to eliminate H_2O_2 (Fig. 1). Therefore, reductant substrates are necessary for i) the catalytic turnover of the enzyme and the consequent elimination of excess H_2O_2 and ii) the protection of the enzyme from inactivation by peroxides (Fig. 2), thus ensuring the protective role of peroxidase.

A similar system has been described in spinach chloroplasts (Hossain and Asada, 1984). Ascorbate peroxidase, an enzyme present in chloroplasts and responsible for eliminating H_2O_2 undergoes an inactivation process when H_2O_2 is added to chloroplast in the dark and also in ascorbate–depleted media. Ascorbate concentrations of less than 20 μM produce a loss of ascorbate peroxidase activity and consequently the chloroplast loses its capacity to eliminate the H_2O_2 generated during photosynthetic processes.

The H_2O_2/ABTS/HRP system described by us simulates that in chloroplasts (H_2O_2/Ascorbate/Ascorbate peroxidase). Therefore, the H_2O_2–eliminating system in chloroplasts recquires a coupled reductant substrate (ascorbate) without which the enzyme would be inactivated. We show in this paper that H_2O_2 itself is the inactivating agent and that the reductant substrate protects the enzymatic activity.

The dark addition of H_2O_2 determines the oxidation of the chloroplast components, due to the fact that the monodehydroascorbate reductase, dehydroascorbate reductase, glutathione reductase systems do not regenerate ascorbate because of the lack of NADPH, thus provoking the inactivation of ascorbate peroxidase. The existence of coupled enzymatic systems which ensure the supply of reductant substrates to the peroxidases is a key factor due to the mechanism of these enzymes. The catalytic cycle of the peroxidases generates free radicals, which unless rapidly reduced, could provoke chain oxidation reactions and consequent damage to the cell structures.

The inactivation process which peroxidases undergo might therefore have some physiological implications. An imbalance between peroxide levels and the availability of specific reductant substrates determines the appearance of a strongly oxidant enzymatic form (Compound I of peroxidase). Compound I reacts with many chemical structures susceptible to oxidation to generate free radicals. One possible way of avoiding the generation of these dangerous radicals would be to control the peroxidase activity by means of enzyme inactivation.

Acknowledgement

M.B.A. has a grant from Instituto de Fomento. Murcia. Spain.

REFERENCES

Acosta, M., del Río, J.A., Arnao, M.B., Sánchez-Bravo, J., Sabater, F., García-Carmona, F. and García-Cánovas, F., 1988, Oxygen consumption and enzyme inactivation in the indolyl-3-acetic acid oxidation catalyzed by peroxidase. Biochim. Biophys. Acta, 955:194-202.

Ator, M.A., David, S.K., and Ortiz de Montellano, P.R., 1987, Structure and catalytic mechanism of horseradish peroxidase, J. Biol. Chem., 262:14954-14960.

Childs, R.E., and Bardsley, W.G., 1975, The steady-state kinetics of peroxidase with 2,2'-azino-di-(3-ethyl benzthiazoline-6-sulphonic acid) as chromogen, Biochem. J., 145:93-103.

Fridovich, I., 1976, Oxygen radicals, hydrogen peroxides and oxygen toxicity, in: "Free Radicals in Biology", W. Pryor, ed., Academic Press, New York.

Hossain, M.A., and Asada, K., 1984, Inactivation of ascorbate peroxidase in spinach chloroplast on dark addition of hydrogen peroxide: Its protection by ascorbate, Plant Cell Physiol., 25:1285-1295.

Rabinowitch, H.D., and Fridovich, I., 1983, Superoxide radicals, superoxide dismutases and oxygen toxicity in plants, Photochem. Photobiol., 37:679-690.

ELECTRON MICROSCOPE OBSERVATIONS AND BA EFFECTS IN APPLE CELLS

Ricardo J. Ordás; Fernández, B. and Rodriguez, R.

Laboratorio de Fisiología Vegetal. Fctad de Biología

Oviedo, Spain

INTRODUCTION

It has been reported that the administration of cytokinins to plants affects nuclear transcription. Other observations suggest that cytokinins also have some effects at the translation level.

Considerable attention has been given to the cytokinin effects on growth kinetics and chemical composition of growing plant cells cultures, but only a few studies have dealt with the morphological changes that accompany their development.

In this paper, the effect of cytokinin treatment on the ultrastructural changes of apple cells are reported.

MATERIAL AND METHODS

A cytokinin-requiring cell line D2,14, cloned from leaf callus of apple (Malus domestica, Borkh) was used for obtaining a cell suspension. Cell culture were routinely subcultured in an agitated liquid medium, M.S. (Murashige and Skoog, 1962) containing 2,4-D (5 uM) and kinetin (0,5 uM). Cells were collected in the early stationary growth phase and washed with cytokinin free control medium.

The washed cells were transferred to fresh medium without cytokinin (N^6-benzyl adenine, BA). Observations were made in medium with BA concentrations of 0.5, 2.5, 5 and 10 uM. Cell suspensions were constantly agitated and maintained at 25ºC. 16 h of white light (1.75 Klux) were provided daily.

For electron microscopy, the cells were fixed in 2,5% glutaraldehyde buffered with 50 uM sodium phosphate (pH 6.8). Before and after osmium-tetroxide (10 g/l) postfixation, the cells were washed in 50 mM sodium phosphate (3 changes of 30 min each). The cells were dehidrated in graded series of acetone and embedded in Spurr's resin (Spurr, 1969). All sections were cut with glass knives and thin sections were stained with uranyl acetate and post-stained with alkaline lead citrate (Reynolds, 1963) and examined under a Zeiss EM-109 electron microscope.

Plant Aging: Basic and Applied Approaches
Edited by R. Rodríguez *et al.*
Plenum Press, New York, 1990

RESULTS

On studying the effects of different BA concentrations on some cell growth parameters (cell density and viability, fresh and dry weight, packed volume, growth rate) an increase of cell proliferation even at low concentrations as 0.5 uM was observed. The maximum response was obtained at a concentration of 5 uM. At concentrations higher than 2-5 uM, however no increase in cell proliferation was observed.

Ultrastructural study showed that the cells (0 days of culture) were polyvacuolized with a rich cytoplasm (fig.1).

They had a number of ribosomes associated in helix and mitochondria with many crests. The Golgi apparatus was formed by a number of nearly dyctiosomes. Each dyctiosome was made up of flat cisterns and a number of vesicles . Furthemore, a large concentration of reserve material, mainly starch, was observed (fig.2).

These cells were then cultivated in BA-containing medium (figs. 3 and 4). In presence of 0.5 uM BA, cells were characterised by a dense cytoplasm rich in organelles and having a number of polysomes and a well developed rRE. Mitochondria were numerous, large accumulation of reserve materials mainly starch in plastids and lipid-like substances, especially in cells cultivated in 5 uM BA were observed. In general, the interphasic nucleus was characterised by having more chromatin than heterochromatin. The nucleolus had a granular and a fibrous zone and (a weakly staining region called) "nucleolar vacuole" was observed in it (fig. 5).

Also, a BA (5 uM)-promoted differentiation to ethioplast from less-developed plastids was observed. However, mature chloroplasts were never observed (fig. 6).

Cells cultivated in medium lacking BA (fig. 7) had a large vacuole, a small developed rRE and free ribosomes (monosomes). They also had fewer mitochondria than BA-treated cells and cup-type mitochondria were often observed (fig 8). Unlike BA-treated cells, untreated cells had no starch in plastids.

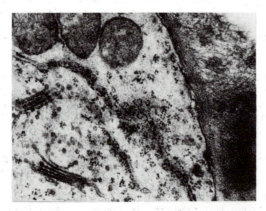

Fig.1 Mitochondria, polysomes and dyctiosomes in an apple cell growing in M.S. medium containing Kn (0,5 µM) and 2, 4-D (5 µM). Spiral and helical polysomes are seen in profusion. Vesicles of various sizes are connected to the dictiosomes (25600 X).

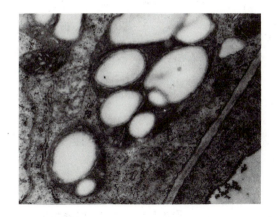

Fig.2 Detail of a plastid containing numerous starch grains (amyloplast) from an apple cell cultured in medium with Kn (0.5 uM) and 2,4-D (5 uM) (25600X).

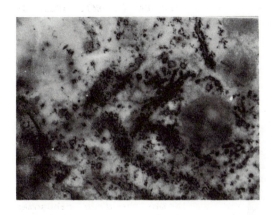

Fig.3 A region of cytoplasm containing mitochondria, polysomes and vacuoles from an apple cell growing in M.S. medium with BA(0.5 uM) and 2,4-D (5 uM), after 10 days of culture (25600X).

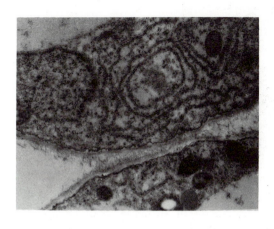

Fig.4 Profiles of rough endoplamic reticulum in an apple cell developing in a M.S. medium containing BA (0.5 uM) and 2,4-D (5 uM), after 10 days of culture (15300X).

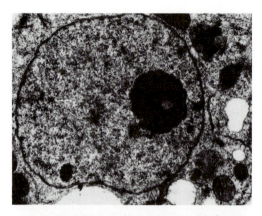

Fig.5 Detail of nuclear region of an apple cell growing in a M.S. medium with 2,4-D (5 uM) and BA (0.5 uM), after 10 days of culture. The convoluted nucleus contains a nucleolus and a nucleolar organizer. The nucleolus contains a central electron transparent region (25300X).

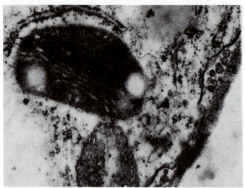

Fig.6 A plastid containing newly formed grana (48000X). Medium containing 2,4-D (5 uM) and BA (5 uM), after 10 days of culture .

Fig.7 Ultrastructural details of apple cell cultivated in a M.S. medium with 2,4-D (5 uM), after 10 days of culture. Noted the free ribosomes (33600X).

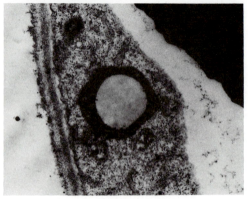

Fig.8 Cup-shaped mitochondrion encircle cytoplasm observed in an apple cell growing in a M.S. medium with 2,4-D (5 uM), after 10 days of culture (47000X).

DISCUSSION

Apple cell line D2,14 is extremely sensible to the presence of BA in culture medium. This regulator stimulates cell growth and division.

Cells in 0 days of culture and BA-treated cells were characterised by a number of polysomes and a well developed rRE. These phenomena are associated with an active protein synthesis (Vujicic et al, 1983). The presence of nucleolar vacuoles in these cells is a sign of high nucleolar activity (Street, 1977; Burgess, 1985) necessary for rRNA synthesis.

Appel cells cultivated in medium lacking cytokinin had a reduced rRE and the ribosomes were free in the cytoplasm as monosomes. Taking account of earlier results obtained in our laboratory (Ordás et al, 1988), the reduced protein synthesis observed in these cells could be explained by an increase in the rate of transformation of polisomes to monosomes. Similar results were obtained in other studies (Muren and Fosket, 1977; Tepfer and Fosket, 1978). They observed cytokinin-induced polyribosome formation in soybean cultures.

Fewer mitochondria and cup-type mitochondria, characteristic of low respiratory activity (Bagashaw et al, 1969; Endress and Sjolund, 1976; Howarth et al, 1983) were observed in untreated cells. Its presence suggest that cells cultivated without BA had a reduced metabolic activity. This hypothesis is supported by the fact that there were a lack of starch in the plastids; Seyer et al (1975) working with tobacco cells, observed that loss of starch in plastids were related to cell senescence. These results show the degenerative process of apple cells lacking exogenous supply of cytokinin. Reduced metabolic activity observed in this study may be, therefore, a consequence of senescent characteristics in the apple cells.

Compared to the untreated BA-treated cells had a well-developed subcellular organization after 10 days. This is result of an active metabolism necessary for sustaining its growth rate. However, the morphological differences between the cells at the start and after 10 days of culture in presence of cytokinins were not remarkable.

REFERENCES

Bagashaw, V., Brow, R. and Yeoman, M.M., 1969, Changes in mithocondrial complex accompanying callus growth. Ann. Bot., 33: 35-44.

Burgess, J., 1985, A Introduction to Plant Cell Development. Cambridge University Press. 10.

Endress, A.G., and Sjolund, R.D., 1976, Ultraestructural cytology of callus cultures of **Streptanthus tortuosus** as affected by temperature. Amer. J. Bot., 63 (9): 1213-1224.

Howarth, M.J., Peterson and Tomes, D.T. 1983. Cellular differentiation in small clumps of **Lotus corniculatus** callus. Can. J. Bot., 61(2): 507-517

Murashige, T. and Skoog, F, 1962, A revised medium for rapid growth and bioassays with tobacco tissue tissue cultures. Physiol. Plant., 15: 473-479

Muren, R.C. and Fosket, F.E., 1977, Cytokinin-mediated translational control of protein synthesis in cultured cells of **Glycine max.** J. Exp. Bot., 28: 775-784.

Ordás, R.J., Rodriguez, R. Fernández, B. and Sanchez, R, 1988, N[6]-Benzyl aminopurine on the translation and transcription activity from apple cells. In: NATO ASI series,a vol. H18: 135-142. Plant Cell Biotechnology, M.S.S. Pais et al, eds., Springer-Verlag. Berlin.

Heidelberg

Reynolds, E.S.,1963, The use of lead citrate at high pH as an electron
    opaque stain in electron microscopy. J. Cell. Biol. 17: 208-212.

Seyer, P., 1975, Effect of cytokinin on chloroplast cyclic
    differentiation in cultured tobacco. Cell Differ., 4: 187-197

Spurr, A.R., 1969, A low-viscosity epoxyresin embedding medium for
    electron microscopy. J. Ultrastruct. Res. 26: 31-43.

Street, H.E., 1977, Plant Tissue and Cell Culture. In: Botanical
    Monographs., vol. 11, H.E. Street, Ed., University of California
    Press. Berkeley. Los Angeles.

Tepfer, D. A. and Fosket, D.E., 1978, Hormone-mediated translational
    control of protein synthesis in cultured cells of **Glycine max.**
    Dev. Biol. 62: 486-497.

Vujicic, Lj., Radojevic and Kovoor, A., 1983, Effect of some nucleic acid
    base analogues and inhibitors of protein synthesis on orderly
    arrangement of ribosomes in callus tissue of **Corylus avellana.**
    Biochem. Physwiol. Pflanzen., 178: 61-66.

POLYAMINE AND ETHYLENE METABOLISMS DURING TOMATO FRUIT RIPENING

M. Acosta, J.L. Casas, J.A. del Río, M.B. Arnao, A. Ortuño and F. Sabater

Department of Plant Biology. University of Murcia. Sto. Cristo, 1. 30001-MURCIA. SPAIN

INTRODUCTION

Polyamines, putrescine (PUT), spermidine (SPD) and spermine (SPM), are relatively small soluble molecules, fully protonated and polycationic at physiological pH.

Many of the biological functions of polyamines appear to be attributable to their cationic nature and are expressed in phases of active growth and differentiation, where increased rates of macromolecular and polyamine synthesis can be found (Slocum et al., 1984). There is now, however, evidence that polyamines might take part in retarding plant senescence by stabilizing cell membranes. The effect of polyamines on delaying the loss of chlorophyll in excised leaves of oat is particularly well-established (Galston and Kaur-Sawhney, 1987).

Fruit ripening can be considered as a feature of plant organ senescence in which a number of physiological and biochemical changes take place to render fruits attractive to a potential consumer. The hormonal control of these changes lies in the plant hormone ethylene which is synthetised from methionine via 1-aminocyclopropane- 1-carboxylic acid (ACC), and S-adenosylmethionine (SAM) (Yang and Hoffman, 1984).

Ethylene is absent in stages of active growth and development where polyamines are predominant, while it appears and suddenly increases at the onset of ripening. In addition to their opposite physiological effects, polyamines and ethylene have been shown to have a common intermediate in their biosynthetic routes, namely SAM. Previous works have revealed the possibility of modulating the flux of SAM towards ethylene or polyamine routes in aged orange peel discs (Even-Chen et al., 1982).

So, the possibility that both ethylene and polyamine metabolisms may be closely linked during fruit ripening cannot be ruled out. If this were the case, the increase in ethylene and ACC which characterizes the ripening process should be paralleled by a decrease in the amines directly competing with ACC for SAM, namely SPD and SPM. The aim of this work is to study the ethylene metabolism and the polyamine level evolution altogether during detached tomato fruit ripening and to establish to what extent the polyamine and ethylene metabolism are related during fruit ripening.

Plant Aging: Basic and Applied Approaches
Edited by R. Rodríguez *et al.*
Plenum Press, New York, 1990

MATERIALS AND METHODS

Plant Material

Tomato fruits (Lycopersicon esculentum Mill. nothovar F1 "Novy") were harvested from a local greenhouse (Francisco López Franco, Cosechero-Exportador, Aguilas, Murcia, Spain) at mature green stage and stored in our laboratory at about 25 °C while ripening progressed.

Ethylene Measurements

Ethylene production by the whole fruits was monitored daily. Fruits were enclosed in glass jars of about 650 ml for 1.5 hours and a 1-ml gas sample was then withdrawn from the inner atmosphere of the jar. The ethylene present in the sample was quantified by gas chromatography.

Lycopene Measurements

Lycopene was determined by extracting the tomato tissue with hexane (5 ml/g) and measuring absorbance at 470 nm. The absorbance spectrum of lycopene in hexane had been previously determined. (Molar absorption coefficient of lycopene ε_{470} = 3450 $M^{-1} cm^{-1}$.

Sample Preparations and Analysis

Three fruits at each ripening stage were individually triturated with an Omnimixer. Three fractions were then separated to determine lycopene, ACC or polyamine content.

ACC was extracted according to Atta-Aly et al. (1987), conjugated ACC was hydrolyzed to free ACC according to Hoffman et al. (1983) and both total and free ACC content was determined by Lizada and Yang's method (1979).

Polyamines were extracted and analyzed by HPLC using the benzoylated method, according to Flores and Galston (1982). Determinations were made using 1,6-diaminohexan as internal standard.

RESULTS AND DISCUSSION

Ripening of the "Novy" tomato fruits was divided into five stages, according to the evolution in lycopene content and the ethylene production pattern (Fig. 1).

The pattern of ACC accumulation is similar to that reported previously (Hoffman and Yang, 1980) (Fig. 2). There is an initial increase in total ACC between mature-green and breaker stages, coinciding with a sudden surge in ethylene production, a decrease in the two following stages and a new increase at the last stage of ripening.

The polyamine levels are shown in Fig. 3. Of the three main plant amines, PUT, SPD and SPM, only PUT and SPD were detected, while no free SPM could be detected at any stage of ripening. Although this same lack in SPM has been reported during "Rutgers" tomato fruits ripening (Dibble et al., 1988), it is not common to other fruits. SPM is effectively found during ripening of pears (Toumadje and Richardson, 1988) and avocadoes (Winer and Apelbaum, 1986), where it is even the major amine in the mature tissue.

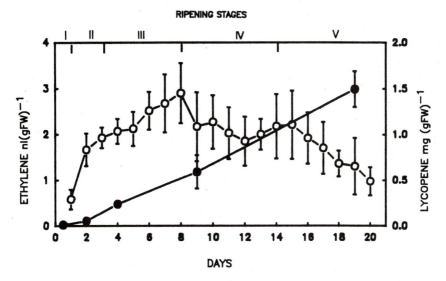

Fig. 1. Ethylene production rate (O) and lycopene content (●) during ripening of "Novy" tomato fruits. Ripening stages are: I.- Mature green; II.- Breaker; III.- Orange; IV.- Ripe; V.- Overripe. Values are the mean of 15 fruits for ethylene and 3 fruits for lycopene ± SD.

In our "Novy" tomato fruits, PUT was the major polyamine in the tissue, coinciding with other results obtained in tomato (Bakanashvili, et al., 1987; Dibble et al., 1988), being between 1.9 and 3.9 times greater than SPD at mature green and overripe stages, respectively. The accumulation pattern of both amines during ripening was similar until the overripe stage (fig. 3), in which PUT underwent a significant increase while SPD maintained a slight decrease.

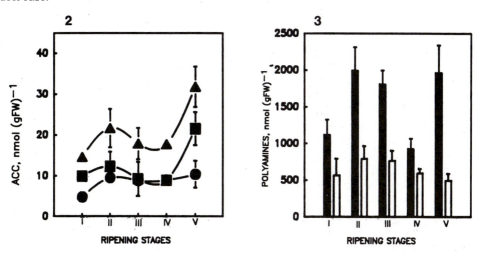

Fig. 2. Content of total (▲), free (●) and conjugated (■) ACC during ripening of "Novy" tomatoes. Values are the mean of three fruits with three replicates. Bars denote ± SD when larger than symbols.

Fig. 3. Evolution of the putrescine (■) and spermidine (□) levels during ripening. Values are the mean of three fruits with three replicates ± SD.

Therefore, during ripening of "Novy" tomato fruits we found the profile of ACC accumulation typical of a clymacteric fruit but we found a high level of PUT and SPD too. Because SPD and ACC has the same precursor, SAM, we can deduce that the increased synthesis of ACC that occurs once ripening is started has no effect on SPD synthesis and so there is no competition between these two substrates for SAM. If we analyze the percentage of SAM which is destined to the ethylene and polyamine biosynthetic routes by means of the ACC/SPD + ACC and the SPD/SPD + ACC ratios, where the term SPD + ACC is an estimate of the total amount of SAM diverted to both routes, we can see that during the whole ripening process, the proportion of SAM employed in SPD biosynthesis is much higher than that employed in ACC synthesis (Table 1). So, although the appearance of ACC in the tissue is a key factor for the initiation of ripening, it does not mean a significant reduction in the SAM-dependent biosynthesis of SPD. Therefore, the initiation of ethylene metabolism does not alter the polyamine levels in the tissue, suggesting an independence between both routes, and here we agree with Kushad et al. (1988).

Table 1. Percentage of SAM destined to ethylene and polyamine biosynthesis during "Novy" tomato ripening.

Ripening Stage	ACC/SPD + ACC	SPD/SPD + ACC
Mature green	2.5	97.5
Breaker	2.6	97.4
Orange	2.3	97.7
Ripe	2.9	97.1
Overripe	6.4	94.0

In consequence, the main role of polyamines in tomato fruit ripening does not seem to be played once ripening is started but rather in the developmental stages immediately prior to ripening, where a drastic reduction in their concentration, mainly of SPD and SPM, might be the "signal" to start ripening. Although there are some experimental results which support this view (Bakanashvili et al., 1987; Dibble et al., 1988) it awaits further investigation.

ACKNOWLEDGMENT

Project CAYCIT N. PA85-0275. M.B.A. has a grant from Instituto de Fomento of Murcia (Spain).

REFERENCES

Atta-Aly, M.A., and Saltveit, M.E.Jr., 1987. Effect of silver ions on ethylene biosynthesis by tomato fruit tissue, Plant Physiol., 83:44-48.
Bakanashvili, M., Barkai-Golan, R., Kopeliovitch, E., and Apelbaum, A., 1987, Polyamine biosynthesis in Rhizopus-infected tomato fruits: possible interaction with ethylene, Physiol. Mol. Plant Pathol., 31: 41-50.
Dibble, A.R.G., Davies, P.J., and Mutschler, M.A., 1988, Polyamine content of long-keeping alcobaca tomato fruit, Plant Physiol., 86:338-340.
Even-Chen, Z., Mattoo, A.K., and Goren, R., 1982, Inhibition of ethylene biosynthesis by aminoethoxyvinylglycine and by polyamines shunts label from 3,4-[14C]Methionine into spermidine in aged orange peel discs, Plant Physiol., 69:385-388.
Flores, H.E., and Galston, A.W., 1982, Analysis of polyamines in higher plants by high performance liquid chromatography, Plant Physiol., 69:701-706

Galston, A.W., and Kaur-Sawhney, R., 1987, Polyamines and senescence in plants, in: "Plant Senescence: Its Biochemistry and Physiology", W.W. Thomson, E.A. Nothnagel, R.C. Huffaker, eds., The American Society of Plant Physiologists.

Galston, A. W., and Kaur-Sawhney, R., 1988 , Polyamines as endogenous growth regulators, in: "Plant Hormones and Their Role in Plant Growth and Development", P.J. Davies, ed., Kluwer Academic Publishers, The Netherlands.

Hoffman, N.E., and Yang, S.F., 1980, Changes of 1-aminocyclopropane-1-carboxylic acid content in ripening fruits in relation to their ethylene evolution rates, J. Am. Soc. Hortic. Sci., 105:492-495.

Hoffman, N.E., Liu, Y., and Yang, S.F., 1983, Changes in 1-(malonylamino) cyclopropane-1-carboxylic acid content in wilted wheat leaves in relation to their ethylene production rates and 1-aminocyclopropane-1-carboxylic acid content, Planta, 157:518-523.

Kushad, M.M., Yelenosky, G., and Knight, R., 1986, Interrelationship of polyamine and ethylene biosynthesis during avocado fruit development and ripening. Plant Physiol., 87:463-467.

Lizada, M.C.C., and Yang, S.F., 1979, A simple and sensitive assay for 1-aminocyclopropane-1-carboxylic acid. Anal. Biochem., 100:140-145.

Slocum, R.D., Kaur-Sawhney, R., and Galston, A.W., 1984, The physiology and biochemistry of polyamines in plants, Arch. Biochem. Biophys., 235:283-303.

Toumadje, A., and Richardson, D., 1988, Endogenous polyamine concentrations during development, storage and ripening of pear fruits, Phytochemistry, 27:335-338.

Winer, L., and Apelbaum, A., 1986, Involvement of polyamines in the development and ripening of avocado fruits, J. Plant Physiol., 126:223-233.

Yang, S.F., and Hoffman, N.E., 1984, Ethylene biosynthesis and its regulation in higher plants, Annu. Rev. Plant Physiol., 35:155-189.

LEVELS OF CYTOKININS IN AGING AND REJUVENATED

<u>Corylus avellana</u> L. TISSUES

B. Fernández, R. Rodríguez, M. J. Cañal, H. Andrés and
A. Rodríguez

Lab. Fisiología Vegetal, Dpto B.O.S.. Fac. Biología
Univ. Oviedo, 33005 Oviedo, Spain

INTRODUCTION

The ability of "in vitro" cloning in many trees is a function of the degree of juvenility of the explanted tissues. Unfortunately, little is known about the physiological and genetic mechanisms that control the differences between the juvenile and mature state of cells, tissues or the plant (Bonga, 1987).

In previous works, we have developed methods to micropropagate <u>Corylus avellana</u> L explants (Perez et al., 1983; 1985). We observed that explant morphogenic capacity decreases as explant age increases. Proliferation of the adult explants needs a longuer culture period and more specific combinations of plant growth regulators that the juvenile embryogenic explants (Rodríguez et al., 1988).

Cytokinins appear to regulate the decline in various macromolecules, protein, nucleic acid, etc., associated with aging in plants (Gilbert et al., 1980) and the loss of membrane permeability (Wittenbach, 1977).

Levels of endogenous cytokinins in <u>Corylus avellana</u> L nodal shoot segments were examined and a comparative study was developed among two different age explants in order to stablish a biochemical expression of rejuvenation.

MATERIAL AND METHODS

Two types of plant material were used: Nodal shoot segments of mature shrubs of <u>Corylus avellana</u> L and nodal shoot segments of forced outgrowth material.

The extraction and purification of basic cytokinins were done using the method described by Horgan and Scott (1985).

Inmediately after collection plant tissues were frozen in liquid nitrogen and allowed to thaw for 18 h at -15ºC in methanol/ chloroform/ 90% formic acid/ water (12:5:1:2 v/v). The solvent was decanted and the tissue was homogenized in chilled methanol/ 90% formic acid/ water

Plant Aging: Basic and Applied Approaches
Edited by R. Rodríguez *et al.*
Plenum Press, New York, 1990

(6:1:4 v/v) and, after 4 h at -15ºC, the blend was vacuum filtered and the filtrates were then combined and reduced to dryness by vacuo at 40ºC and redissolved in 20 ml of water, adjusted to pH 3 with acetic acid and percolated through a column of cellulose phosphate (NH) equilibrated to this pH. After washing the column with water (pH 3), cationic compounds may be eluted with 2M ammonium hydroxide. The eluates were reduced to dryness by vacuo at 40ºC, redissolved in pure methanol and passed through a 0.45 um filter, reduced to dryness with N_2 and redissolved in 1 ml of 5% acetonitrile in water (pH 7.0 with TEAB).

The separation of the cytokinins was done using a Varian Model 5000 liquid chromatograph connected to a Varichron detector at 254 nm. The separation column (150 x 4.6 mm) was a 5 um Spherisorb ODS-2. Elution was achieved with a 40 min linear gradient of 5-20% acetonitrile in water (pH 7.0 with TEAB) at a flow rate of 2 ml x min^{-1}. Fractions of 1 ml were collected and freeze dried. Each fraction was assayed by ELISA or by Amaranthus bioassay.

Bioassay. Isopentenyladenine activity was assessed using the Amaranthus caudatus bioassay (Biddington and Thomas, 1973). Betacyanin was extracted by mean four cycles of freezing and thawing and the quantity determined by calculating the difference between the optical densities at 542 nm and 620 nm. A series of bencyladenine (BA) standars from 10^{-9} M to 10^{-5} M was used to quantify the response.

Enzyme-linked immunoabsorbent assay. The ELISA test for Zeatin riboside (t-ZR) and zeatin (Z) were carried out with a Phytodetek Immunoassay test kit. Absorbance at 405 nm was measured in a Perkin Elmer Lambda Reader.

To obtain a linear standard curve the logit-transformation is used for transformation of B-values. The ZR concentration in the samples was calculated from the standard curve expressed in pmol per 100 ul. The Zeatin is expressed as pmol ZR equivalentes in 100 ul.

RESULTS AND DISCUSSION

The separation by HPLC of the extracted cytokinins, showed a totally different profile in mature nodal shoots and forced nodal shoots segments (fig. 1).

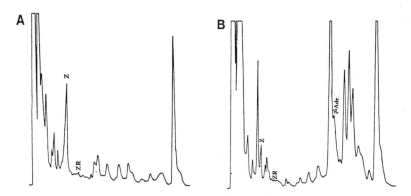

Fig. 1. Reverse -phase HPLC analysis of a extract of: A) mature nodal shoots (5 g of fresh wheight), B) nodal shoot segments (3 g of fresh weight) of forced outgrowth material.

The most remarkable data was the detection of isopentenyladenine only in the extracts of the nodal shoots segments of forced outgrowth plant tissues. The other cytokinins studied (zeatin and zeatin riboside) were present in both kinds of explants.

The figure 2 shows the levels of the three cytokinins studied (Z, ZR and I^{6}Ade) in nodal shoot segments of mature and forced outgrowth material.

The E.L.I.S.A. test for zeatin and zeatin riboside showed that these cytokinins were present both in mature and forced nodal shoots, although the highest levels of the two compounds were found in nodal shoots segments of forced outgrouwth material.

Isopentenyladenine, as quantified by the Amaranthus caudatus bioassay, was detected only in the forced nodal shoots segments, being its levels in the tissues substantially important (0.54 ng g fresh weight^{-1}.

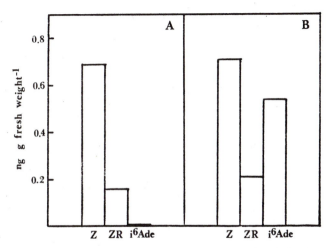

Fig. 2. Zeatin riboside, zeatin and isopentenyladenine levels in nodal shoot segments of: A) mature and B) forced outgrowth material.
ZR is expressed as ng g fresh weight^{-1}.
Z$_6$is expressed as ng ZR equivalents g fresh weight^{-1}.
i^6 Ade is expressed as ng BA equivalents g fresh weight^{-1}.

Since the nodal shoots segments of forced outgrouwth material had showed the highest morphogenic capacity, our results suggest a relationship between the morphogenic capacity of the explants and the levels of some individual cytokinin. Now we are studing the levels of this phytohormones in embryogenic tissues and material from "in vitro" culture, in order to relate aging and cytokinin content with morphogenic capacity.

REFERENCES

Biddington, N.L. and Thomas, T.L.H., 1973. A modified Amaranthus betacyanin Bioassay for the rapid determination of cytokinins in plant extracts. **Planta**, 111: 183–186.

Bonga, J.M. 1987. Clonal propagation of mature trees: Problems and possible solutions. In **Cell and Tissue Culture in Forestry** (Vol. 1). pp. 249-271. J.M. Bonga and D.J. Durzan eds. Martinus Nijhoff Publishers. Holanda.

Gilbert, M.L., Thompo, J.E., Dumbroff, E.B. 1980. Delayed cotyledon senescence following treatment with a cytokinin; and effect at the level of membranes. **Can. J. Bot.**, 58: 1797-1803.

Horgan, R. and Scott, I.M. 1987. Cytokinins. In **The Principles and Practice of Plant Hormone Analysis**. pp. 303-365. Academic Press. London.

Perez, C., Fernández, B. and Rodríguez, R. 1983. "In vitro" plantlet regeneration through asexual embryogenesis in cotyledonary segments of <u>Corylus avellana</u> L. **Plant Cell Reports**, 2: 226-229.

Perez, C. Rodríguez, R. and S. Tamés, R. 1985. "In vitro" filbert (<u>C. avellana</u>) micropropagation from shoot and cotyledonary segments. **Plant Cell Reports**, 4: 137-140.

Rodríguez, R. Diaz-Sala, C. and Ancora , G. 1988. Sequential culture of explants taken from adult <u>Corylus avellana</u> L. **Acta Horticulture**, 227: 460-462.

Wittenbach, V.A. 1977. Induced senescence of intact wheat seedlings and its reversibility. **Plant Physiol.**, 59: 1039-1042.

CHANGES IN POLYAMINES RELATED WITH PRUNING

AS A METHOD FOR REJUVENATION IN FILBERT

Manuel Rey, Roberto Astorga, Carmen Díaz-Sala,
Antonio F. Tiburcio* and Roberto Rodríguez

Lab. Fisiología Vegetal, Dpt. B.O.S.,
Univ. Oviedo, 33005 Oviedo, SPAIN
(*) Lab. Fisiología Vegetal, Fac. Farmacia,
Univ. Barcelona, 08028 Barcelona, SPAIN

ABSTRACT

Polyamine titer in leaves taken from mature and severely pruned trees (two successive treatments) of filbert (Corylus avellana L. cvs Gironella and Negreta) was investigated in order to determine the possible relationship between polyamine levels and the degree of juvenility of the plant material achieved by pruning. The results revealed different patterns in the polyamine metabolism, probably related to the effectiveness of the pruning, based on the increase of rooting as an indication of rejuvenation.

INTRODUCTION

Changes in polyamine levels have been observed during plant growth and development, acting as modulators of some cellular and physiological processes (Galston, 1983). They have been related with aging and senescence in plant tissues (Kaur-Sawhney and Galston, 1979; Kaur-Sawhney et al., 1982). Also, high levels of polyamines have been associated with rapidly growing and young tissues (Palavan and Galston, 1982; Shen and Galston, 1985). Morphogenic processes, both in vivo (Friedman et al., 1982; Jarvis et al., 1983) and in vitro (Feirer et al., 1984; Kaur-Sawhney et al., 1988; Tiburcio et al., 1988), are also associated with polyamine metabolism.

Filbert micropropagation success depends on the physiological changes related to ontogenical and chronological ages of the trees used as a source of explants (Díaz-Sala et al., in this volume). Decline of the rooting ability was also observed in relation to these changes (Astorga and Rodríguez, unpublished data), suggesting that some metabolic changes occurred leading to the loss of morphogenic potentiality.

We used severe pruning of filbert trees in order to obtain a good source of explants for micropropagation purposes. Severe pruning is a well known method for rejuvenation in woody species (Mott, 1981; Evers, 1987). It allows the partial regaining of juvenile characteristics in the newly formed shoots (Mott, 1981). Thus, morphogenic capacities can be recovered, resulting in an increase in micropropagation success.

The purpose of this study was to determine if severe pruning was related with changes in polyamine metabolism, thus perhaps reflecting the degree of juvenility of the plant material subjected to this treatment, allowing a true interpretation of the results obtained during filbert vegetative and in vitro propagation.

MATERIAL AND METHODS

Pruning treatments were imposed on shoots (at the soil level) of 5-year-old filbert (Corylus avellana L. cvs. Gironella and Negreta). Also, the pruned trees (once-) were pruned again the following year (twice-pruned trees). The treatments were carried out in a filbert experimental field-plantation of the Estación Experimental Agraria de Villaviciosa (Asturias, Spain).

Polyamines were determined in the two first sub-apical leaves of active-growing shoots of adult (non-pruned), once, and twice pruned trees, according to the method described by Tiburcio et al. (1985). The leaves were homogenized in pre-chilled mortars with 5% perchloric acid (300 mg f.w./ml). The extracts were centrifuged at 27000 g for 20 min. Alicuots of the supernatant and the pellet were hydrolyzed overnight in 12 N HCl at 100-110°C and then filtered through glass wool.

Alicuots of the crude supernatant (S fraction), which contained the free polyamines, the hydrolyzed supernatant (SH fraction), which contained the polyamines bound to cytoplasmic soluble compounds, and the hydrolyzed pellet (PH fraction), which contained the insoluble polyamines, bound to macromolecular compounds in the plant cell, were all dansylated overnight in alkaline conditions. Pure standards of the polyamines were dansylated overnight. Dansyl chloride (5-dimethylamino-1-naphtalenesulfonyl chloride) and polyamine standards were purchased from Sigma Chem. Co. (St. Louis, MO, USA). Proline was added to remove the excess of dansyl chloride.

The dansylated polyamines and standards were extracted with benzene and analyzed by thin layer chromatography (TLC) using silica gel 60A plates (Whatman Int. Ltd., Maidstone, England) with a concentrating zone. The solvent used was chloroform/triethylamine (25:2, v/v). The spots were identified under ultraviolet light and eluted with ethylacetate. Fluorescence was measured with a Perkin Elmer Luminiscence Spectrometer (excitation 350 nm, emission 495 nm), and the results were compared with the dansylated standards. The experiments were carried out twice, except where indicated, and the mean ± Standard Error (S.E.) was calculated.

RESULTS AND DISCUSSION

In this species, there is little research on polyamine metabolism. In reviewing the literature, only one paper concerning this topic was found (Meurer et al., 1986), reporting the identification of spermidine conjugates from pollen on the basis of nuclear magnetic resonance and mass spectral data. Our procedure for polyamine determinations was performed for the first time in filbert, that was adapted from that used with tobacco callus (Tiburcio et al., 1985), changing the solvent mixture used for running the TLC plates. Chloroform/triethylamine (25:2, v/v) was the solvent mixture that yielded the best separation on the plates (Fig. 1).

The supernatant fraction obtained after centrifugation of the plant extracts presented a gradual pattern of colour. This could reflect metabolic differences between the three plant materials tested, perhaps in the anthocyanin metabolism, or other metabolites of low molecular weight, because the youngest material yielded a red-coloured supernatant, in contrast with the oldest material (yellow-coloured one).

The qualitative pattern of polyamines was very similar among the three materials; remarkably, diaminopropane (DAP), a product of the oxidation of the major polyamines, was detected. However, the posible enzyme for this reaction, polyamine oxidase (PAO, EC 1.4.3.4), was found only in the Gramineae (Smith, 1985).

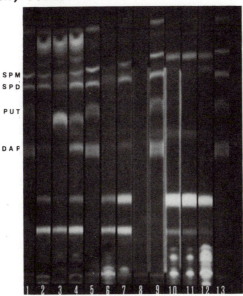

Fig. 1. TLC of Dansyl-polyamines from filbert leaves. Lanes 1,5,9 and 13, standards (2 nmol each). Lanes 2,3 and 4, non-pruned, 6,7,8, once-pruned, and 10,11,12, twice-pruned filberts. S fraction: lanes 2,6,10. SH: lanes 3,7,11. PH: lanes 4,8,12.

Table 1. Polyamines in leaves of adult (non-pruned) and severely pruned filbert (Corylus avellana L.) trees.

		cv. Gironella			cv. Negreta		
		NON-PRUNED	ONCE PRUNED	TWICE PRUNED	NON-PRUNED	ONCE PRUNED	TWICE PRUNED
Put		N.D.*	N.D.	N.D.	N.D.	N.D.	N.D.
Spd	S	173.5 ±2.9	91.2 ±13.5	312.2 ±7.06	118.8 a	79.0 a	181.9 a
	SH	33.2 ±7.7	89.9 ±3.6	N.D.	N.D.	59.4 a	N.D.
	PH	40.5 ±4.7	35.8 ±2.6	74.6 ±3.02	37.7 ±1.3	35.0 ±1.2	36.0 ±0.2
Spm	S	114.2 ±11.1	87.7 ±36.1	130.5 ±0.6	88.9 a	97.8 a	143.6 a
	SH	36.2 ±0.04	63.3 ±4.06	N.D.	N.D.	49.9 a	N.D.
	PH	146.4 ±6.5	198.0 ±24.4	137.7 ±7.9	140.2 ±4.08	150.5 ±1.9	104.4 ±3.2
DAP	S	47.4 ±5.6	30.9 ±1.7	114.8 ±6.3	45.0 ±1.1	51.1 ±0.5	81.3 ±0.2
	SH	25.2 ±0.2	33.9 ±4.2	N.D.	N.D.	38.8 ±1.1	N.D.
	PH	N.D.	N.D.	N.D.	23.5 ±0.1	25.2 ±0.7	24.9 ±0.4
DAP/PAs	S	0.164	0.173	0.259	0.217	0.289	0.250
	PH	---	---	---	0.132	0.136	0.177

Values are means ±S.E.   (*) N.D.: Not detected.
nmol/g. fresh weight   (a) Data from a single experiment.

The fluorimetric measurements showed great quantitative differences in relation to the pruning treatment (Table 1). For Gironella filberts, the highest amount of both Spermidine (Spd), Spermine (Spm), and DAP was obtained in the youngest trees, mainly Spd; these data are in agreement with those obtained for young and old leaves of bean (Palavan and Galston, 1982). On the other hand, these levels were unexpectedly lowest in the once-pruned trees; this result, although supporting the effectiveness of the pruning, is very difficult to discuss, but additional experiments are in progress. However, the DAP/Spd+Spm ratio (in the S fraction) was higher as the age of the trees decreased. These results were only confirmed in part with the Negreta trees, perhaps reflecting some cultivar differences.

Putrescine (Put) was not detected in these experiments. Palavan and Galston (1982) reported that in young and old leaves of bean Put was not detectable either. However, in these leaves arginine decarboxylase (ADC) and ornithine decarboxylase (ODC) activities were present and to a greater degree in young leaves. We did not determine these Put biosynthetic activities, but they may also be present in filbert leaves. One possible explanation is that Put is synthesized, perhaps at a higher rate in young plant material than in old, and it is rapidly metabolized into the polyamines. This was also supported by the fact that Put was also undetectable in the acid-insoluble fractions.

These experiments indicate that pruning was very effective in changing the physiological status of the treated trees, as compared with the non-treated ones. The polyamine level could be higher as a result of the pruning, and then fluctuating around a certain value, characteristic of the mature trees. This hypothesis may be tested by a comparison between the three materials tested here and the results of another pruning treatment (three-times pruned). On the other hand, polyamines appear to be a useful marker for monitoring the progress of rejuvenation techniques and mechanisms controlling maturation, leading to a more effective development of micropropagation programmes.

Financial support is acknowledged from the spanish CAYCIT (Grant no. 860-84)

REFERENCES

Evers, P. W., 1987, Correlations within the tree, in: "Cell and Tissue Culture in Forestry", J. M. Bonga and D. J. Durzan, eds., Martinus Nijhoff Pub., Dordrecht.

Feirer, R. P., Mignon, G., and Litvay, J. D., 1984, Arginine decarboxylase and polyamines required for embryogenesis in the wild carrot, Science 223:1433.

Friedman, R., Altman, A., and Bachrach, U., 1982, Polyamines and root formation in mung bean hypocotyl cuttings. I. Effects of exogenous compounds and changes in endogenous polyamine content, Plant Physiol. 70:844.

Galston, A. W., 1983, Polyamines as modulators of plant development, BioScience 33:382.

Jarvis, B. C., Shannon, P. R. M., and Yasmin, S., 1983, Involvement of polyamines with adventitious root development in stem cuttings of mung bean, Plant Cell Physiol. 24:677.

Kaur-Sawhney, R., and Galston, A.W., 1979, Interaction of polyamines and light on biochemical processes involved in leaf senescence, Plant Cell Environ. 2:189.

Kaur-Sawhney, R., Shih, L., Flores, H. E., and Galston, A. W., 1982, Relation of polyamine synthesis and titer to aging and senescence in oat leaves, Plant Physiol. 69:405.

Kaur-Sawhney, R., Tiburcio, A. F., and Galston, A. W., 1988, Spermidine and flower-bud differentiation in thin-layer explants of tobacco, Planta 173:282.

Meurer, B., Wray, V., Grotjahn, L., Wiermann, R., and Strack, D., 1986, Hydroxycinnamic acid spermidine amides from pollen of Corylus avellana L., Phytochem. 25:433.

Mott, R. L., 1981, Trees, in: "Cloning Agricultural Plants via In Vitro Techniques", B. V. Conger, ed., CRC Press Inc., Boca Raton, Florida.

Palavan, N., and Galston, A. W., 1982, Polyamine biosynthesis and titer during various developmental stages of Phaseolus vulgaris, Physiol. Plant. 55:438.

Shen, H., and Galston, A. W., 1985, Correlations between ratios and growth patterns in seedling roots, Plant Growth Regul. 3:353.

Smith, T. A., 1985, Polyamines, Ann. Rev. Plant Physiol. 36:117.

Tiburcio, A. F., Kaur-Sawhney, R., Ingersoll, R. B., and Galston, A. W., 1985, Correlation between polyamines and pyrrolidine alkaloids in developing tobacco callus, Plant Physiol. 78:323.

Tiburcio, A. F., Kaur-Sawhney, R., and Galston, A. W., 1988, Polyamine biosynthesis during vegetative- and floral-bud differentiation in thin-layer tobacco tissue cultures, Plant Cell Physiol. 29:1241.

EARLY FLOWERING IN SEEDLINGS OF ASPARAGUS PROMOTED BY DIURON AND ATRAZINE

Mª Luisa González Castañón

S.I.A. - D.G.A.

Apartado 727

50080 Zaragoza, Spain

ABSTRACT

Seeds of asparagus treated with different solutions of atrazine (2-chloro-4-ethylamino-6-isopropylamine-s-triazine) or diuron (3-(3,4-Dichlorophenyl)-1,1 dimethylurea) and shaken for a week increased flower formation from 2.5 % in controls to 22 % in concentrations of 0.20 mM diuron or 19 % in 0.22 mM atrazine, at an early stage of development.

The first flowering usually takes place in Asparagus officinalis one to three years after seed germination. However, flowering at seedling stage permits separating male plants from female ones in a few weeks.

The treatment with diuron at 0.4 mM did not affect germination rate but plant growth decreased significantly as measured by shoots and roots weight.

INTRODUCTION

Asparagus officinalis is a dioecious and perennial crop. Flowering does not occur until one to three years after sowing the seeds. It is impossible to know the sex of the plants before flowering.

For selection and breeding it is interesting to know the sex of the plant at an early development stage in order to reduce the cycles of selection.

Abe and Kameya (1986) found that flower formation in 25 days old seedlings was increased by soaking the seeds for 12 days in a solution of atrazine or diuron from 4 % in controls to 37 % in treatments with 0.4 mM of atrazine or around 0.1 mM diuron.

In order to test the effects of these compounds, experiments have been carried out on seeds of Asparagus officinalis L. cv Largo, used as parentals in our breeding programme. The application of treatments has been reduced to seven days to assess whether a reduction in the treatment duration affects its effectivity.

KEYWORDS: Asparagus, Flowering induction, Atrazine, Diuron.

Plant Aging: Basic and Applied Approaches
Edited by R. Rodríguez *et al.*
Plenum Press, New York, 1990

445

MATERIAL AND METHODS

Experiment 1. Lots of seeds from Asparagus officinalis L. cv Largo were treated with 0.1 or 0.22 mM of atrazine and 0.05, 0.1 or 0.2 mM diuron solutions.

Lots of 25 seeds were put in Erlenmeyer flasks with 25 ml solution of each concentration. Flasks were maintained at room temperature and shaken for seven days, then the seeds were transferred to Petri dishes with water until germination. The germinated seeds were put in trays containing soil and vermiculite. The seedlings were maintained at 25 ± 2° C with sixteen light hours and irrigated with Hoagland solution at two days intervals. After three weeks the plants with apical flowers were recorded.

Experiment 2. Ten lots of 25 seeds were treated with a solution of 0.4 mM diuron for 14 days and two other lots with water as controls. The same procedures from experiment 1 for manipulation of seeds and seedlings were followed.

Two months after germination the weight of shoots and roots was determined.

Each experiment was repeated twice.

RESULTS AND DISCUSSION

Experiment 1. Three weeks after germination some seedlings showed an apical flower, these flowers were male and female although male flowers were predominant.

The higher value for seedlings with an apical flower was 19 % and 22 % when the seeds were treated with 0.22 mM atrazine and 0.2 mM diuron respectively while in controls only 2.5 % was recorded.

The results obtained with atrazine treatment were similar to Abe and Kameya. However, with diuron treatments we have found a lower rate of flowering 17% - 22 % versus 36 % according to their results. With the lower concentration, 0.05 mM diuron, we have found a higher response 8 % plants with tip-flower versus 3 % found by Abe and Kameya.

Increasing the concentration corresponded to an increase of flowering. Analysis of regression resulted linearly positive (Table 1).

The time of seeds treatment does not seem to be the cause of these differences, unless there could be a relation between the time of treatment and concentrations higher than 0.1 mM in the case of diuron.

Experiment 2. The germination rate (Table 2) with herbicide solution (76 %) was not altered as compared with the control (84 %).

The treatment with diuron caused a decrease of the growth (Table 2). The weight of the shoots and roots from the seedlings treated were 29 % and 25 % of the control plants. The treatment affected to a greater extent the root system than the aerial part of the sedlings.

In soybean seeds Reider and Buchholtz (1970) determined the uptake of atrazine by the seed and they found a direct relationship between uptake and concentration.

Table 1. Effect of seed treatment with diuron and atrazine on flowering in asparagus seedlings. Treatment was for 7 days. Linear regression of % of plants with flower on product concentrations.

Treatment	Plants with flowers (%)
Control (water)	2.5
Diuron (mM)	
0.05	8.0
0.10	17.0
0.20	22.0
Linear regression	$P \leq 0.001$ (r = 0.9427)
Atrazine	
0.10	12.0
0.22	19.0
Linear regression	$P \leq 0.001$ (r = 0.9651)

Table 2. Effect of seed treatment with diuron on weight of shoots and roots in asparagus seedlings. Treatment was for 14 days. Data were recorded two months after germination.

Treatment	Germination	Weight (mg/plant) shoots	roots
Control (water)	84	166.7*	183.3*
Diuron (0.4 mM)	76	48.5	46.9

* Values significantly different at $P \leq 0.01$

In our case, the herbicide also seems to have passed to seeds as proved later on by the weight decrease of seedlings.

Atrazine and diuron used as herbicides are known to affect electron flow in photosystem II (van Reusen, 1982). The inhibition of photosynthesis by diuron increases flowering in plants (Posner et al., 1977).

The induction of flowering in asparagus seedlings promoted by diuron and atrazine has not been proved to be due to an effect produced by the inhibition of photosynthesis because when we treated the seed the photosynthesis had not yet started.

The treatment with these herbicides induced flowering in seedlings and could be used as a method for determining the sex of asparagus plants in an early stage of development.

REFERENCES

Abe, T., and Kameya, T., 1986, Promotion of flower formation by atrazine and diuron in seedling of asparagus. Planta 169: 289-291.
Posner, H.B., Posner, R.S., and Gower, R.A., 1977, Effects of DCMV on long-day flowering of Lenna per pusilla 6746 and photosynthetic mutant strain 1073, Plant Cell Physiol. 18: 1301-1307.
Reider, G. and Buchholtz, 1970, Uptake of herbicides by soybean seed. Weed Sci. 18: 101-105.
van Reusen, J.J.S., 1982, Molecular mechanisms of herbicide action near photosystem II, Physiol. Plant, 54: 515-521.

EPILOGUE

There is a lack of clear definitions of such terms as Ageing, Maturation and Senescence. Thus it was necessary for the participants to arrive at mutually acceptable terminology. It is admited that definitions arrived at under such circumstances will be subject to severe limitations. However the following definitions provided an adequate working framework for the meeting.

Ageing; Increase in disorders caused by the environment of the cell, organ or plant, probably due to oxidative stress.

Maturation; Genetically programmed developmental processes inducing morphological and physiological changes which lead to the reproductive state.

Senescence; Genetically programmed developmental processes leading to the death of the plant or its parts (of specific organs.)

It is clear that as new information about these processes is obtained the above definitions will need to be modified and in some circumstances redefined in completely new terms. It is also possible that the use of terminolgy from other scientific disciplines will help to clarify such definitions.

Future research into the above processes will certainly include important contributions from other fields. Detailed mathematical modelling may provid insights into the hierarchical structure of key processes and indicate important switching and control points for development.However it is accepted that very complex models will be needed and that these can only be constructed on the basis of more precise and detailed experimental data. To adquire such data, new experimental techniques will be needed. These will certainly involve the application of novel techniques from the physical sciences. In particular the use of new developments in spectroscopy are envisaged together with advanced techniques for the computerised analysis of data. Such approaches have been discussed at this meeting. On the theoretical side, the application of fractal mathematics, which has proved so useful in describing complex and apparently chaotic physical processes such as liquid flow, will certainly help to shed light on certain phenomena in plant biology.

The continuing development of existing techniques such as tissue culture, genetic manipulation and recombinant DNA methods will be essential to provide new and exciting insights into developmental processes.

It is envisaged that the marriage of novel and existing techniques, together with new theoretical approaches will herald a new era of plant physiological research. This will undoubtedly lead, in a few years time, to an enormously increased depth of understanding of plant growth and development, and assist greatly in our efforts to manipulate these processes.

INDEX